Heaven is Everywhere

Jefferson Glassie

Heaven Is Everywhere
Jefferson Glassie

Published by:
Peace Evolutions, LLC
Post Office Box 458-51
Glen Echo, MD 20812-0458

Order books from: www.heaveniseverywhere.com | www.peace-evolutions.com
Free digital download of book to purchasers at website http://heaveniseverywhere.com/ebook/ with password 'peace4us'.

Copyright © 2014 Jefferson Glassie

All rights reserved. No part of this book may be reproduced or transmitted in any form or by any means, electronic or mechanical, including photocopying, recording, or by any information storage and retrieval system, without written permission from the author, except for brief quotations for purposes of a book review.

The quotes and citations in this book constitute fair use under the Copyright Act of the United States as criticism, comment, and education and we thank all sources very much; if anyone has a question about this, please contact jeff@peace-evolutions.com.

Printed in the United States.

Cover photos by NASA
Cover design and book layout by Kent Fackenthall
Back cover text: Tracey Wright

Publisher's Cataloging-in-Publication
(Provided by Quality Books, Inc.)

   Glassie, Jefferson Caffery.
   Heaven is everywhere / by Jefferson Glassie.
   p. cm.
   LCCN 2012931771
   ISBN-13: 978-0-9753837-9-7
   ISBN-10: 0-9753837-9-5

    1. Peace of mind. 2. Separation (Psychology)
    3. Heaven. I. Title.

BF637.P3G53 2012   158.1
        QBI12-600036

"The people who are crazy enough to think they can change the world are the ones who do."

Apple's "Think Different" commercial, 1997,
from *Steve Jobs*, by Walter Isaacson

Tidbits from *Heaven is Everywhere*:

There were too many of these mythical entities with the same characteristics for Jesus to have been the only real one. The Jesus Christ of the Bible must have been a myth. All the supernatural things the Bible said he did – perform miracles, walk on water, turn water into wine, cure the sick, raise the dead, and rise from the dead himself – were impossible, but millions of us believed that stuff. It was making too much sense that the Jesus of my youth was just a story.

. . .

I feel my heart swell with the thought that everything is love. Or, that I could at least think of everything as love! Wouldn't that be a good perspective?

. . .

Wait a minute. … If we assume there's only love and fear, and oneness is love, does that mean that the opposite, separateness, is fear? Well, it wouldn't just be separateness, because there is no such thing. If everything is one, nothing is separate, right? So, would fear just be the perception of separation?

. . .

If fear causes anger, violence, and murder, we need to understand fear, fully. I don't think any of these truly unfortunate people were evil. I just don't believe it. But they do sound scared, frightened, and fearful, and that's exactly the point.

. . .

[W]hat humans believe and think creates the social world we live in. What people think determines how people act. And if their thinking is fear-based, then we have a fearful, murderous, hateful world. Just like the elephants that were never hunted, or the animals on the Galapagos, things would be entirely different if we didn't fear one another.

. . .

In our roles as observers of the universe, we help create reality, as quantum physics teaches, so it's critical that we don't look at things in an unhelpful way. Viewing life as awful and humans as inherently vile does not promote peace. Observing life as perfect does.

. . .

As my friend Peter Ainslie says, conservative thoughts are Us versus Them, and liberals are about All of Us. The perception of separation versus oneness; fear versus love. Not right or wrong. When thinking where to stand on political principles, if one wants to cast a ballot for separation, vote Republican.

· · ·

Friends ask what quantum physics can mean to us in our daily lives. (Well, not that many friends, but you know, a couple people did recently.) I think it's this; if according to science, there isn't really a past or future, then letting fears from the past and anxieties of the future determine our state of mind is silly. If the past and future aren't really in the past or future, then they're illusions. There's no separation even of time and space. Seeing anything other than the now as everything seems to deviate from an accurate perspective of the truth.

· · ·

If I only had 24 hours to live, I'd want to be with my family and friends, and be able to see outside. Look at some trees. Even if it was raining, as it is the moment I'm typing these words. I look out at the rain, and a solitary tulip in front of my window, pink petals open to the sky. When thinking of life this way, as a series of perfect moments, all thoughts of money dissolve. When in the Now and feeling the oneness, there's no need or room for money. The concept of money is incompatible with the Now. Money is only about not having enough or using it to get stuff. The absence of money-consciousness when in the Now, to me, is proof of its fear roots.

· · ·

Did you ever think of the earth as alive? We talked about the Hubble images; in a funny way, don't those galaxies and star clusters look alive? For that matter, doesn't the earth look like it's a living entity in those iconic photos from space?

· · ·

To me, the ideal human relationship is not monogamy, promiscuity, or a harem. Those all involve fear-based concepts. Marriage is based on being exclusive, not inclusive, and the fear of losing your spouse, afraid of being alone again. Promiscuity isn't fearful in itself, but is based on fear of commitment and lack of trust. Harems are based on control; clearly fear-based from what we've learned. My guess is you'll find many of these relationships where people are very happy. That's great. My only point is that polyamory is another option based on love, trust, and inclusiveness. So, it seems more love-based to me. It's not afraid of changing partners. It teaches great lessons of forgiveness and empathy. And I think that humans should celebrate their sexuality more, like bonobos. That wouldn't be awful.

· · ·

Ever have a scar heal? How about a sore knee or tendonitis? What about a headache? We take healing so for granted, that we don't realize we can truly fix ourselves. So, we can heal our minds, too.

Dedicated to

My Mother and Father

Table of Contents

Foreword 9
One | **My Old Time Religion** 11
Two | **Historical or Mythical?** 27
Three | **Religion, Atheism, the Virus and Metaphysical Naturalism** 47
Four | **Learning about Love** 73
Five | **Spitting Blood** 87
Six | **Post Dog Bite Disorder** 101
Seven | **Random Paranoias** 117
Eight | **Who's Afraid of Daddy?** 141
Nine | **A Dog's Life** 157
Ten | **Elephants, Slime Mold, and Worms - Oh, My!** 169
Eleven | **People Are Naked** 187
Twelve | **Stardust** 201
Thirteen | **Let It Be** 217
Fourteen | **Now, My Personal Allegory** 235
Fifteen | **Is it supposed to be?** 253
Sixteen | **Politics, as Usual** 277
Seventeen | **"Money, it's a crime..."** 305
Eighteen | **NWAL Becomes Intuitive** 327
Nineteen | **Nothin' but High and Low Water** 347
Twenty | **Bigger Love** 381
Twenty One | **Of Course We Can** 411

Foreword

I've been writing this book for decades. Seems like, anyway. From when I first sat around on Jefferson Island in 1976 writing what I thought about life, my path has been toward this. I decided I want to bring peace to the world, and this is how my soul wants me to do it.

After I self-published *Peace and Forgiveness* in 2004 my wife Julie said to me, "Maybe you should write another book and explain it more." I took that to heart. In about 2006, after fonging around for a while, I started collecting articles from newspapers and magazines to provide some support for what I was thinking. I also culled quotes from many of the books I had read during my 'mid-life crisis.'

I began writing this book in June of 2009. I finished the first draft in December 2012. The book named itself. I gave out electronic prepublication copies to friends and held a series of book talks. I continued to collect articles and quotes for the next year and half, and then incorporated more current and new information into the book. Some of the things I learned during that time were really important, and so I'm glad I waited to finish it.

I only know a handful who've read it – admittedly, it's a bear to finish - but a few did. Some really liked it. One new friend said, "*Heaven is Everywhere* has changed my life." I guess that makes it all worthwhile, don't you think?

Ultimately, my message is: we can change our minds from fear to love, and then we really can have peace of mind and peace on earth. I'm thinking it's worth believing that. If we don't think we can, how will we ever do it?

Jefferson Glassie
Bethesda, Maryland
2014

One | **My Old Time Religion**

My mother was excommunicated by the Catholic Church for marrying my father. It was a mortal sin to marry a divorced man. I'm not sure there was an official letter from the Pope, or anything. I never asked.

She went to Mass every Sunday my entire childhood, even though she wasn't allowed to receive Holy Communion. I remember seeing her sitting in the pew, wearing a black veil. She didn't tell us anything about it at the time, but did make sure we were educated at Catholic schools.

I went to Blessed Sacrament School for eight years (the school was called "BS") and was taught by the Sisters of the Holy Cross. Then, it was on to Georgetown Prep for four years under the tutelage of the Jesuits. My brother and sister had twelve years of Catholic education, too.

We never thought about doing anything different. Life was good in Chevy Chase, Maryland, living in an old 1912 house my father refurbished. He'd been born in Washington, DC, and grew up on Bradley Lane, like me. His family raised chickens. He had a pony cart to take his aunt to Washington Senators games at Griffith Stadium. He'd been married, had two children, made a lot of money during World War II, but fell in love with my mother. Divorce, excommunication, then me.

When I was in my early twenties, a priest named Father Persig came to see Mom. I didn't know him really, but he seemed like a nice man because he forgave her. Well, actually, he gave her confession and I guess God forgave her. It seems the Church had somehow changed its mind about excommunicating those who'd married divorced men or women. I didn't understand, but everything was OK, spiritually, after that. She could go to communion at Mass and receive the host, the body of Jesus. Sometimes they'd have wine, and she could drink his blood, too. I think she was happy about that, you know, to finally be in God's good graces again.

I was a good Catholic boy. I did my homework and went to Mass. I didn't talk back to the nuns. Didn't chew gum in class. I went to confession regularly like I was supposed to.

"Bless me father, for I have sinned. It has been __ weeks since my last confession." The next part was the hard part. You had to tell the priest your sins. I wasn't embarrassed about it, and was glad to tell him. But it wasn't easy figuring out my sins. I think every once in a while I might've stolen some kid's pencil at school. Then I'd have something to confess. But usually, it was just: "I got mad at my mother and was mean to my little sister." When I got older, I could say: "I used curse words __ times." You had to do something bad or the priests would be disappointed in you.

After you told the priest your sins, he'd forgive them and give you a penance. The typical penance was saying three Hail Mary's. Sometimes, you'd get three Hail Mary's and three Our Father's. Every once in a while a priest would tell you that you had to say the *whole Rosary*. That was over *fifty* prayers, in a row. I think those were just the angry priests; it wasn't so much about what you'd done.

I was also an altar boy. In those days, they said the Mass in Latin. The priest wore jeweled robes and we altar boys sported black cassocks and white surplices on top. On feast days, we wore red cassocks. The priest in charge of altar boys at BS was pretty particular. We had to have our shoes shined, black sox, dark pants and, when we sat, hold our hands out straight on our knees. Slouching was not allowed. Fingernails *must* be clean.

We had to have the Latin down cold. In those days, the priest faced the altar away from the people. So, most of the time, no one in the congregation could hear us reciting the prayers anyway; they knew the responses by heart.

We poured water over the priest's fingers into a little bowl at the washing of the hands. That was symbolic of washing away the sins of those attending Mass. We also had to pour the wine (with a little bit of water) into the chalice for the transubstantiation. The priest would raise the chalice up slightly to tell you when to stop pouring. You had to be careful; they didn't like it if you poured too much. It was pretty important being an altar boy. Girls weren't allowed to do it back then, but the Church since changed the rules and girls can now be altar boys.

There were also bells we had to ring at just the right times. The most important ringing happened when the priest changed the bread and then the wine into, respectively, the body and blood of Christ. I had to ring the bell three times; when the priest genuflected, raised the host/chalice over his head to God, and then genuflected again. I also remember being on my knees and bowing really low during the Confiteor, and having to turn my head toward the priest beating my chest in unison. "Mea culpa. Mea culpa. Mea maxima culpa." ("My fault. My fault. My most grievous fault.")

We had a lot of rules. Men couldn't wear a hat in church, but women had to wear a hat. Many women wore veils to cover their head; little black or white lacy things that seemed to count as a hat. We couldn't eat meat on Fridays (but it was OK to eat fish). We had to go to Mass on Sunday and on "Holy Days of Obligation." We were taught these rules and all about our religion at Blessed Sacrament School by the nuns in their black and white habits. Each classroom had fifty kids in a class; ten each in five rows. No one made a sound or acted up, because the nuns might whack you on the wrist with a ruler if you did.

I recently found a 1964 copy of the old *Baltimore Catechism* book like the one used as a textbook in religion class. Millions of young Catholic girls and boys were taught the fundamental principles of the faith using the *Baltimore Catechism*. Here are some of the lessons we learned.

Lesson 1 in the *Baltimore Catechism* explains *The Purpose of Man's Existence*. "God made us to live with Him in His happy home in heaven. So He sent his Son, Jesus Christ, to lead us to heaven. Jesus leads us with love and with protection as a good shepherd leads his flock."

That was comforting. Jesus was the actual Son of God, and he takes care of us just as a shepherd cares for his little sheep. I'd never known a shepherd, of course, or any sheep. But it made you feel safe. I used to pray to Jesus so he could lead me to heaven where I could be with him after I died.

Point 6 in the first Lesson asks, "Where do we find the chief truths taught by Jesus Christ through the Catholic Church?" The chief truths of life are laid out right there in the *Baltimore Catechism*, which I probably studied when I was about ten years old. The answer is, "We find the chief truths taught by Jesus Christ through the Catholic Church in the Apostle's Creed." It also was referred to as the Nicene Creed, and spelled out these truths:

> *I believe in God, the Father Almighty, Creator of heaven and earth;*
> *and in Jesus Christ, His only Son, Our Lord; Who was conceived*
> *by the Holy Spirit, born of the Virgin Mary, suffered under Pontius*

> *Pilate, was crucified, died, and was buried. He descended into hell; the third day He rose again from the dead; He ascended into heaven, sitteth at the right hand of God, the Father Almighty; from thence He shall come to judge the living and the dead. I believe in the Holy Spirit, the Holy Catholic Church, the communion of saints, the forgiveness of sins, the resurrection of the body, and life everlasting. Amen.*

So, when I was little, these were the most important things I was taught to believe. It was a bit mysterious, though. It wasn't easy to think of Jesus being "conceived by the Holy Spirit, born of the Virgin Mary." What did "conceived" mean? Plus, the idea of Jesus rising from the dead, going down to hell, and then up to heaven was hard to grasp. But, hey, these were the *chief truths*, so they were OK by me.

Lesson 2 was titled *God and His Perfections*. There was a picture of Jesus in a boat calming the storm, while his disciples were scared witless in the stern. We were told:

> *By calming the storm at sea, Jesus showed us, first, that the Father knows all things. Even though he was asleep, He knew there was a storm. God is **all-knowing**. [No emphasis added.]*
>
> *Second, Jesus showed us that God can do all things. Science tells us that a storm has the power of many atom bombs. Yet, with only a word [actually three words, "Peace, be still."], Our Lord calmed the storm. God is **almighty**.*
>
> *Third, Jesus showed us that God is **all-good**. He calmed the storm so that no harm would come to those He loved.*

These examples showed God was perfect. We were told to "praise God and His perfections by going to Mass as often as you can." We also were told in Lesson 2 that, "God is right in this room, even though we cannot see Him. He is always with us to help and protect us. … He knows all things, past, present, and future, even our most secret thoughts, words and actions."

As a young lad, this was *very* impressive. Jesus could stop a storm. He was much more powerful than Superman. All I could think was, "Wow, Jesus was so almighty and good, he makes me seem so puny." If He was in the room with me all the time, that meant you had to be really good. And if he knew what I was thinking, I could be in big trouble, particularly as I got older.

Since Jesus was all-knowing, I figured he could help me with my career path. I was in about 4th grade when I began wondering what I was going to be when I grew up. Well, of course when I was really little and rode around in a red fire truck, I was going to be a fireman. Later, when I was in my cowboy and Indian phase, I wanted to be a cowboy. I had several sets of cowboy boots that I grew out of, and a dashing cowboy outfit with a badge, hat, and holster. I played with tiny cowboy and Indian figures and shot up all the Indians. But for some reason I began thinking about this seriously in 4th grade and figured it was time to ask God.

Do you remember those label makers we had back then? They looked a little bit like guns, with a trigger. You put this plastic tape in the back of the gun, selected the letter or number on the dial, and then pulled the trigger hard. The letter or number would imprint on the tape. You could use the tape to identify lockers or other things you owned. I was considering at the time whether I should be a priest, policeman, or baseball player. Actually, I don't remember exactly what the third option was. It might have been army man, fireman, but probably not lawyer, author, or musician.

Anyway, I typed out three pieces of tape with different letters on each. '"P" for priest; "C" for cop (I couldn't use another "P" for police, or it could've been confusing): and "B" for baseball player. I didn't peel off the back and just left them on my dresser one night. My idea was that God would take the backing off the tape with the letter of my career choice, then stick it on the dresser so I could see it in the morning. That would be a clear direction from God, or maybe Jesus, as what to do.

Unfortunately, when I woke up, none of the letters was stuck to the dresser! "Huh. God didn't tell me what to do." I really thought he would. He knew the past and future, and *all* my thoughts. He could stop storms with the power of atomic bombs, so I was sure he could remove the back of the tape and stick it to my dresser. Now, I'd have to figure it out on my own, without help from God. It was sorta disappointing, but I still believed and all that. I was fortunate, though, in that I hadn't considered my mother's reaction if the tape had been stuck to the dresser. I don't imagine she'd have bought my story that, "God did it."

Lesson 3 was about *The Unity and Trinity of God*. We reviewed these questions:

> *Is there only one God?*
> *Yes, there is only one God.*

> *How many persons are there in God?*
> *In God there are three Divine Persons – the Father, the Son, and the Holy Spirit.*

That was a little hard to understand; three persons in one God. But Lesson 3 provided an explanation:

> *How there are three Persons in one God is a mystery. It is something we cannot understand. But we believe it because Jesus told us. We will not understand it until we reach heaven. We call this mystery of three Persons in ONE God the 'BLESSED TRINITY.'*

We actually called them the Holy Trinity, but that's just semantics to explain the three-persons-in-one-God concept. It was useful for them to have a name like that, especially when it's a *mystery*.

I remember being interested in the next Lesson about *Creation and the Angels*. I never saw any angels, though I've known people who said they've seen 'em. The *Catechism* says, "The chief creatures of God are angels and men. [I assume that meant the human species, rather than leaving women out of the 'chief creatures' category altogether.] Angles are "created spirits, without bodies. The Angels can't be seen or heard or touched. They have much greater power than man has and they know much more."

So, here's a question; if they can't be seen, heard, or touched, how does anyone know about them? And what did those people who saw angels really see?

Here's another important fact about angels; there are good angels and bad angels. "Some angels were not faithful to God. They would not obey Him. They were cast out of heaven and are called 'devils.'" These bad angels tempt us into sin, but we have the good angels on our side. They remained faithful to God and are in eternal happiness in heaven. "The good angels are always before the throne of God. They love and adore Him and do what he asks of them."

God also gives each one of us a "guardian angel" to help us ward off the temptation of the devils. I wasn't sure exactly how we got tempted by the bad angels and helped by the good angels, since we couldn't see, hear, or touch them. But I did pray to my guardian angel for help, as Lesson 4 told me. Sometimes, I would try and listen for the good angel on one shoulder, and then for the bad angel on the other shoulder, but I never did hear anything from the angels.

Lesson 5 was very important, and explained about *Creation and Fall of Man*. As related in the Book of Genesis in the Bible, God made Adam and Eve first. You've probably heard the story. God told Adam and Eve they had to make a choice; did they want God or themselves to be first in their lives? He told them not to eat the fruit of a certain tree in the Garden of Paradise. (I always wondered what kind of fruit; it musta been good.) But the devil tempted Eve, who convinced Adam with her womanly power of seduction to eat the fruit.

According to the *Catechism*, this meant that Adam and Eve loved themselves more than God. So, they "lost sanctifying grace and the right to heaven, and were driven from the Garden of Paradise. ... On account of the sin of Adam we come into the world without grace, and we inherit his punishment. ... This sin in us is called original sin." Because of it, we are "filled with selfishness."

I didn't feel all full of selfishness, but it is very sad that we enter this world with no grace. We Christians have Baptism, which does give our grace back, though I wasn't sure about the other religions. I remember being told that Jewish people (I didn't really know any back then) couldn't go to heaven because they didn't get baptized. That never seemed fair to me, but God was the decider. Not *my* fault.

Another point we got from Lesson 5 came from this question and answer:

> *Was any human person ever free from original sin?*
> *The Blessed Virgin Mary was free from original sin, and this favor is called her Immaculate Conception.*

I thought for a long time (when I got older) that the Immaculate Conception referred to Mary conceiving Jesus without having sex. But even though that was supposedly true, it wasn't the Immaculate Conception. It might be called 'coitus notus,' I suppose. I did wonder how anyone *really knew* that Mary was free from original sin. That would also have meant God let her be born without original sin. So, indeed, God had the power to allow people be born with, or without, original sin. I wondered then why Jesus really had to be born and die for our sins, as we were taught, because God coulda just wiped that original sin away when we were born.

I'm not going to go over every Lesson in the *Baltimore Catechism*, in case you were getting worried, but there are a few more important concepts. Lesson 6 is about

Actual Sin. See, original sin was not the only kind of sin; there's another kind called actual sin, which "is any willful thought, desire, word, action, or omission forbidden by the law of God." Sounds kinda legalistic, doesn't it?

There are two kinds of actual sin: mortal sin and venial sin. "Mortal sin is a grievous offense against the law of God [that] takes away the life of the soul." It basically, "drives Our Lord out of the life of the one who commits it." I was afraid of mortal sins, because I didn't want my soul to die and have a big black mark on it. The *Catechism* did say that, "little boys and girls are not usually in danger of committing mortal sins," so that was good. But we were warned against venial sins because we'd be more likely to sin as adults. I do remember thinking of how my soul would look with a venial sin. I imagined an ugly blackish blotch, like this little boy in Lesson 6 must have had:

> *Suppose a boy says "no" to his mother without thinking when she asks him to help with the dishes in the middle of his favorite TV show. But afterward he thinks of how Our Lord says "Yes" to His Father, even when His Father asked Him to die on the Cross for us. If that boy is sorry for saying "no" and is willing to say "yes" to his mother the next time, no matter how good the program is, the Our Lord is pleased.*

I wanted the Lord to be pleased with me (though I did understand about not wanting to do the dishes in the middle of my favorite TV show). As a practice tip in Lesson 6, we were advised to, "Look at a crucifix each day and think how much Jesus suffered for our sins." I did look at the crucifix a lot. They were in each room at school, often near the American flag.

The next two Lessons were *The Incarnation* and *The Redemption*. The first of these discussed how Mary was asked to be the Mother of the Son of God by an Angel, and said, "Yes." So, the Holy Spirit got her pregnant. (I guess that technically wouldn't have been "sexual relations.") God had not abandoned humans after the fall from grace caused by Adam, so God "sent His Son to earth to save men from their sins. ... The Savior of all men is Jesus Christ ... born of the Blessed Virgin Mary on Christmas Day, in Bethlehem." That was pretty cool.

The Lesson on redemption taught us how Jesus willingly went to Jerusalem with the Apostles "to suffer and die and rise again." You probably know about this, too, but Jesus was tortured by the Romans and then crucified on a cross. He died on Good Friday, though it probably wasn't that good for him.

It was very important for us, though, because:

> *He took all the sins of the word on Himself and died that they might be destroyed. ... The sins of the world had offended God very deeply. To make up for sin it was necessary to give God something that pleased Him more than sin displeased Him. Jesus our Good Shepherd did this for us by giving His Father the love of His Sacred Heart. When Our Lord died on the Cross the love in His Heart pleased His Father more than all the sins of the world displeased Him. ... To show He was pleased, the Father raised the body of His Son from the dead and took him to heaven with Himself. ... Christ rose from the dead, glorious and immortal, on Easter Sunday, the third day after His death.*

This was *big stuff!* And a lotta mystery going on, which was way past my mental capacity at the time (or even now for that matter). We were just told to love Jesus, and to bear our small crosses by offering them up to God like Jesus did on the Cross. There's a picture of a little girl looking at a crucifix and saying, "This homework is a cross for me, Jesus, but I will do it gladly for you." That's the way we were taught to think about life; it was a big cross we should bear for Jesus. And we were taught Jesus sits at the right hand of the Father in heaven. The Catechism explains this, "When we say Christ 'sits at the right hand of the God' we do not mean that He is in heaven doing nothing. From heaven Christ rules over all men. He is our King."

I hadn't thought he was doing *nothing* in heaven, and understood he ruled over us and that would probably make him pretty darn busy. He knew all our thoughts and actions. I mean, Jesus was truly amazing. He was different than us, separate from us, and way better than us. As God, he could do anything and had saved us all from original sin. Most of us wore chains and medals with pictures or engravings of Jesus or Mary. We went to Mass basically every Sunday, confession every few weeks on Saturday afternoon, and said our prayers every night. I liked the "Hail Mary" better than the "Our Father." Here's how the "Hail Mary" went, for those who might not know it:

> *Hail Mary, full of grace, the Lord is with thee. Blessed art though amongst women, and blessed is the fruit of my womb Jesus. Holy*

Mary, Mother of God, pray for us sinners, now and at the time of our death. Amen.

A rosary is basically five sets of ten Hail Mary's with an Our Father thrown in between. When I later went through my "midlife crisis," I had little finger rosaries made out of silver that I used to say rosaries with. I carried them around in my pocket for several years. It was always comforting to say Hail Mary's. I never really felt like a sinner, but sure did call myself a sinner thousands of times by saying that prayer.

That was a major characteristic of Catholics. We *were* sinners. We were always told we were sinners. All our prayers confirmed how bad we were. The Mass reminded us we were unworthy. There's a part before communion when we all said together, "Lord, I am not worthy, but just say the word, and my soul shall be healed." We were taught to fear the wrath of God and not being able to go to heaven with him. If we couldn't go to heaven, you know what that meant. We were going to hell. And that meant eternity without God; eternal damnation. *The ultimate fear.* I don't think we walked around like scaredy cats all the time, but there was always this underlying fear. A lot of fear. Every time one says, "I'm afraid," that really means, "I am in fear" of something. That perception of being separate from God was the big fear as a Catholic.

But things could be alright, you know. We could go to heaven, if we just *believed* and didn't piss God off. Lesson 14 on *The Resurrection and Life Everlasting* told us, "at the end of the world the bodies of all men will rise from the earth and be united again to their souls, nevermore to be separated." Not to be separate from God again, ever. *That would be awesome!* There was only one person whose body "was raised from the dead and taken into heaven" directly, and that was the BVM. She had been born without original sin and went right to heaven. I'm not sure where she sits up there, on the right or the left, but she's the mother figure who intercedes for us with God.

Anyway, after the general resurrection, judgment is passed on all men (presumably women, too). The possible reward or punishment you could get was heaven, purgatory, or hell. The Catechism didn't mention limbo, which is where little babies went who died before they were baptized, according to what I was taught. I recall reading somewhere recently that limbo isn't officially considered a place any more under Church doctrine.

Skipping ahead again to Lesson 16 (I hope I'm not going to fast for you), we learned about the *First Commandment of God*. I was sort of surprised when I saw this, because I guess I'd forgotten the First Commandment. My bad. Here it is:

I am the Lord thy God; thou shall not have strange gods before Me.

Well, that's pretty clear. No one was supposed to worship false gods. God definitely did *not* like that, because he made it his *First Commandment*. That Lesson also told us how we should worship God, which was "by acts of faith, hope, and charity and by adoring Him and praying to Him. ... A Catholic sins against faith by not believing what God has revealed and taking part in non-Catholic worship. ... The sins against hope are presumption and despair. ... The chief sins against charity are hatred of God and of our neighbor, envy, sloth, and scandal."

As a young boy growing up, I'm sure I didn't hate God or my neighbor, and didn't understand what envy, sloth, and scandal were, so I don't think I committed those sins. I also didn't take part in non-Catholic worship. I don't ever remember going to a Protestant, Lutheran, Seventh Day Adventist, or Jewish ceremony. I suppose we were rather sheltered.

The Second Commandment of God is:

Thou shalt love thy neighbor as thyself.

You know, I got that one. I thought everyone should love one another and didn't think anyone needed to fight. I knew my mother loved me, and my father, too. I think my sister and brother loved me. I didn't hate anyone. And how did the *Catechism* answer the question of what we had to do to love God, our neighbor, and ourselves?

To love God, our neighbor, and ourselves, we must keep the commandments of God and of the Church

That didn't give me a lot of guidance on how to love others. Maybe some of the other Commandments would do that. Let's see:

3. *Remember thou keep holy the Lord's day.* (We went to Mass.)
4. *Honor thy father and thy mother.* (Was that honor, or love?)
5. *Thou shalt not kill.* (OK, understood.)
6. *Thou shalt not commit adultery.* (What? Not be an adult?)
7. *Thou shalt not steal.* (Oops.)

8. *Thou shalt not bear false witness against thy neighbor. (?)*
9. *Thou shalt not covet thy neighbor's wife. (Covet?)*
10. *Thou shalt not covet thy neighbor's goods. (Covet??)*

These prohibitions on coveting had me confused. The *Catechism* said that "covet" means "to wish to get a thing unjustly." How does that apply to Mrs. McKay down the street? "By the ninth commandment we are commanded to be pure in thought and in desire." When I was ten, this was a very strange commandment.

The "Oops" mentioned above at the 7th Commandment refers to a big sin of mine I just remembered. I stole money from my dad's wallet.

You see, my friend Tom and I were in about third grade and played "army" all the time. His father had been in the army, and I watched a lot of World War II shows. A favorite of mine was called, "Combat" with Vic Morrow as the Sarge. Tom and I would roam the alleys of northwest DC in green army uniforms and our plastic machine guns. We shot and killed lots of imaginary "Krauts" and "Japs." I don't think I could have killed a real person.

Remember, this was about 1963, less than twenty years after World War II. I was a baby boomer, and people thought differently back then. There were the communist threats, and soon to be the "Gooks" we fought in Viet Nam. It seemed we had to be afraid of and fighting some other people all the time; Indians, Germans, Japanese, Italians, Communists.

One day, Tom showed me he had two dollars. "I took these out of my father's wallet," he said. I thought, "That's a pretty good way to get money." My father kept his clothes and stuff in my room (the one with the aforementioned dresser). So, he left his wallet there at night. I wasn't thinking about stealing or the Commandments or anything. I wasn't afraid or feel bad that I didn't have any money. I just wondered what it would be like to take some. I swiped a ten-dollar bill and put it in my wallet. I didn't get caught.

So, next I took a twenty dollar bill. Nothing happened. No one said anything or treated me differently. I didn't hear my guardian angel and didn't see any black splotches on my soul. I ended up taking about seventy bucks and stuffing it into my wallet, which otherwise had nothing in it except some sort of ID and probably a Jesus or Mary holy card.

One day, my father stopped me in my room. He said my mother had found my wallet and was stunned to find all that money. He said he remembered missing a bunch of money one day. I didn't say anything, but I guess I outwitted him. I hadn't taken it all at once; I'd spread my larceny out over several days.

He told me stealing was wrong. He was not a Catholic, so didn't say that Jesus was disappointed with me or anything like that. I admitted I'd taken the money. He grabbed a plastic ruler and told me to lay on the bed. He spanked me with the ruler several times. I don't remember crying, but imagine I must have for effect. He actually broke the ruler, which I thought was funny, but didn't laugh. He told me not to steal anymore.

I wasn't afraid. I knew I'd just taken the money to see what it was like. Tom had done it, so why couldn't I? I didn't mean to hurt anyone. I didn't want to make God angry or the BVM sad for me. My father reacted as I would've anticipated. I shouldn't steal. That was a good practical lesson. But I didn't feel like God was going to strike me down or send me to hell. I must have told the priest about this in confession; that would have been a good sin to tell.

In addition to the Ten Commandments, the *Baltimore Catechism* also had Lessons on the seven sacraments, which are Baptism, Confirmation, Holy Eucharist, Penance, Anointing of the Sick, Holy Orders, and Matrimony. I want to mention a bit more about the Holy Eucharist, also called Holy Communion, because I referenced "transubstantiation" earlier.

"When Our Lord [at the Last Supper] said 'This is My body,' the bread was changed into His body; and when He said, 'This is My blood,' the wine was changed into His blood." We were told that, at Mass, "after the bread and wine had been changed into Our Lord's body and blood [by the priest], there remained only the appearances of bread and wine. ... The appearances in the Holy Eucharist are what look like bread and wine. The *substance* is Our Lord Himself under these appearances."

That was hard to get your head around. It was another one of those *mysteries*. When the priest came around to give you communion, he said, "The body of Christ," and we said "Amen." That meant, "So be it." "Yes, it's the body of Christ." One time, years later, I received Communion for the first time in a long time, and had forgotten you were supposed to say "Amen." I just said, "Thanks, Father."

When I was an altar boy, the parishioners would kneel at the altar rail to receive Holy Communion, but weren't allowed to touch the host. The priest put the host right on the person's tongue. The altar boy carried a little brass plate with a handle, called a "patent," and put it right under the person's chin in case the host dropped out. I caught a few hosts with my patent, I must say. It was a big deal though if the host fell *onto the floor.* The priest would immediately get down and pick it up off the floor, kiss the floor, and then I think he ate the host. It caused quite a commotion. Later on, the Church decided it was OK to touch the host, so the priests started placing the host in the hands of the persons receiving Communion, and then they put it into their own mouths. That probably was a better idea, from a logistical standpoint. And altar boys and girls didn't have to use the patent anymore.

As a side note, they didn't usually give us wine at Mass. Sometimes, on special occasions, there was wine. You would get in line for the host, and there'd be a separate wine line. You took a little sip out of the golden chalice, and the person who gave it to you (they let lay people do this after a while) wiped off the place on the chalice where you'd touched your lips to it. You know, being antiseptic and all.

I'd also like to talk briefly about *Confession*, from Lesson 31. This is where you had your sins washed away. Pretty much no matter what your sins were, you could get rid of them and erase them from your soul. This was good. How is confession defined? "Confession is telling our sins to a priest to obtain forgiveness. We go to the priest who takes Christ's place and Christ, through the priest, forgives our sins. … It is necessary to confess every mortal sin which has not yet been confessed and forgiven; it is not necessary to confess our venial sins, but it is better to do so."

You could ingest the Lord and if you had any sins, you could get them forgiven. I was glad to be a Catholic.

The *Baltimore Catechism* also had an appendix, which was stated as being useful for the instruction of adult converts to the faith. Here are a couple of the questions and answers I find interesting.

How can we prove that there is a God?

We can prove that there is a God because this vast universe could not have come into existence, nor be so beautiful and orderly, except by the almighty power and wisdom of an eternal and intelligent being.

How can we prove that the soul of man is immortal?

We can prove that the soul of man is immortal because man's acts of intelligence are spiritual; therefore, his soul must be a spiritual being, not dependent on matter, and hence not subject to decay or death.

How can we prove that all men are obliged to practice religion?

We can prove that all men are obliged to practice religion because all men are entirely dependent on God, and must recognize that dependence by honoring Him and praying to Him.

Whence do we chiefly derive our historical knowledge of Jesus Christ, His life and teaching, and of the Church He established?

We derive our historical knowledge of Jesus Christ, His life and teachings, and of the Church He established chiefly from the books of the Bible, which can be proved to be reliable historical records.

What else are the books of the Bible besides being reliable historical records?

Besides being reliable historical records, the books of the Bible are the inspired word of God, that is, written by men with such direct assistance from the Holy Ghost as to make God their true Author.

This all was pretty persuasive to a young boy, especially because practically everyone I knew accepted the same dogma. We all believed the same thing.

Two | Historical or Mythical?

Fast forward, to 2005. Learning doesn't happen chronologically. Thoughts and events from various points in the past lay dormant until unlocked. My father used to say learning occurs in steps. You go along doing something the same way until, at one point, you started doing it better.

Tennis is a good example. I could practice my serve, and it'd still be pretty lame. But if I kept practicing, taking lessons, thinking about it, and working on it, eventually my serve would get better. It's the same as trying to learn the best outlook on life for yourself. Really.

A lot had intervened since I was an altar boy at BS. I'd been married, went to law school, had three kids, got divorced, and married again. I had grown away from the Catholic Church, but went back when we had the kids. They all went to Little Flower Catholic Church, and Catholic high school; both sons went to Georgetown Prep like me. I wasn't practicing any religion in 2005, but had learned tons during my midlife crisis; that is, my divorce and the aftermath. I was reading a lot and developing new beliefs.

In 2004, I wrote and published *Peace and Forgiveness*, which explained my beliefs about life, love, and faith. I also wrote a companion book of poems, called *Poems of Peace and Forgiveness*, and produced a CD of acoustic blues music called *Songs of Peace and Forgiveness*. I've been on sort of a spiritual quest for a while now. Very rewarding. But let me first tell you a little about Peter.

I met Peter Ainslie in 2005 in our little enclave of Brookmont. He was the minister of the Church. Peter's had a bit of trouble walking; he was seventy-something. He got married for the first time in 2008 to Sharon, his soul mate. I saw them walking downtown one Sunday afternoon. She had a cane and held onto Peter's arm. They were jaywalking.

They got married at the Brookmont Church in June, and Peter asked me to light the candles on the altar. I had lots of experience using those brass candle lighter/snuffer-outers in my youth. I lit all the candles for Peter's wedding, wearing a seersucker suit rather than a red cassock, with all the poise and posture of a fifty-year-old altar boy. Unfortunately, one candle went out after I'd left the sanctuary. Several members of the congregation were kind enough to point out the situation, so I went back up there and re-lit the one candle. It musta been a breeze.

Peter is a minister of the Disciples of Christ, but he delves more deeply into life. He reads lots of spiritual and metaphysical books. He told us about his ventures with Buddhism and other religions.

He and I used to find ourselves talking about all sorts of spiritual topics. Then, at some point, he gave me a one-page bibliography he'd prepared listing books on the topic of the historical versus mythical Jesus. My understanding of faith has never been the same.

The name of the first book on the list was: *The Christ Conspiracy, The Greatest Story Ever Sold*, by Acharya S. What kind of name was that? And there actually is no period after the S, like Harry S Truman. That was weird. Later I found out Acharya S is the pen name for D.M. Murdock. She's classically educated and has written several other books, among them *Suns of God, Krishna, Buddha, and Christ Unveiled*, and *Who Was Jesus? Fingerprints of The Christ*. What was all this about; Christ conspiracy? Suns of God? Her name freaked me out, so I picked another book on the list to read first.

I started reading The *Jesus Mysteries: Religious Lies and Gnostic Wisdom*, by Timothy Freke and Peter Gandy, two Brits. I read it when Julie and I went on our delayed honeymoon to Costa Rica in November 2005. I was enthralled. The dedication page says, "This book is dedicated to all those who love their enemies." I always *thought* you should love your enemies (in spite of my aforementioned "army" days in grade school). The first line of the book is "Wake up!" and then, "See through the illusion of separateness and recognize that we are all essentially one." This didn't sound like the *Baltimore Catechism*.

The authors ask, "Have you ever seen a picture of Jesus laughing? Probably not, because we have inherited a distorted form of Christianity created by the Roman Church in the fourth century, which focuses exclusively on Jesus the 'man of sorrows.'"

Historical or Mythical?

Come to think of it, I don't think I ever saw Jesus laughing; he smiled sometimes, but usually was pretty serious, even stern or in agony. They also said:

> *The image that has dominated our culture is that of a man being tortured to death on a cross. But the original Christians didn't see Jesus as an historical man who had 'suffered for our sins.' They viewed Jesus as the mythical hero of a symbolic teaching story, which represents the spiritual journey leading to the experience of awakening that they call 'gnosis', or 'knowing'.*

Huh?

It's true the cross is ubiquitous. Not just the huge crucifixes in every Christian church, or the semi-naked Jesus with a bloody crown of thorns nailed to the wood on schoolroom walls, but also the silver or gold crosses women and men now wear around their necks as jewelry.

What was all this about Jesus not being an historical man, but a myth? How could that be? All my life I believed that Jesus lived on earth and was some sort of a God. Everyone I knew growing up believed he was born of a virgin, died for our sins, rose from the dead, and went to heaven, where we wanted to be with him. Most of the people I've known who weren't Catholics at least believed there was a Jesus who lived and preached about 2,000 years ago. But Freke and Gandy say:

> *Jesus was obviously a myth. ... There is no more evidence for the existence of Jesus than there is for Moses, Joshua, David, Solomon and all the rest [of the Biblical figures]. All are Jewish literary creations. The Jesus story is a symbolic allegory based on ancient Pagan myths. ... Scholars have demonstrated that not one of the [] gospels is an eyewitness account of the life of Jesus.*

This was huge; and I was interested. It had never crossed my mind that Jesus never lived, or that there was no Moses. But then again, whenever I did try to read the Bible, it didn't make a whole lot of sense to me.

Wait a second. Who are these guys again? I'd never heard of Freke and Gandy, but the Bible is the most popular book of all time and leads the world in book sales every year. Well, according to the book jacket, Freke has a degree in philosophy and has authored more than twenty books on spirituality. Gandy has a masters degree in classical civilization, and has written seven books with Freke. So apparently they aren't crackpots.

They go on to explain the Biblical chronology about Abraham was not consistent with history, the story of Moses was likely a myth, and there's no evidence for either David or Solomon's existence. They said stories were told throughout history about men who were god-on-earth.

> *The earliest religious texts in the world come from ancient Egypt and tell the story of Osiris. Osiris is a god who became man and wandered through Egypt teaching the people about religion and the right way to live. He was put to death by the forces of evil, but was magically restored to life and ascended into Heaven to become the judge of souls in the afterlife.*

Similar stories were found in other cultures. In Greece, this "godman" is called *Dionysus*, in Asia Minor he is *Attis*, in Syria *Adonis*, in Persi *Mithras*, just to name a few. Freke and Gandy refer to these as "mystery" religions, also referred to as "pagan" beliefs. They list some of the common elements of the stories about these godmen:

- *His father is God and his mother is a virgin girl.*
- *He is hailed by his followers as the saviour, God made flesh the Son of God.*
- *He is born in a cave or humble cowshed on the twenty-fifth of December in front of shepherds.*
- *He surrounds himself with twelve disciples. He offers his followers the chance to be born again through the rites of baptism.*
- *He miraculously turns water into wine at a marriage ceremony.*
- *He rides triumphantly into town on a donkey while people wave palm leaves to honour him.*
- *He attacks the religious authorities who set out to destroy him.*
- *He dies at Easter time as a sacrifice for the sins of the world, sometimes through crucifixion. On the third day he rises from the dead and ascends to Heaven in glory.*
- *His followers await his return as the judge during the Last Days.*

- *His death and resurrection are celebrated by a ritual meal of bread and wine, which symbolise his body and blood.*
- *By symbolically sharing in the suffering and death of the God-man, initiates of the mysteries believed they would also share in his spiritual resurrection and know eternal life.*

This was scary. Of course, these are some of the important aspects of the teachings about Jesus. I'd never heard of these other "godmen." But if they really did have similar characteristics, and the others were clearly myths, where does that leave us with Jesus?

I remember as a kid reading about the Greek gods, you know, Zeus, Apollo, Poseidon, Aphrodite, etc. I heard stories about the Cyclops, the Argonauts, and the Nymphs. Those were all just stories, of course, and I figured the Greeks were pretty dumb to have believed all that stuff. Now I wondered, were the teachings of the Catholic Church also based on myths?

The authors said that the Emperor Constantine adopted Christianity as the official religion of the Roman Empire and used it to help control the people. He called the Council of Nicea in the Fourth Century to lay out the tenets of the new religion. Freke and Gandy say, "The Nicean Creed was designed by a despotic Roman emperor and imposed on Christianity by force. Yet, incredibly, it is still repeated in churches throughout the world today." Gulp.

The problem, they said, is that the people interpret these teachings literally. They call them the "Literalists." That's what Peter Ainslie calls them, too. Freke and Gandy say:

> *Literalism is stuck in the past. ... The very nature of life is that things constantly change, but Literalists don't like change. You can tell when a spiritual tradition has succumbed to Literalism because it stops changing. People repeat the same old creeds, perform the same old rituals, and wear the same old costumes, although they no longer have any idea why.*

Another gulp.

> *Religion is obviously crazy. ... The time has come to announce that*

> the emperor has no clothes. Literalist religion deserves to be ridiculed, not respected. It is irrational, immoral, and outmoded. And hysterically funny. ... Literalism is bonkers! If we told you that a friend of ours had been born of a virgin, could walk on water and had come back from the dead, who but the truly demented would believe us. Yet billions happily believe this based on a bizarre old book. ... The irony is that most extreme Fundamentalists are actually only one small step away from waking up. They are already completely convinced that everyone else's religion is utter nonsense. All they need to do now is realise that so is their own.

I must say, this was challenging. I hadn't thought much about whether Catholicism was true or made sense. I just believed it and went to Mass because *I always had*. Now, I was being told that everything I'd believed, and that my devout mother - who *knows* God had His hand on her shoulder - also believed, was just baloney. How does one accept this?

I picked up another book by Freke and Gandy, called *The Jesus Mysteries; Was the Original Jesus a Pagan God?* They said, "It is easy to believe that something must be true because everyone else believes it. But the truth often only comes to light by daring to question the unquestionable, by doubting notions which are so commonly believed that they are taken for granted." How many people could do that? Could I do that? Could you do that? How hard is it to think outside of the box and determine what you believe, not what others believe?

I then decided to try Acharya S. What the heck. In the introduction, she says:

> Although many people believe religion is a good and necessary thing, no ideology is more divisive than religion, ... which is dependent on division, because it requires an enemy, whether it be earthly or in another dimension. Religion dictates that some people are special or chosen while others are immoral and evil, and it too often insists that it is the duty of the "chosen" to destroy the others. ... More horrors have been caused in the name of God and religion than can be chronicled ... .

Historical or Mythical?

I wonder how many people who tried to read her books would be turned off by that first page and put it down? Probably a lot. But I was interested in her rationale.

Her book is well-documented, with *lots of footnotes*. She continues along the same lines as Freke and Gandy, saying it's clear that "the story of Jesus Christ was invented and did not depict a real person." The Bible is "riddled with inconsistencies, contradictions, errors and yarns that stretch the credulity to the point of non-existence." Many people believe the Bible is true because of the prophecies that have come true, but she says much Biblical prophecy was actually written after the fact and that very few of the prophecies actually did come true. She cites the Epistles of Paul and states that Paul never actually refers to a historical Jesus, though referencing the Christ spirit "within" in Gnostic manner.

She says, "[T]here are basically *no non-biblical references to a historical Jesus* by any known historian at the time during and after Jesus's purported advent." She goes to great length referring to many of the known writers of the day, such as Flavius, Josephus, Pliny the Younger, Tacitus, etc., none of whom make any reliable reference to a man who was said to have done so many miracles and drawn huge crowds.

She delves into significant detail when explaining, like Freke and Gandy, that the Christ character was a "compilation" of the various gods worshipped long before Jesus supposedly lived. "The legend of Jesus nearly identically parallels the story of Krishna, for example, even in detail, with the Indian myth dating to at least as far back as 1400 BCE." She echoes what Freke and Gandy say about the similar aspects to the mythical godmen stories.

> [W]e find the same tales around the world about a variety of godmen and sons of God, a number of whom also had virgin births or were of divine origin; were born on or near December 25th in a cave or underground; were baptized; worked miracles and marvels; held high morals, were compassionate, toiled for humanity and healed the sick; were the basis of soul-salvation and/or were called 'Savior, Redeemer, Deliverer;' had Eucharists; vanquished darkness; were hung on trees or crucified; and were resurrected to heaven, whence they came.

On page 106, she has a list of names, I swear, with *over fifty* of these godmen! They include Adonis, Apollo, Dionysus, Hercules, Zeus, Attis, Buddha, Krishna, Horus,

Osiris, Issa, Jupiter, Mithra, Prometheus, Thor, and Zoroaster, from different regions and cultures around the ancient world.

She discusses several of the more prominent godmen in detail, but let's look at one as an example. Horus of Egypt, the son or renewed incarnation of Osiris:

- *Horus was born of the virgin Isis-Meri on December 25th in a cave/manger with his birth being announced by a star in the East and attended by three wise men.*
- *His earthly father was named "Seb" (Joseph)....*
- *At age 12, he was a child teacher in the Temple, and at 30, he was baptized, having disappeared for 18 years.*
- *Horus was baptized in the river Eridanus or Iarutana (Jordan) by "Anup the Baptizer" ("John the Baptist"), who was decapitated.*
- *He had 12 disciples, two of whom were his 'witnesses' and were named "Anup" and "Aan" (the two "Johns").*
- *He performed miracles, exorcised demons and raised El-Azarus ("El-Osiris"), from the dead.*
- *Horus walked on water.*
- *His personal epithet was "Iusa," the "ever-becoming son" of "Ptah," the "Father." He was thus called the "Holy Child."*
- *He delivered a "Sermon on the Mount" and his followers recounted the "Sayings of Iusa."*
- *Horus was transfigured on the Mount.*
- *He was crucified between two thieves, buried for three days in a tomb, and resurrected.*
- *He was also the "Way, the Truth, the Light," "Messiah," "God's Anointed Son," the "Son of Man," the "Good Shepherd," the "Lamb of God," the "Word made flesh," the "Word of Truth," etc.*
- *He was "the Fisher" and was associated with the Fish ("Ichthus"), Lamb and Lion.*
- *He came to fulfill the Law.*
- *Horus was called "the KRST," or "Annointed One."*
- *Like Jesus, "Horus was supposed to reign one thousand years."*

This wasn't funny anymore. There were too many of these mythical entities with the same characteristics for Jesus to have been the only real one. The Jesus Christ of the Bible *must* have been a myth. All the supernatural things the Bible said he did – perform miracles, walk on water, turn water into wine, cure the sick, raise the dead, and rise from the dead himself – were impossible, but millions of us believed that stuff. It was making too much sense that the Jesus of my youth was just a story. What was I to make of that?

Acharya did not relent. In a chapter titled "Astrology and the Bible," she says the Bible is actually an "astrotheological text," and allegorically includes many passages referring to the heavenly bodies. She explains that astrology was a key element of early Hebrew beliefs. The Israelites worshipped many gods, including the sun, moon, and stars, she says, and the Bible includes much "astrological imagery." The sun and the moon are given human emotions in the Psalms and the book of Isaiah, for example. The book of Job is full of astrological images, including references to the constellations Pleiades and Orion and the 'Mazzaroth,' which is the zodiac.

In the Psalms, there are references to things like the heavens telling the story of God, and the stars have names given to them by the Lord. Psalm 84:11 states "For the Lord God is a sun." The sun and moon had personalities in Isaiah. She notes that the Hebrews were also moon worshippers, with feasts and holidays based around the phases of the moon, as is true today. I knew that the date we celebrated Easter, the resurrection of Jesus, was based on the moon. Easter Sunday is the first Sunday after the first full moon, after the vernal equinox. I'd wondered about that.

She gives a number of examples of the astrological features of the Old Testament, including Jacob's twelve tribes being symbolic of the twelve houses of the zodiac. The seventy two angels ascending and descending Jacob's ladder represent the twelve decans, i.e., five degree portions of the zodiac.

What really struck me was her explanation of the astrological aspects of Jesus. First, the sun is often worshipped as god. After all, it is the bringer of light that warms us and grows the plants. Without the sun, we could never exist in our current forms. In the book of Job 38, the stars are called the "sons of God." "Hence, one star would be a 'son of God,' as well as the son of the Sun. Thus, the sun of God is the son of God." She explains the reason so many of the stories of the godmen are similar is that they're based on the ancient knowledge of the movement of the sun through the heavens. Jesus and the other godmen are personifications of the sun. For example, as the source of the resurrection story, the ancients knew the sun annually goes southward until December 21, the winter solstice, when it "dies," and seems not to move for three days, and then "rises again" and comes back to bring life. This event was generally celebrated on the day equivalent to December 25th.

For thousands of years, people had been looking up at the sun, moon, and stars. There weren't electric lights or Hubble telescopes. They must have sat around the fire, looked at the sky, and told stories. Just like we watch movies now. Their lives much more than ours revolved around the sky and the light from it. They hardly knew how to read and write, most of them. Yet, they must have wondered, just as we do, about where they came from and where they'd go. The sun as the dominant heavenly body, the most "visible proxy of the divine," was featured in countless stories through the ages. These stories were the way they passed down history and other information from generation to generation. They weren't necessarily true, but had meaning.

Here are some other interesting perspectives on the astrological origins of the Christ story, according to Acharya:

- *The sun of god was born of a virgin, which refers to the new or virgin moon or the constellation of Virgo;*
- *The sun is at its zenith at noon, and then is "most high;"*
- *The sun's birth is attended by three kings, the three stars on Orion's belt;*
- *The sun enters each sign of the zodiac at 30 degrees, i.e, at the "age" of 30;*
- *The sun's followers are the 12 signs of the zodiac, the 12 "disciples;"*
- *The sun "changes water into wine," by creating rain, which grows the grapes for wine;*
- *The sun "dies in winter," the vegetation dies, and is resurrected in the spring;*
- *The sun "walks on water," referring to its reflection;*
- *The sun rising in the morning is the "saviour of mankind;"*
- *The sun wears a corona, "crown of thorns," or halo; and*
- *The sun is the "light of the world."*

All these sayings and images were so familiar from the religion taught to me as a child. Jesus walked on the water, was most high, had twelve disciples, was the light

Historical or Mythical?

of the world, and the savior of mankind. I wasn't an expert in astrology, the Bible, or any of this, but all of a sudden it made sense. Yes, they were *just stories!* Acharya goes into great and extensive detail about the Bible stories, and their origins in pagan and mythological beliefs based on study of the heavens.

But I still wanted to learn more, so I got another book on Peter's list and read it. *The Jesus Puzzle: Did Christianity Begin with a Mythical Christ?* by Earl Doherty. Just from the title, I could tell where he was headed with this book, too. According to the jacket, he is "an historian and classic scholar living in Canada." He is published in several magazines and journals and, "Since 1996 gained world-wide attention with his groundbreaking website" www.jesuspuzzle.com. I checked it out, and there is a lot of substance there; the kind of substantial information and research I was coming to expect from these authors.

Doherty begins his book with what he calls the "Twelve Pieces of the Jesus Puzzle." I'll give you just a few:

1. *The Jesus story is not found in any Christian writings earlier than the Gospels, the first of which was in the late first century.*
2. *No non-Christian record of Jesus before the second century.*
3. *The early epistles, such as Paul, only speak of Christ as a spiritual being, not as an historical person.*
4. *All the gospels come from one source; whoever wrote the Gospel of Mark. The Acts of the Apostles are purely myth.*
5. *The Gospels are not historical accounts, but constructed by "midrash," the Jewish re-working of old Biblical passages to reflect new beliefs.*

I'll say Doherty's book is arcane; I had a hard time following it. But the same concept kept being hammered; these old Bible stories were *mythical*. He questions the historical nature of Moses, Abraham, Confucius, and others, comparing them to William Tell, the reported founder of the Swiss Federation, but who also likely never existed. Dougherty acknowledges that modern humans might have trouble understanding the mythical thinking common throughout the ages, but contends there was just too much of it to give any credence to the existence of these figures.

I then turned to the font of all knowledge – the Internet – for more information. There are a lot of people out there writing about this issue. Who knew? A guy named

Frank Zindler wrote a story at www.atheists.org called *Did Jesus Exist?* In a well-researched article, he cites over forty known writers within a century of the time Jesus allegedly lived, but *not one of them mentioned Jesus Christ*. Some of the writers were Plutarch, Pliny the Younger, Quintilian, Ptolemy, and others I hadn't heard of, though Acharya cited some of them. I don't suppose there were many published writers back then; the printing press wasn't even invented until about 1440 CE. (I finally learned what that means; "CE" stands for "Common Era," what we used to call "A.D.," or "Anno Domini," i.e., after the year of the Lord. BCE is "Before the Common Era," what we used to call "B.C.," or "Before Christ." It's the modern way of referring to years.)

There were two writers who apparently are generally cited as referring to Jesus Christ: Josephus, a Pharisee, and Tacitus, a pagan, according to Zindler. He explained, as had Freke and Gandy, Acharya, and Dougherty, that their writings allegedly about Jesus were either forgeries or completely unreliable. You would think, if Jesus was drawing multitudes to hear him speak and see him perform miracles, that someone other than the authors of the Gospels would have written about him.

And no one even knows who wrote the Gospels! This is according to another writer, Marshall Gauvin in an article titled *Did Jesus Really Live?* http://www.infidels.org/library/historical/marshall_gauvin/did_jesus_really_live.html. He says the first Gospel of Mark was written in about 70 CE, Luke and Matthew (which were largely taken from Mark) in about 110 and 130 CE, respectively, and John in about 140 CE. The first historical mention of the Gospels was by a Christian Father, Irenaeus, in about 190 CE. The Gospels were all written in Greek, too, not Aramaic, the language in Palestine at the time. Those original Gospels no longer exist, so no one really knows what they said; the oldest available manuscripts of the New Testament were from about the Fifth Century.

The people back then didn't know the world was round or understand any of the scientific facts we know about disease, the atmosphere, etc. Gauvin says there's no proof that the city of Nazareth even existed at the alleged time of Jesus; and we'd always called him "Jesus of Nazareth." Gauvin says there's no reliable historical evidence that Paul existed either, though if he did, his writings never talk about a historical Jesus or his life. Rather, the Epistles of Paul only refer to "the Christ," so that Paul, if he existed, was likely discussing the inner Christ that arises in the context of the old Mystery religions. Gauvin says, "The Jesus Christ of the Gospels could not possibly have been a real person."

I found many sources and books, even beyond Peter's list, that said the same thing – no historical Jesus.

Even that modern bastion of current information, *Newsweek*, as reported in an article on December 17, 2012, titled, "The Myths of Jesus," by Bart D. Ehrman, pretty much called it out:

> As Christians around the world now prepare to celebrate Jesus' birth, it is worth considering that much of the "common knowledge" about the babe in Bethlehem cannot be found in any scriptural authority, but is either a modern myth or based on Gospel accounts from outside of Christian Scripture. Some obvious examples; nowhere does the Bible indicate what year Jesus came into the world, or that he was born on December 25; it does not place an ox and an ass in his manger; it does not say it was 3 (as opposed to 7 or 12) wise men who visited him. ... But Christians throughout the Middle Ages were rarely interested in historical accuracy; ...
>
> We have good documentation of Cesar Augustus, and there never was a census of his entire empire. Let alone one in which people had to register in their ancestral home. In this account Joseph and Mary need to register in Bethlehem (which is why Jesus was born there) because Joseph is descended from King David, who came from Bethlehem. But David lived a thousand years earlier. Is everyone in the entire Roman Empire returning to their ancestral home from a thousand years earlier? Imagine the massive migrations for this census. And no historian from the time thought it was worth mentioning? This is not a story based on historical fact.

I also found a movie, *The God Who Wasn't There*, by Brian Flemming. It's sort of a documentary, and uses clips from movies about the story of the life of Jesus. He goes through it all; virgin birth, Jesus teaching in the Temple at age twelve, then doing miracles, raising Lazarus from the dead, being crucified, then rising from the dead, and sitting at the right hand of the father. The movie also contends that Paul didn't know anything about a historical Jesus, Mary, Joseph, etc. It notes that allegorical religion was very popular back then.

He interviews a number Christians and, the funny thing is, they don't know anything about how Christianity began or how it spread. They didn't know about Horus, Osiris, or the other godmen. I hadn't either. Flemming raises the now familiar list of these mythical figures with the litany of similar features. He says that humans have been obsessed thousands of years back in history with blood sacrifices, so he asks if there's any wonder the Jesus myth involves a crucified savior whose blood they can drink and whose body they eat?

Then, he equates this lust for blood with Mel Gibon's movie, *The Passion of the Christ*, which he says is a favorite of Christians. There are hundreds of violent scenes from the movie that are absolutely gory. Why were the Christian viewers so taken with this bloodbath of a film?

What I remember hearing about the film was people saying, "It's amazing how much Jesus loved us and suffered for us. We have to strengthen our faith in him and resolve to love him more each day." That sounded like guilt to me. It was the same feeling the *Baltimore Catechism* tried to impose on me.

The movie also devotes a lot of attention to the "rapture." I didn't know much about the rapture when I was little; maybe Catholics weren't into it. I remember about Jesus coming back to judge the living and the dead at the end of the world, but I never it assumed it would be in my lifetime. Apparently, there are millions of Christians who *do* believe that Jesus will come back to earth in about twenty years or so!

Flemming admits he used to believe the same thing, as a fundamentalist Christian growing up. He'd been told the *only* chance to be saved from the fires of hell was by accepting Jesus Christ as his personal savior. He learned hell is an actual place people really do go if they don't accept Jesus. He was told Jesus would forgive anyone of anything, except not believing in the Holy Spirit; it's the one unforgiveable sin that results in damnation. If he doubted the Holy Spirit, he was afraid he'd go to hell. It's a scary thing to put on a kid.

I decided to read still another of the books on Peter's list, this time *The Pagan Christ; Recovering the Lost Light*, by Tom Harpur. Peter told me that he'd emailed Harpur, who also lived in Canada. He asked Tom, why don't people understand the mythical and allegorical nature of the ancient religious stories? Tom replied that, in Canada, many do. But for some reason, in the United States and other areas, such as Islamic countries, the Literalists hold sway. In his book, he echoes the understanding in the other works just described. There's nothing in the Bible about the life of Jesus, he says, that isn't found in earlier religious teaching. Apart from the Gospels and Epistles, he says there's no hard evidence for Jesus at all from the first century. The

Literalists not only hang on to the *literal word* of these writings, but through the centuries "did everything in their power, through forgery and other fraud, book burning, character assassination, and murder itself, to destroy the crucial evidence of what had happened." Yikes!

He focuses on similar concepts as Acharya and Gnostic understandings like Freke and Gandy. Harpur says religion is about the, "incarnation of the divine in the human. ... [W]hile the sun is the source of all that is in our solar system, it is also by its light alone that we are able to see and know everything that exists, and for that reason, it was a natural symbol in antiquity for the ultimate being, for God."

Ancient religions were solar and lunar based, he says. "It is called sun worship, but it was much more spiritual than this suggests or sounds." The radiant sun gods of ancient times, from Hercules to Horus to Jesus, represented both the divine and the human simultaneously. Science tells us that we're all, "literally shot through with particles of the sun." When the ancient people extolled the virtues of the sun, they saw it as symbolic of their *own* divinity as well. They saw the one truth, "that we are embodied souls or spirits destined, through the love of God, for eternity. This is a spirituality full of hope and power. ... The sole and crucial difference between the ancient myths and Christianity is that Christianity eventually concentrated this universal concept into a single historical figure."

Harpur explained the similarities of Jesus with other avatars and saviors. Harpur said the ancient people who told the stories of the godmen didn't believe them. They believed the *principles* behind them. In fact, he says, the *only* way to adequately express the divine nature of our world, the "workings of the spirit in the human heart or the cosmos," is through "myth, allegory, imagery, parable, and metaphor."

That was a new concept for me. Really, was that true? The *only way* to describe the divinity of creation is through story? The nature of God and the universe had been taught to me in stories of ancient figures, and we believed them to be *literally true*. Now, it seemed the *meaning* behind them was the key. Thinking about it, in what ways have I felt closest to the spirit of creation? Through music. Art. Movies. These seemed like efforts to capture some element of "god" in human terms. Harpur says the plays of William Shakespeare are true, of course, when "it comes to matters of the human heart and spirit." But they are clearly fiction from a historical perspective. Good point!

As humans spend their lives crafting poems, films, books, songs, paintings, etc., are those efforts just to describe divinity? If the religious myths are attempts to understand God, then might they be platforms for perspective? So, is *my* challenge to create an individual allegory so that I can understand creation? I felt I needed to ask Peter more about this myth stuff. If all those tales in the Bible meant something, I wanted Peter to explain them so I could understand their meaning.

So, I talked with Peter and asked him what he thought about some of the old parables from the Bible. He looks at all of the Biblical tales, and for that matter the stories of all religions, as allegories. They are mythical stories that are a way of explaining spiritual concepts. Peter said that Christianity started by the sharing of mythical stories but then, for a number of reasons, the stories became thought of as having actually happened. While the original allegories can help reveal important precepts, making the stories historical destroys their ability to teach. He said the original myths were not for everyone, only those who were interested in esoteric spiritual meanings.

For example, the story of the Israelites fleeing Egypt is really about all of us, and our being tied into material and unpleasant things in our own lives. We can escape them by leaving behind the thoughts that bind us. The story of Job was about us, Peter said, and how we deal with our daily struggles. There was no person named Job who was an historical figure. The story of the virgin birth of Jesus was about how we can have new life within us.

Remember the story I mentioned from the Baltimore Catechism out of the Bible where Jesus was in the fishing boat with his disciples? How he calmed the storm? The lesson for us then was how powerful he was; he could stop a storm that had the power of an atomic bomb. Peter said that's really just about how a person can quell their inner troubles by remaining calm and peaceful.

Of course, the biggest story of all is the death and resurrection of Jesus Christ. Mel Gibson made the passion of the Christ into a very real, historical, disturbing image. Peter says the resurrection is only about how we all have the Christ spirit of love and life inside us, and how we can see ourselves united with God in a new way. It's a present experience of inclusiveness, about loving everyone, not a past experience. If we make these stories about the past or future, we deny the present and destroy the true meaning of life; we make our lives separate and exclusive of others. Everyone has the Christ spirit inside them; the people of all religions have the spark of life in them. Peter said no one should take any of the religious stories of the Bible, the Torah, the Bhagavad-Gita, Book of Mormon, or any of the old religious texts literally.

This put faith and religion in a new light for me. The references to the Gnostics by Freke and Gandy made more sense now. The Gnostics used stories as allegories to demonstrate in a more understandable way how a human could come to better understand the spirit and divinity within.

At the same time, while I got what Peter was saying, these old ancient Biblical stories didn't really do it for me. I still felt – what was it? – a *chasm* between my conscious awareness and the inner life those allegories were trying to reveal. But I was beginning to wonder how people really believed them.

I have a cousin who lived near us growing up. She was very attractive, about my age, and a terrific figure (if you can say that about a cousin). She moved away years ago, got married, etc. I connected with her by email a few years ago, and we somehow conversed about faith. I must have been in my mid-life crisis (more on that later). I've always had a feeling that all animals had some sort of soul, though that concept couldn't be reconciled with my religion.

Anyway, in one of our metaphysical email exchanges, I asked, "Don't you think a spider has a soul?" She replied, "A spider is a mean, soulless, gutless insect that no way has a soul." I thought that was a bit harsh, but she could sense my "faith" was shaky. She recommended a book that she said proved Jesus Christ was God. I tend to be an eclectic reader, and I want to find the truth, so I read it. Or at least I tried.

The name was *Evidence that Demands a Verdict*, by Josh McDowell. The book is a text that studiously references Bible passages and claims all the prophecies of the Bible had come true. He said that the Bible was the *inerrant word of God*. He said the Christian faith is objective, and the stories of Jesus from writers who "either wrote as eyewitnesses of the events they described or recorded eyewitness firsthand accounts of the events."

Based on what I read in the other books noted above, with the passage of decades separating the Gospels, this didn't ring true for me. McDowell says:

> *They certainly knew the difference between myth, legend, and reality. A professor of a world literature class in which I was speaking asked the question, 'What do you think of Greek mythology?' I answered with another question, "Do you mean, were the events of the life of Jesus, the resurrection, virgin birth, etc. just myth?' He said, 'Yes.' I replied that there was one obvious difference between these things applied to Christ and these things applied to Greek mythology that is usually overlooked. The similar events, such as the resurrection, etc., of Greek mythology, were not applied to real, flesh and blood indi-*

> viduals, but rather to mythological characters. But when it comes to Christianity, these events are attached to a person the writers knew in time-space dimension history, the historic Jesus of Nazareth whom they knew personally.

I thought, Josh, no. It's fact that the Gospel writers did *not* know Jesus personally.

I read as much of the book as I could, but it was *not* convincing. Attempting to reconcile the inconsistent passages of the Bible with one another to prove the existence of Jesus doesn't work. Two particular sections of the book undermined his "evidence" for me. In a portion titled, "He Changed My Life," McDowell says, "Jesus Christ is alive. The fact that I'm alive and doing the things I do is evidence that Jesus Christ is raised from the dead."

Huh? No way; what did that mean? He tells the story of how he'd rejected religion, but felt bad; he wasn't happy. But when he accepted Jesus Christ into his life, he became happy.

I was happy for him, but seriously, this was *not* evidence. When he placed all his trust in Jesus Christ, he was able to let go of his unhappiness. Maybe happiness had something to do with letting go?

The other part of the book that struck me was in the back, after the index, entitled, "Have You Heard of the Four Spiritual Laws?" Frankly, I hadn't. I knew of the three persons in one God, and the Ten Commandments, but not the four spiritual laws that govern our relationship with God. Here they are:

> *Law One – God loves you, and offers a wonderful plan for your life.* (Nice.)

> *Law Two – Man is* **Sinful** *and* **Separated** *from God. Therefore, he cannot know and experience God's love and plan for his life.* (Emphasis not added.)

> *Law Three – Jesus Christ is God's* **only** *provision for Man's Sin. Through him you can know and experience God's love and plan for your life.* (Again, emphasis not added.)

> *Law Four – We must individually* **receive** *Jesus Christ as savior and*

lord; then we can know and experience God's love and plan for our lives. (Ditto.)

Dang, that sure sounds like my old *Baltimore Catechism*. And then there was this diagram, which is meant to show God and man were separate, and the *only way* to bridge the gulf between God and man was through Jesus Christ.

<u>God</u>
I
Jesus
I
Man

The visual representation was so clear. In this view of life, just like the Catholic view, humans are sinful, unworthy, *and separate* from God. Only through Jesus can they have a chance of being saved from everlasting separation. That was the ultimate fear, that separation from God; that separation from the ultimate love. It seemed to drive everything.

I didn't believe a spider had no soul. I couldn't believe the only way to know or be with god was through Jesus Christ. God would *not* have condemned human beings to eternal separation just because they'd never heard of Jesus. And, as I'd been learning, *what if there was no Jesus?* What would humans do or think then? Would they be screwed with no "faith" to fall back on? Religion must do some good, right?

Three | Religion, Atheism, the Virus and Metaphysical Naturalism

My son, when he was about 12 or 13, said to me, "Going to church makes me feel bad." I'd been puzzled by that. It had never made me feel bad. I liked the music. I liked being on my knees during the quiet part of Mass and just thinking. It was peaceful; I thought about deeper things than normal.

I hadn't focused on the separation part of it. I realized I was a sinner and all that, but felt I was going to be OK. My parents loved me and God must have loved me too (whatever that meant). I thought religion was good. Everyone determined what was morally correct based on religious tenets. That seemed to make sense. Religion was the core of goodness in our lives. Right?

Sure, religions do a lot of wonderful things. We always think of the missionaries helping the poor people, taking care of the lepers, and building wells in African countries. Religions give a lot of money to the poor, run shelters, and take care of young pregnant women. According to the book *Religious Faith and Charitable Giving*, by Arthur C. Brooks, religious people give more to charity than secular folks, and even give more to nonreligious charities. So, they're pretty generous people.

Pretty happy, too. Studies show that religious people tend to be happier than others. For example, a study by Professor Andrew Clark of the Paris School of Economics and Dr. Orsolya Lelkes of the European Centre for Social Welfare Policy and Research, reported in the *BBC News* on March 18, 2008, found that life satisfaction seems to be higher among the religious population (http://news.bbc.co.uk/2/hi/health/7302609.stm). Professor Clark said: "We originally started the research to work out why some European countries had more generous unemployment benefits than others, but our analysis suggested that religious people suffered less psychological harm from unemployment than the non-religious." The authors of the study acknowledged that other factors could be at play, such as family life and community.

But in another study, twice as many religious people say that they are very happy with their lives, while the secular are twice as likely to say that they feel like failures. In the article in *Slate* on November 7, 2008, "Does Religion Make You Nice? Does atheism make you mean?" Paul Bloom argues that *community* is the most important aspect, "rather than a belief in constant surveillance by a higher power. Humans are social beings, and we are happier, and better, when connected to others."

I remember being happy as a kid. We played sports in the Catholic Youth Organization, hung out at the playground, and went to dances. Fifty kids in each classroom made us rather chummy. There seemed to be a general respect for religious morals and the clergy. I'm not going to dwell here on the positive aspects of religion. Certainly most religious people think religion is good. I did too. But is that an accurate assessment?

Not according to Freke and Gandy. In *The Jesus Mysteries; Was the "Original Jesus a Pagan God?*, they say that "Literalist" religion is responsible for most of the great wars and atrocities in the history of humans on the planet.

> *Let's not forget that the Catholic Church created the Inquisition in order to ethnically cleanse Spain of hundreds of thousands of Muslims and Jews, which it did with unimaginable brutality. ... When the Fascists and Nazis started persecuting the Jews by humiliating them, depriving them of all rights, herding them into ghettos, killing and burning them, they were only doing what the Church had done for centuries. ... Literalist religion is the greatest threat to world peace in the twenty-first century.*

OMG! Who is taught about the Inquisition anymore? Could it possibly be true that the *Christians* in Italy and Germany were responsible for the Holocaust? Is religion – which always seemed to be based on love – really the biggest threat to world peace?

I had to think about this. I couldn't conceive of my Catholic friends and me killing and torturing Jewish people, or *any* people. But then again, as a ten-year-old boy I ran through the alleys near Blessed Sacrament School, blowing away make-believe Krauts and Japs.

Religion, Atheism, the Virus and Metaphysical Naturalism

I've often been led on my life journey by happenstance. It occurs often with books. I'll be browsing in an airport store, and a book will call out to me. The shout-out this time came from a book titled *The End of Faith*, by Sam Harris. In the first chapter, "Reason in Exile," he says:

> *A glance at history, or at the pages of any newspaper, reveals that ideas which divide one group of human beings from another, only to unite them in slaughter, generally have their roots in religion. ... Our situation is this: most of the people in this world believe that the Creator of the universe has written a book.*

Never thought of it that way. People do claim, "the Bible is the inerrant word of God." We assumed the Bible was divinely inspired. I guess that means God wrote it. If you added up all the Christians, Muslims, and Jews, they'd all say that God was responsible for writing the Bible, Koran, and Torah, respectively. Harris says they each make "an exclusive claim as to infallibility." If they were all about love, that would seem to be something infallible, but that's not what Harris means. "The central tenet of every religious tradition is that all others are mere repositories of error, or, at best, dangerously incomplete. Intolerance is thus intrinsic to every creed."

The Inquisition was about intolerance, big time. There were actually several; the medieval Inquisition in the 12th and 13 centuries, the Spanish Inquisition from 1478–1834, the Portuguese Inquisition from 1536–1821, and the Roman Inquisition from 1542-1860. That wasn't too long ago. It's not known how many people were killed in the Inquisitions, but certainly many, many thousands. But did you know the Inquisitions were all ordered *by the Popes* to fight heresy, which had been punishable by death in the Catholic Church since the 5th Century? Pope Gregory IX was the reported "Father of the Inquisition," but Pope Innocent III (cool name, huh?) followed him, and had inquisitors who answered directly to him. Apparently, the Church confiscated the property of those who were prosecuted, which is one way it became wealthy.

And this actually could be traced to the Bible. Deuteronomy 13: 7-11 requires the killing of those who would serve other gods:

> *If your brother, the son of your father or of your mother, or your son or daughter, ... tries to secretly seduce you, saying, "Let us go and*

> serve other gods," ... you must kill him .... You must stone him to
> death since he has tried to divert you from Yahweh your God.

John 15:6 says:

> If anyone does not abide in me, he shall be cast outside as the branch
> and whither; and they shall gather them up and cast them into the
> fire, and they shall burn.

In Luke 19:27, Jesus says:

> Those enemies of mine who did not want me to be king over them,
> bring them here and kill them in front of me.

Sam Harris explains what a person might expect if maintaining his or her innocence during the Biblically inspired trials and witch hunts of the Inquisition:

> You may be imprisoned for months or years at a time, repeatedly
> beaten and starved, or stretched upon the rack. Thumbscrews may
> be applied, or toe screws, or a pear-shaped vise may be inserted into
> your mouth, vagina, or anus, and forced open until your misery
> admits of no possible increase. You may be hoisted to the ceiling on a
> strappado (with your arms bound behind your back and attached to
> a pulley, and weights tied to your feed, dislocating your shoulders). To
> this torment squassation might be added, which, being often sufficient
> to cause your death, may yet spare you the agony of the stake.
> If you are unlucky enough to be in Spain, where judicial torture had
> achieved a transcendent level of cruelty, you may be placed in the
> "Spanish Chair:" a throne of iron, complete with iron stocks to secure
> your neck and limbs. In the interest of saving your soul, a coal brazier
> will be placed beneath your bare feet, slowly roasting them. Because
> the strain of heresy runs deep, your flesh will be continually larded
> with fat to keep it from burning too quickly. Or you may be bound to
> a bench with a cauldron filled with mice placed upside-down upon
> your bare abdomen. With the requisite application of heat to the iron,
> the mice will begin to burrow into your belly in search of an exit.

That doesn't sound like something a church should do. I mean, I thought three Hail Mary's and three Our Father's was a lot of punishment! I never got thumbscrews as a penance, even after I stole that money from my father's wallet. Seriously, did Christians really do that to people? Surfing around on the Internet explained other things, such as, "The Catholic Church learned a human being could live until the skin was peeled down to the waist when skinned alive. Often the rippers were heated to red hot and used on women's breast and in the genitalia of both sexes." A picture of a "breast ripper" was on the same page. http://www.exposingchristianity.com/Inquisition.html. Other devices used in the Inquisition included, the rack, skull crusher, the wheel, and, of course water torture like we did to some of our prisoners in the "war on terror."

You've heard that Galileo, the Italian astronomer, had made the outrageous case that the Earth was *not* at the center of the universe. The Catholic inquisitors tried him for heresy and he was forced to recant, but lived his life under house arrest. (Thankfully, the Church officially pardoned him in 1993.) Between the years 1623 and 1633, the Jesuits and other inquisitors in Wurzburg, Germany were said to have killed some 900 witches, most burned at the stake, for having intercourse with the devil, including many children as young as two or three.

OK, OK, this was getting depressing. What causes all this insanity? Surely, all church-goin' types aren't this crazy?

Harris also claims the Nazis were agents of religion. He says that though their atrocities were generally secular, the Nazis hated the "impure elements – homosexuals, invalids, Gypsies, and, above all, Jews – [who they believed] posed a threat to the fatherland." That threat, says Harris, "was a direct inheritance from medieval Christianity."

I've never gotten into the holocaust; never been to the holocaust museum in downtown DC. I think it's terrible what happened to the Jews during World War II. I feel sadness about that, and about the Inquisition, and about all of the mean and nasty things humans have done throughout history. I only went on here about the atrocities of the Inquisition, because the level of violence associated with my religion surprised me. I believe it's important to look at one's faith directly and see what it really means. I didn't do that most of my life, but I do now. Searching for the kernel of truth that can help me understand life and figure out how to deal with it. Understanding the infirmities of religion should help determine whether one continues to believe. So, perspectives like those of the writers I've mentioned help me open my eyes and think for myself.

Harris says, "the history of Christianity is principally a story of mankind's misery and ignorance rather than of its requited love of God." He points out that 44 percent of American are either certain or believe that Jesus will return to earth in the next fifty years. He says, "Only 28 percent of Americans believe in evolution; 72 percent believe in angels."

Harris contends, however, that we face even more serious problems today from Islam, another religion of the dark ages whose proponents now carry modern weapons. He explains that, while many believe *jihad* (the struggle or striving) is a movement in defense of Islam, it's really a call to conquest. He says that devout Muslims, "have no doubt about the reality of paradise or about the efficacy of martyrdom as a means of getting there." Muslims believe, as many Christians do about the Bible, that the book their creator wrote, the Koran, is the inerrant word of God.

The separation and distinction of Muslim believers from others is apparent in the religion. Non-believers are condemned in the Koran, and there is no ambiguity about it, he says. "The only future devout Muslims can envisage – *as Muslims* – is one in which all infidels have been converted to Islam, subjugated, or killed," says Harris. In fact, converting from Islam to another religion *is punishable by death.*

Harris then goes on in five pages to recite the exact translated quotes from the Koran that basically say, "Kill all the non-Muslims." I thought the Bible was pretty outrageous in calling for sinners and nonbelievers to be murdered, but the Koran could not be clearer. Harris says there's no choice ultimately for Muslims if they live by the Koran; they must believe that everyone who's not a believer in the Islamic God Allah should be killed or damned forever. Here are a few sample quotes from the Koran, courtesy of Sam Harris:

> *God is the enemy of the unbelievers. ... Slay them wherever you find them. Fight against them until idolatry is no more and God's religion reigns supreme. ... As for the unbelievers, neither their riches nor their children will in the least save them from God's judgment. They shall become fuel for the fire. ... We will put terrors into the hearts of the unbelievers. ... If you should die or be slain in the cause of God, God's forgiveness and His mercy would surely be better than all the riches they amass. ... Do not be deceived by the fortunes of the unbe-*

> lievers in the land. Their prosperity is brief. Hell shall be their home, a dismal resting place. ... Those that disbelieve and deny our revelations shall become the inmates of Hell. ... They shall drink scalding water and be sternly punished for their unbelief.

According to Harris, "On almost every page, the Koran instructs observant Muslims to despise nonbelievers. Islam, more than any other religion human beings have devised, has all the makings of a thoroughgoing cult of death."

So, my guess was Sam Harris mainly was condemning religions as having no value for society. It's true, he does contend it's a major problem that so many humans are, "motivated not by what we know but by what we are content merely to imagine. Many are still eager to sacrifice happiness, compassion, and justice in this world for a fantasy of a world to come."

He surprised me though, by making the argument that one of the biggest problems we have is with the *moderates* who think all people should be able to believe what they want; that we owe deference to faith, in spite of the fact that much about religion is shibboleth. He contends religious faith is not based on reason, but on mirage. One of the central beliefs of his book is:

> [T]hat religious moderates are themselves the bearers of a terrible dogma; they imagine that the path to peace will be paved once each of us has learned to respect the unjustified beliefs of others. ... [T]he very idea of religious tolerance – born of the notion that every human being should be able to believe whatever he wants about God – is one of the principal forces driving us toward the abyss. The problem that religious moderation poses for all of us is that it does not permit anything very critical to be said about religious literalism.

Isn't it important, though, to have a basis in religion? When my three children were small and we were considering schools for them, their mother and I agreed it would be best to bring them up Catholic, as we had been. In fact, our oldest son Jay went to kindergarten at – you guessed it – Blessed Sacrament, the same school I had gone to decades before. There weren't fifty kids in a class, closer to fifteen or twenty. (No one could even imagine today sending their children to a school that had more than that; it would be child abuse.)

Jay learned to read in kindergarten at good old BS, because they had dedicated teachers and the classes were reasonably disciplined. There weren't many nuns there anymore, and most didn't wear those old Sally Field Flying Nun habits. The Sisters of the Holy Cross at Blessed Sacrament wore black gowns with white halos around their heads and white bib-like cowls over their chests.

You know, that was a funny thing about nuns. When I was growing up and learning about the birds and the bees, we boys noticed that all the nuns looked flat chested. Other women didn't look that way; why nuns? I never saw a stacked nun. So, it made sense when someone said they had to cut off their breasts before becoming nuns. You know, giving them up for God.

Seriously, I believed that. As kids, we also believed in Santa Claus and the Easter Bunny. Certainly, there was a Tooth Fairy who came around and took the teeth we lost. I mean, *who didn't believe in the Tooth Fairy?* When you went to bed, the tooth was carefully placed right there under your pillow, and then in the morning, it was gone and a quarter would be in its place! What more proof did one need than that?

Eventually, we came to understand those tales weren't true. But we kept believing in the Bible and Christ stories. We baptized the children as prescribed by the rules, you know, to be sure they could go to heaven. We watched them receive their First Holy Communion when they were about seven or eight. We enjoyed taking them to their weekly discipline lesson – Mass. All of our children, Jay, Anne, and Max (we had three under the age of four, like good Catholic parents), fell asleep on my shoulder as infants during Mass. We (sort of) sang the songs, kneeled and sat, stood and kneeled, and shook our pew neighbors' hands at the designated portion of the Mass. I believed it gave us strength.

We weren't conservative Catholics. We were fairly liberal. We believed in a woman's right to choose, for example. And we definitely believed in tolerance for other religions (even though the poor non-Christian bastards theologically speaking would be going to hell).

Now, this Sam Harris guy is saying that *we* were part of a major problem in the world. We didn't think of other faiths as wrong or bad. We accepted that they believed different things. Wouldn't condemning other religions violate the freedom of religion protected by the U.S. Constitution? Didn't freedom of speech also apply here? As long as you didn't shout "Fire!" in a crowded theater, didn't one have the right to speak his or her mind?

Time magazine ran a photo story on February 23, 2009 called, *How the World*

Heals, illuminating a number of practices by religious believers. For example, there's a church north of Sante Fe, New Mexico that has *holy dirt*. This dirt is believed to have healing powers, so visitors kiss the ground, rub the dirt on their body, or eat it. If a sick family member can't travel to the church, rubbing the dirt on his or her picture is one technique to implement the dirt's healing properties.

I knew about Lourdes, the site in France where the Blessed Virgin Mary was said to have appeared to a girl in 1858. Water from a spring near the sighting is believed to have healing powers, so people drink it hoping for a miracle cure. (Hmm, you could probably make some mud with the Sante Fe dirt and it would be pretty powerful stuff, though I wouldn't want to eat it.)

Shamans in Siberia summon invisible spirits with a bodhran drum, so they can communicate with the spirits and "persuade them to provide a better life for humans." You know, I never disbelieved in spirits. There could be spirit energies out there (or somewhere). My father's spirit could be around me in the ether. People claim they can communicate with spirits, and maybe they can. Anything is possible. I haven't experienced any spirits, myself. I tend to not think about them, and for sure not when I go into the basement alone at night. I don't want to be scared of something that might not exist. I suppose that might also apply to hell.

In Lac, Albania, there's a holy rock called Shna Ndo. People touch it to heal sickness. There's a picture in *Time* of a little boy being held up to the rock. I suppose if you're sick, anything you can do to get better is fine. I do think people are very powerful and can often heal themselves.

The *Time* article also explained about some sects of Islam believing that certain verses of the Koran have healing powers, and there's a picture of a man apparently suffering from temporary blindness being prayed over. I wonder if the people who believe in the power of Koranic verses also would believe in the powers of the Lourdes water, and vice versa? Or do you think they each believe the others are crazy?

In Havana, Cuba, religious practices blending precepts from Native American and Roman Catholic beliefs lead to the worship of Santos, "an amalgam of African gods and Christian saints." The magazine has a picture of a High Priest about to sacrifice a rooster to help Fidel Castro overcome his illness.

So, I suppose you might say peoples' religious beliefs even today can be quite arcane. Another article in *Time* on November 26, 2007, titled, "One Day in America," by Nancy Gibbs, stated: "The vast majority of Americans believe in God, and more than

90% own a Bible, but only half can name a single Gospel, and 10% think Joan of Ark was Noah's wife." LOL!

According to an article by Sylvia Moreno and Adam Kilgore, "Late-Night Call Revealed Secret World," in the *Washington Post* on April 9, 2008, members of the Fundamentalist Church of Jesus Christ of Latter Day Saints, which broke away from the Mormon Church in 1890 when it banned polygamy, "believe that having multiple wives gives them access to the highest level in heaven, the Celestial Kingdom." The group was busted that month for polygamy after a complaint from a young girl. "Many of the female inhabitants [of the compound] did not know how to spell their last name and many could not state their birth date."

I'm not sure I buy the argument about getting to the highest level of heaven with multiple wives, but there sure does seem to be something unusual going on when it comes to religion and sex. I mean, most people like sex. It's inappropriate to take advantage of others for sex, especially if they're young and if people in religious orders are doing it. I'm sure you know about the sexual abuse scandal in the Catholic Church for decades in the Twentieth Century. Turns out there were nearly 5,000 priests accused of sexual abuse, mostly in the United States and mainly involving young boys.

An Associated Press report on May 20, 2009, by Shawn Pogatchnik, titled "Thousands Beaten, Raped in Irish Reform Schools," also disclosed a horrific situation in Ireland in which thirty thousand boys and girls considered to be "petty thieves, truants, or from dysfunctional families" were routinely beaten, molested, and raped" by the Christian Brothers. Sexual abuse was "endemic" said the report of the Irish Commission reviewing the matter and:

> *In some schools, a high level of ritualized beating was routine. Girls supervised by orders of nuns, chiefly the Sisters of Mercy, suffered much less sexual abuse but frequent assaults and humiliation designed to make them feel worthless. ... In some schools a high level of ritualized beating was routine. ... Girls were struck with implements designed to maximize pain and were struck on all parts of the body. ... Personal and family denigration was widespread.*

Sisters of Mercy; ironic, huh? Sounds like a mini-version of the Inquisition, with the religious leaders condemning others using tools of punishment far worse than the alleged "crimes." Similar abuse took place around the globe; in the United States, of course, but also in Belgium, where at least 13 children were driven to suicide according to Reuters, as reported in the *Washington Post* on September 11, 2010, in an article titled, "Belgium Report: Abuse was Rampant in Church."

I had a lot of interaction with priests and nuns growing up. I can honestly say that I never felt sexual pressure. The clergy was usually very stern, but respectful. The coaches on the football team would yell at you, but you knew they just wanted you to play better in a positive way. I assumed those vows of chastity were strictly observed by the religious.

Why did so many have sex problems? Maybe it was the devil in them? We certainly learned about all the temptations the devil could place in front of you, though I don't remember young boys ever being an example. The Mormons believe in Satan, according to an article by Annalise Shumway, "LDC believe Satan is Real," in a copy of *The Salt Lake Tribune* that I picked up on July 2, 2005.

> *For Mormons, the devil is a fallen angel. He rebelled against God and Christ in a pre-Earth life and persuaded other spirits in heaven to join with him. As fallen angels, they didn't get bodies. Now they want desperately to have one, so they try in every way they can to jump into someone else's skin. [The Mormons] believe that missionaries are particularly vulnerable. Thus, the army of men and women serving the church on [] missions are forbidden to swim, which some LDS members attribute to the belief that Satan controls the water.*

I don't remember hearing that Satan ruled the seas; I thought that was Neptune? In any event, the devil played a key role in the founding of the Mormon Church, in that its founder Joseph Smith had a wrestling match with the devil in 1838 in a grove of trees. Smith also was visited a number of times, beginning in 1823, by the angel Moroni, who reportedly still has possession of the golden plates that were the source of the Book of Mormon. That's why no one can find the plates, because Moroni has them.

It's not only the Christians who have such interesting stories and perspectives. I'm not much of a Hindu scholar, but an article titled "A Daily Round of Rituals for Boys Becoming Priests," by Ram Lakshmi, in the *Washington Post* on October 18, 2007, talks about a school in New Delhi where boys study was dictated by the 3,500 year old tradition of learning to become a Hindu priest. They rise at 4:45 am, these teenage males, where after eight years of focused training in classical Sanskrit they can "perform prayer rituals involving fire worship and rhythmic incantations. ... The boys follow a grueling routine of do's and don't's – they cook and eat only vegetarian food, wash their own clothes by hand, cannot call or visit their families, cannot take medicine except for a physical injury, and cannot watch television. Parents cannot bring any gifts." They cannot go out, they are kept away from the "world of illusions and desires. They lead pure, austere lives." Their teacher says, "The knowledge of the Vedas is the only education that cannot be erased. Everything else is impermanent in this world."

Well, there's another take on religious teaching. Some folks think the Bible is the inerrant word of God, but others think the Vedas are the only permanent teachings. I'm beginning to see a pattern here.

There's much in the news today about Islam, so I've cut out a number of articles. First, an article from January 25, 2008 in my trusty *Washington Post*, by Nora Boustany, "Afghan Reporter's Death Sentence Draws Wide Condemnation." A reporter for a daily newspaper and a journalism student at Balkh University in Mazare, Afghanistan was sentenced to death in a five minute proceeding in front of three judges, but with no lawyer of his own. What was his crime? "His alleged offense was distributing to classmates a report, printed from a Web site, commenting on a Muslim woman's right to multiple marriages. The article, written in Farsi ... questioned why men are allowed to have four spouses in Islam while women are denied the same right."

Sounds like a good question to me. They really sentenced him to death for handing out an article from the Internet? Sounds like the Inquisition, again. The poor young fellow was tortured so that he would confess. His sentence was later commuted to 20 years, and finally he was secretly pardoned by Afghani President Hamid Karzai and allowed to leave the country. I sure hope he learned his lesson about downloading things off the web.

He got off easy. Ask the couple who tried to elope in Afghanistan; the Taliban caught them and executed them in public. The couple, reportedly in either their teens or early twenties, were shot dead by a firing squad using AK-47's in front of a mosque for their actions, which were considered an "insult to Islam." The girl apparently had been unhappy about an arranged marriage. Islamic religious leaders officially sanctioned the killings. Stuart Whatley, "Taliban Execute Eloping Afghan Couple," *Huffington Post*, April 14, 2009. I don't understand why killing people like that could *ever* be countenanced by a religion.

Another article from the *Washington Post* on October 27, 2009, was titled "Saudi

King Abdullah Blocks Flogging of Female Journalist Yami," by Abeer Allam. Apparently, Rozana al-Yami, age 22, was "sentenced to 60 lashes by a Jiddah court for working for an unlicensed [television] network." Whoa; you can get whipped for that? "The network, known as LBC, has generated controversy in Saudi Arabia because it recently aired a program that featured discussion about sex." What is it with religion and sex? The network was from Lebanon and the show was named, 'Bold Red Line' in which a Saudi man, "boasted of his sexual exploits. ... Mazen Abdul Jawad, the man in the LBC show, was sentenced ... to five years in prison and 1,000 lashes for promoting debauchery. Three friends who appeared on the show with him were given two-year terms and 300 lashes each, while a cameraman was sentenced to two months in jail and 300 lashes."

An article appearing in the *Huffington Post* on September 14, 2009, "Indonesia: Adulterers Can Now Be Stoned To Death in Aceh Province," by Fakhrurradzie Gade, told how a new law was adopted there in accordance with strict sharia law permitting adulterers to be stoned to death. Of course, I'd heard of people in the ancient world being stoned to death, *in the Bible*. I didn't know how they actually did that, until I saw a trailer for a movie about a stoning. They generally don't just smash the person's head with a big stone; they wrap the person in a white sheet, put them in a hole in the ground waist deep, and then throw stones at them.

In November 2009, a Somali man was stoned to death by members of the group Hizbul Islam for adultery, with hundreds of villagers forced to watch. I saw pictures on the *Huffington Post* from December 15, 2009: http://www.huffingtonpost.com/2009/12/15/somali-man-stoned-to-deat_n_392503.html. The man was buried half way in the dirt, but they didn't cover him with a sheet. Then masked men threw white rocks at him, until they dragged him from the hole bloody and dead. The girl with whom he had the alleged adulterous affair received only 100 lashes, because she wasn't married at the time (huh?). Does that mean they would have stoned Tiger Woods to death and whipped all those women?

Stoning actually is still permitted in Afghanistan, Pakistan, Saudi Arabia, the United Arab Emirates, and parts of Nigeria. They certainly seem judgmental over there. In the United States, criminals aren't put to death until they go through years of appeals in the courts, and then they are killed by lethal injection or the electric chair, which seem much more humane than stoning (?).

In Pakistan, according to a *Washington Post* article, "Pakistani Cleric Calls for Islamic Law Nationwide," by Pamela Constable, on April 18, 2009, a Muslim cleric has been urging followers to adopt more dogmatic Islamic laws. Maulana Abdul Aziz has "advocated the strict separation of men and women in accordance with rules that he said are set forth in Islamic law. He once issued a fatwa, or religious edict, against a female government official for publicly hugging a man who was not her husband." He and his followers supported the Taliban's rise to power in the region. In fact, "a growing number of religious groups in Pakistan seek to make sharia the exclusive form of national law, asserting that it provides swifter justice than government courts and protects public morality from vulgar Western influences."

Seriously. Vulgar Western influences? Excuse me, I don't think I've ever heard of a stoning here in Maryland. I agree with Sam Harris on this. Some religious beliefs are *not* conducive to a peaceful society. Imagine how you'd feel if you could get stoned to death for hugging a neighbor. Man, that's *some kind of fear* to deal with all the time.

In that *Post* article, Noor Mohammed, 15, recounts his allegiance to Aziz and celebrates how he, "will lead us on a path to glory and the spreading of an Islamic way across the country. ... I am ready to lay down every sacrifice for sharia, including martyrdom."

OK, that's scary. A fifteen-year-old kid is ready to be killed for the right to stone people to death. And we're supposed to accept these "beliefs?" Which of these religious precepts makes the most, or any, sense?

Did you know the Inquisition is alive and well today in ... Papua New Guinea? A story in the Australiasia *Independent News* on May 8, 2009, began:

> *Nearly all of the residents of Koge watched as Julianna Gene and Kopaku Konia were dragged from their homes, to be hung from trees and tortured for several hours with bush knives. No one came forward to help. In the eyes of the villagers, the women were witches. They deserved to die. "They used their powers to bewitch a man to death," said Kindsley Sineman, a community leader. "We had to get rid of them, as they could have killed others. We had to protect our village."*
>
> *Fear of the supernatural and the stigma of being branded a witch is so great that around 30 of the victims' relatives were chased out of the village ... . Although there are no official statistics on sorcery killings, more than 50 were reported to the police in just two Highland*

Religion, Atheism, the Virus and Metaphysical Naturalism

> *provinces last year. ... Belief in black magic is so ingrained that the government legally recognizes sorcery, under the 1976 Sorcery Act.*

This ain't Harry Potter. Those people really believe that. Just like they believed in the righteousness of the Inquisition back in the Middle Ages.

The Pew Forum reported on December 17, 2009, on a new study showing that *two-thirds of the world* lives under religious limits:

> *Global Restrictions on Religion, a new study by the Pew Research Center's Forum on Religion & Public Life, finds that 64 nations - about one-third of the countries in the world - have high or very high restrictions on religion. But because some of the most restrictive countries are very populous, nearly 70 percent of the world's 6.8 billion people live in countries with high restrictions on religion, the brunt of which often falls on religious minorities. Some restrictions result from government actions, policies, and laws. Others result from hostile acts by private individuals, organizations, and social groups. The highest overall levels of restrictions are found in countries such as Saudi Arabia, Pakistan, and Iran, where both the government and society at large impose numerous limits on religious beliefs and practices.* http://pewforum.org/Government/Global-Restrictions-on-Religion.aspx.

If religion can be so problematic, what's the alternative? Is atheism any better? If everyone believed there was no God, no heaven or hell, and no life after death, could that make for a more peaceful society? Doesn't religion bring morality? I decided to check out some atheistic writings.

Richard Dawkins in *The God Delusion* describes an atheist as "somebody who believes there is nothing beyond the natural physical world, no *super*natural creative intelligence lurking behind the observable universe, no soul that outlasts the body and no miracles – except in the sense of natural phenomena that we don't yet understand." Well, that sounds reasonable.

I used to think atheists were craven lunatics, unprincipled chaps that just didn't understand how wonderful God was. I assumed they'd got off at the wrong stop. But Dawkins definition was somehow comforting. We don't have to worry what God is thinking about us all the time; we didn't have to be afraid of going to hell. Hmm.

Heaven is Everywhere

God, says Dawkins, in the sense of being a "superhuman, supernatural intelligence who deliberately designed and created the universe and everything in it, including us … is a delusion, … a pernicious delusion." The Gospel writers, he says, can't be trusted. They certainly never met Jesus, if he did exist, and simply "rehashed" the Old Testament so that he would meet all of the prophecies. He says, "The only difference between *The Da Vinci Code* and the gospels is that the gospels are ancient fiction and *The Da Vinci Code* is modern fiction."

Dawkins says, "The God of the Old Testament is arguably the most unpleasant character of all fiction: jealous and proud of it; a petty, unjust, unforgiving controlfreak; a vindictive, bloodthirsty, racist, infanticidal, genocidal, filicidal, pestilential, megalomaniacal, sadomasochistic, capriciously malevolent bully." He wonders if the people who believe in this God actually know what's written in the Bible.

For example, he says that in Leviticus chapter 20, the following offenses merit the death penalty: "cursing your parent; committing adultery; making love to your stepmother or your daughter-in-law; homosexuality; marrying a woman and her daughter; bestiality; … . You also get executed for working on the Sabbath; the point is made again and again throughout the Old Testament." And he gives the example of a man in Numbers 15 who God commands be stoned to death for picking up sticks on the Sabbath. I'da been stoned last Saturday in my yard.

Dawkins quotes Thomas Jefferson, who in writing to John Adams, said, "The day will come when the mystical veneration of Jesus, by the Supreme Being as his father, in the womb of a virgin, will be classed with the fable of the generation of Minerva in the brain of Jupiter."

Dawkins doesn't believe there's a supreme being, but recognizes the wonders of the world in a similar way that Albert Einstein does, who he quotes as saying, "I am a deeply religious nonbeliever." I guess people in some faiths would want to kill him for that. "The idea of a personal God is quite alien to me and seems even naïve," said Einstein. Dawkins says that Christianity and Islam, "teaches children that unquestioned faith is a virtue," which makes it easier for people to believe in a personal God and all sorts of other preposterous "memes" (Dawkins' term for an idea or behavior that spreads from person to person within a culture).

Other memes that support religions include: that we will survive our own death, martyrs will go to a wonderful paradise and enjoy seventy two virgins, heretics and blasphemers must be killed or ostracized, everyone must respect religious beliefs

Religion, Atheism, the Virus and Metaphysical Naturalism

more so than other beliefs, and we're not meant to understand "some weird things" such as the Holy Trinity and transubstantiation.

What I get from Dawkins is a cerebral, yet sometimes arcane, pooh-poohing of religious beliefs in an anthropomorphic supreme being. What I don't get is: *what should I believe in?* I want something to believe *in*.

Atheism can't just be *all* anti-god, I thought. So, I checked out some more of this thought train. I tried another famous atheist, Christopher Hitchens, and his book *God is Not Great; How Religion Poisons Everything*. He talks about:

> [F]our irreducible objections to religious faith: that it wholly misrepresents the origins of man and the cosmos, that because of this original error it manages to combine the maximum of servility with the maximum of solipsism [according to dictionary.com, the theory or view that the self is the only reality], that it is both the result and cause of dangerous sexual repression, and that it is ultimately grounded on wish-thinking.

I'm not sure I would've said it that way, or even if I'm smart enough to understand what he means by the second one, in particular. Both Dawkins and Hitchens are so intelligent I often don't understand what the hell they're talking about. But, as I study this issue, I'm coming to the conclusion that his first point is true. The idea that this incredibly complex beautiful universe, which we understand more every day, is run by some supreme guy who gets mad at you no longer makes sense to me. The religion of my youth begins to look downright sophomoric. Why do people gravitate to such thinking? And how did I think like that for so long? I don't know, but I want to find out.

His third point is interesting, and we've seen some of that in the examples I've noted above. But is religion both the *result* <u>and</u> *cause* of dangerous sexual repression? I can understand one might think religion causes such repression, but is religion *caused by* sexual repression? Why are people sexually repressed in the first place? Need to think about that too.

Finally, he says, religion is based in wishful thinking. I suppose that means a belief in a friendly big-brother sort of a God who'd be really cool and wonderful to hang out with for the rest of time. Maybe he means that people believe they're actually more special than other humans, or animals, and deserve a revered place in heaven.

He ascribes these faults at least in some part to the fact that:

Heaven is Everywhere

> *Religion comes from the period of human prehistory where nobody*
> *– not even the mighty Democritus who concluded that all matter*
> *was made from atoms – had the smallest idea what was going on.*
> *... Today, the least educated of my children knows much more about*
> *the natural order than any of the founders of religion, and one would*
> *like to think – though the connection is not a fully demonstrable one*
> *– that this is why they seem so uninterested in sending fellow humans*
> *to hell.*

I think back on Christopher Columbus, and how people thought he'd fall off the edge of the flat earth. I remember reading 18th Century Europeans thought washing was a terrible thing to do and was bad for you (or the Christians thought was immoral because of the nudity involved). Letting in the "ill humors" of the night air would make you sick. The "humoral" theory of disease was handed down from ancient Greeks and Egyptians and believed in Europe for centuries. The four humors were cold, hot, dray, and moist. Bleeding, blistering, purging, and induced vomiting were regular treatments up to the 19th Century. When George Washington came down with a bad cold in 1799, his physicians prescribed a bloodletting of over half a gallon. *A half a gallon of blood* was let out of cuts in his arms in attempts to cure him! He died, of course.

So, I take Hitchens' point. About two thousand years ago when the Gospels were written, how much do you think those folks knew about life? Enough to craft a religion that humans still believe today? I guess so.

Hitchens also talks about the ways that religion is immoral, and identifies several basic precepts that support this notion. He says religion presents a false picture of the world to the innocent and credulous, imposes harmful doctrines of blood sacrifice and eternal reward/punishment, and attempts to enforce impossible tasks and rules. And he concludes that teaching religious doctrine to boys and girls is essentially child abuse, primarily because it focuses on sin and warped sexual values engendering guilt that can last a lifetime.

Hitchens concludes: "Religion has run out of justifications. Thanks to the telescope and microscope, it no longer offers an explanation of anything important." What we need, he says, is a *New Enlightenment*, where scientific inquiry can lead everyone, not just the gifted few, to a more rationale worldview. The electronic revolution will help foment this change. He says that humans should finally be able to divorce sexual life from fear, disease, and tyranny, but only if, "we banish all religions from the discourse."

Another writer, who follows from the theories of those like Dawkins and Hitchens, is Darrel Ray. He's a really interesting guy. He'd grown up in a fundamentalist Christian household, but since changed his views and has a book published entitled, *The God Virus, How Religion Infects Our Lives and Culture*. He attempts to show how religious beliefs are experienced in everyday life, by demonstrating the "remarkable parallels between the propagation methods of some biological systems and the strategies of religion. ... Richard Dawkins and others have noted the similarities of religions to parasitic behavioral control in certain animals."

Ray, who is a psychologist, essentially analyzes religious behavior in people using the metaphor of viruses. He refers to these concepts as viruses, rather than memes, because he feels the concept is more understandable.

He looks to some examples of viruses and parasites in nature, like the rabies virus, which takes over the brain of the host without regard for its wellbeing and then dies. With religious people, particularly those who convert, "it is difficult to have a rational conversation about the irrational aspects of religion. It is as though something invaded the person and took over a part of his personality." Just like a virus.

Ray says religion has five basic properties like viruses, including the ability to:

1. *Infect people.*
2. *Create antibodies or defenses against other viruses (i.e., religions).*
3. *Take over certain mental AND physical functions and hide itself within the individual in such a way that it is not detectable by the individual.*
4. *Use specific methods for spreading the virus.*
5. *Program the host to replicate the virus.*

He proceeds to explain these points in an engaging and interesting way, but I will only focus on a few of the more relevant aspects (obviously, you can read the whole book yourself, or check out any of the sources noted here; this is all about learning).

He focuses on guilt and fear as central facets of religion. "Guilt has become one of the greatest tools of western religions. Of course, guilt existed long before Judaism, Christianity, or Islam, but these religions have developed highly effective, guilt inducing techniques. ... You are never good enough." Ray says without any shame:

> *From the time I was 11 years old through early adulthood, I prayed fervently for god to take away my urge to masturbate. Parents, Church, Sunday School teachers, even gym teachers, implied that masturbation was sinful and that you would go to hell for doing it. As much as I prayed, I did it even more. Even at church camp! God surely would send me to hell, if not kill me outright for masturbating at church camp! What is more, I was masturbating to the thought of the sexy preacher's daughter at the camp. That had to be a deadly sin.*
>
> *Not until I came across Masters and Johnson's Human Sexual Response (1966) and later Alex Comfort's The Joy of Sex (1972) did I realize that my religion was totally wrong about masturbation. As I quickly lost my guilt, I soon began asking, "What else could be wrong with my religion?"*

I guess this would be an example of what Dawkins and Hitchens were saying. Ray talks further about sexual issues with religion, noting the ways many religious folks are programmed just like cult members to believe homosexuality is a sin and that same sex relationships and premarital relationships are against God's law.

> *All major relations and most minor ones sanction women who express their sexuality in ways proscribed by religion. ... [S]exual control is a key religious strategy. ... The Catholic virus has ... celibacy, abstinence, no abortion, no contraceptives, Mary was a virgin, Jesus was not married and non-sexual ... . Other forms of control can be seen in dress, especially that of women. It is interesting that the habit of Catholic nuns resemble the burka of Islam. Both are symbolic of sexual control and surrender to a male figure – husband, father, pope, Jesus, bishop, etc.*

I have to admit, that's the way it seems to me. Ray also devotes attention to fear in religion:

> *A key strategy of most religions is fear. ... If you are involved with a deeply infected person, spouse, parent, child, neighbor, or friend, recognize that they are fundamentally afraid. They are experiencing a*

> *profound and unsettling fear that they have to deal with almost daily, like a disease that can only be managed, never cured. ... Fear of death keeps people close to religion. Constant focus on an afterlife creates a phobic condition – fear of death, hell or the unknown. Chronic fear of death is the condition of many people infected by the virus and can lead to neurosis, anxiety attacks, or depression. ... Death is a very important concept to the Christian, Jewish and Islamic god viruses. Much of the focus on death in these Scriptures relates to fear, terror, punishment, guilt, and damnation. Infected persons have had the notion 'death = possible damnation' drilled into their head since childhood."*

I look back at my Catholic upbringing and recall that guilt and fear, like the statements in the *Baltimore Catechism*. Where do they come from? What is the relationship between guilt and fear? The deep linking between sex and religion hadn't really occurred to me before. The male hierarchy seemed normal and natural. God was the father; my father was the head of the household. But in my youth, neither really had anything to do with sex. Sex was not in our lives at all, except every little boy somehow does seem to find out about masturbation (I can't exactly speak for little girls). When I became a young man, it can make you a little frustrated and angry. How come you just can't have sex whenever you want?

Ray also expounds on anger in religion. He says many religious sermons seethe with anger and fighting words. "Anger is an underlying condition of fundamentalism." Many of the preachers are angry about the secular world, and how religious people are treated. While many people call atheists angry, Ray doesn't see it that way. "Nonbelievers don't tend to burn people at the stake, persecute, fail to hire or discriminate based on belief. Religions that promise salvation from eternal damnation may tend to more easily hook fearful or worrying types of people. Fundamentalist churches seem to attract people who have a strong need for right/wrong, good/bad, black/white approaches, with little tolerance for ambiguity."

On the other hand, atheists and non-believers are proportionately underrepresented in prison versus believers, and constitute a very high percentage of scientists and highly educated folks. He concludes that religion also inhibits intelligence. He urges people to stand up for a rational view of the world and reject what he calls the immoral religious values and training that over the years have been "racist, sexist, insulting and cruel to other groups, and misleading if not downright wrong about human relationships, sexuality, science, education, other religions and many other things. … Religion provides little or no guidance that a rational person could not determine without religion."

I felt somewhat overwhelmed by these multi-pronged attacks against religion (as some of you may feel now), but the idea that one doesn't need religion to figure things out makes sense. However, I still wasn't finding something to believe *in*. Ray also didn't say *what* the virus actually is, so I have to think about that now, too.

I tried another book, *Sense and Goodness Without God, A Defense of Metaphysical Naturalism*, by Richard Carrier. The title was attractive because it seemed like it would offer a way to be good without a faith in God, though I didn't know what "metaphysical naturalism" was.

Carrier's book explains his journey to develop a personal worldview, i.e., what he thinks is true. He studied many religions and philosophies, having spent time studying Taoism and Eastern spirituality. Tao taught him a "simple truth: that my humanity was a good and natural thing." He strives to use reason and science to develop his "truths," and argues there is "probably only a physical natural world without gods or spirits, but that we can still live a life of love, meaning, and joy." His metaphysical naturalism is basically the mind-frame that everything can be explained or interpreted without any supernatural source, without need of further explanation. This was sounding interesting, and reminded me of my own journey.

Carrier comes to similar conclusions about religion as the other atheists we've discussed. He talks about reading the Bible from cover to cover and concluding it was completely intelligible. The Bible features an immoral, angry, and repugnant God, for example, willing to order Abraham to kill his son. He cites numerous passages in the Bible, such as those condoning child murder, genocide, sexism, and child abuse. He basically says the Bible is a "pile of baloney." He points out specifically, for example, that the Bible condones, which it does - I checked - slavery (see Luke 12:47 and 1 Timothy 6:1-2.). He says the superstition of the Bible is so antihuman that it "would be immoral not to fight it."

Rather than only castigating religion, however, Carrier argues that each of us should develop our *own* worldview, and opines the best way to develop one is just to look at what we actually *know*, even if we can't know it all.

Carrier starts by basing his inquiry on science, which he contends can actually explain the foundation of our world. You know, the Big Bang, astronomy, and all that.

Religion, Atheism, the Virus and Metaphysical Naturalism

Science tells us our universe started some fourteen billion years ago, when the Bangster created jillions of stars and galaxies. He says it appears from scientific study that the purpose of the universe may be to create black holes! These are "stars that have been crushed by gravity so much that even light can't escape them."

Wow! I'd heard of black holes, but I don't think I ever saw one. Carrier says there are jillions of those, and that *we* are much more of an aberration than black holes in the universe. He surmises based on the so-called "Smolin theory" that there may be new Big Bangs and universes waiting to happen inside each of the black holes! The idea is that there are *many universes going on at the same time*. I don't think they talked about *that* in the *Baltimore Catechism*.

"Metaphysical naturalists conclude that the 'ultimate being' is this ensemble of universes. We have no evidence that the ultimate being consists of any more than that .... In fact, the multiverse is a simpler explanation than god, because it has all those attributes of god sufficient to ground its own being and cause this universe to exist, minus all that stuff about intelligence, knowledge, desires, or omnipotent powers. So it does the same work as god with less baggage." He believes that a "mindless multiverse exists, has always existed, and exists by nature."

These concepts strike me as more reflective of the enormity of the universe than putting a man-like entity in charge. There's something comforting about not thinking there's a Supreme Dude up there getting mad at us. Carrier says there's no reason for us to believe the life we see has a cause, just because our experience seems to indicate everything else has a cause.

What there is, of course, is *energy*, according to Carrier. Scientists have "demonstrated beyond all reasonable doubt that matter is simply another form of energy, one among many. In fact, there is really only one fundamental substance out of which every material thing is made; energy. Light is energy. Heat is energy. A magnetic field is energy. Gravity is energy. An atom is energy, as is a proton or a quark. A Dodge Caravan is energy. So is a pencil."

Who thinks of things from this perspective? If everything is energy, why doesn't everyone understand that? Why doesn't the Gospel or the Koran talk about the energy, or gravity, or multiverses? And, from the perspective of these metaphysical naturalists, what are we?

By this time, I'm guessing Richard Carrier is going to tell me what he thinks we are. He says:

Heaven is Everywhere

> *Scientists have come to realize there is no empirical distinction between what people popularly call a 'soul' and what scientists and philosophers have long called a 'mind.' ... The evidence seems clear; our mind, hence our very existence, depends entirely on the brain. ... So when our brains die, and eventually disintegrate, we cease to exist, never to return. This means we will never meet our deceased heroes or loved ones again, that our own lives are of only limited duration and that 'justice' in the sense of fitting rewards or punishments cannot be dealt to the dead. Naturally, this is all terribly annoying.*

Yuh, huh. Annoying, to say the least. If that's right, I have (hopefully) about 14,000 more days left, and then – *I'm no longer me*. I will never be again, and will never be able to see the sunrise, watch the river flow by, or talk with my daughter. Ouch! This metaphysical naturalism stuff doesn't have the heavenly rewards promised by my Catholic religion. I've been growing increasing suspect of the religious mumbo-jumbo, but at least you have a chance to go on somewhere after this, even if it's just the fires of hell. Wouldn't Hades be better than nothingness, because at least you would *know* you were suffering? What would Carrier say about this sad state of affairs?

> *Yet our lives are a joyous occasion. ... Meaning can be found in our own existence, the here and now, but also in our hopes and dreams for the future, and not just our future but the future of others. In the simplest terms, the meaning of life is a healthy mind in a healthy body, pursuing and manifesting what it can most deeply love: the creation of good works, and the society of good people, in a well-tended world. ... Our being here, to acknowledge it, to study it, to know it, and to love it, gives the universe meaning, not the other way around. ... But it does not matter if our life is brief, for merely the opportunity itself is priceless. ... And for us, the sages have said it for millennia; it really is love that is key – love of learning, love of doing, love of others, love of ideals, love of country or cause, anything, everything, is the foundation of meaning.*
>
> *Immortality is inconsequential in this equation. We have no reason to fear death. Why fear the end of fear itself? We live for only one reason: because we love life, all of it, any of it. And if it disappoints us*

that there is not enough happiness in the world, not enough goodness, we can contribute to rectifying that. ... By making the universe that little bit brighter and more meaningful, my own life had value and meaning as a consequence, and that addition to the world remains even when I do not.

Wow, again. This was a new concept, but one I now *feel in my bones*. Everything I've learned and can remember seems to flood toward this type of perspective.

I'm disappointed there's not enough happiness in the world. It pisses me off and makes me sad that humans kill one another. If these metaphysical naturalist characters like Richard Carrier are right, we the people by our collective consciousness of fear, hate, and murder are wiping out the one-in-an-infinity chance of others even to *be*.

Can I contribute to rectifying that? Can you?

I say we try.

Four | Learning about Love

How do I get started rectifying the problems of the world? What's the first step in developing my own worldview? If all of a sudden you realized everything you assumed to be true in your life *was not*, where'd you go from there?

I'll begin with a hug I had one Labor Day weekend in about 1997. Our family goes to an island in the Potomac River. We often took friends as guests. One of the other moms and I had a nice connection. Something had been starting between us, but I hadn't realized it. One night, groups of people were milling around outside under the moonlit sky, and we were alone.

I said, "Mary, you all are just so much fun to be with; how about a hug to say thanks for being such good friends." We hugged, and the *stars exploded*.

I don't remember what else we said, but we both felt it. Indelibly changed forever. Later, when I read *One Day My Soul Just Opened Up*, by Iyanla Vanzant, I understood what that meant.

But Mary and I were in a pickle. We both couldn't stop wanting to be together. We talked on the phone surreptitiously. We met a few times; I sat in the backseat of my car with her and cried a whole packet of Kleenex. We didn't have sex; that's not what it was about. I didn't want to leave my marriage, wife, or family, but *my soul was in charge*. I consulted with a Jesuit priest and a counselor and tried to figure out what the hell was going on.

We were found out a couple months later; busted. I wanted to do what I needed to save my marriage. My wife was as good as she could be, and I really tried. I had anxiety like crazy, constantly. No contact with Mary, and all that. But like I said, my soul was in the driver's seat. I finally went to a healer after my wife set up an appointment with him. ("You set up an appointment with some healer who lays hands on people without telling me?") I decided I should go ahead and try; didn't want to be accused of not trying. The first thing I said to Avery Kanfer - aforesaid healer - was, "I'm tired of fighting myself."

He replied, "Why are you fighting yourself?"

That was another of many meaningful twists in my life plot. Avery helped me understand what was going on. "Just witness your emotions," he said.

"What if my kids hate me?"

"Your children will love you even if you're on the moon. You have to do what you have to do, for your own peace of mind," he said.

I later moved out; it was the hardest thing I ever did. I was on autopilot, and there wasn't any choice. It had been a couple years since the hug, and another few years before the divorce. It affected my relationship with my kids, and seems like it still does. I'm very sorry about all that, but I don't believe in regrets anymore.

My soul somehow knew I had a lot to learn about life. The period after the hug was like a life-education rocket ship. Not that I hadn't learned a lot before, but no period in my life was as intensely focused on trying to learn *why* as my "mid-life crisis." It's not really a crisis, though. I now define that transitional time in nearly everyone's life – which can be triggered by the death of a friend, spouse, marriage, or anything for that matter – as going from *who you've become* to *who you are*. If you don't heed the call from your soul, you can basically die inside and frustrate that inner Christ.

My ex-wife played an instrumental role in my learning. I'd met her in 8th grade at a dance at BS. We talked on the phone for hours, though I don't know what about. I invited her to Hot Shoppes for lunch, but her strict Catholic mother wouldn't let her go. We lost track of one another until I was about 26 years old. I was walking east on M Street, crossing 15th, and she was headed west. It was like one of those cartoons where Wile E. Coyote's head stops to look at the camera and his body keeps going. Some energy field threw us together. We talked for a few minutes. She was with a girl friend headed somewhere. They'd both just come back from Europe where they'd been doing TM and following Maharishi Mahesh Yogi. Her eyes sparkled. She was living at home and I still remembered her phone number. As we walked away, I thought to myself, "OK, now that I've found myself a woman, … ."

We got engaged six months later and married when we were both 28. When we turned 30, she threw away the birth control stuff and we had three kids in four years. It's been really busy ever since.

Learning about Love

One reason I was attracted to Missy, my first wife, was because of TM. That's Transcendental Meditation, not trademark. The theory of existence taught by Maharishi was totally new for me and very attractive. They said TM didn't replace one's religion, but basically TM seemed like Hinduism with a mantra. Everything came out of the *being*. The more you meditated and dipped into the being, like a shirt that was dyed yellow again and again, the more you'd reflect the love and peace of the being; the brighter you'd be. The goal was to be *enlightened*.

I remember talking with her about reincarnation, how the TMers believed you could fly by using a super-duper mantra, and lots of intellectually invigorating stuff. She taught me how to meditate; gave me my mantra. I also took a course to learn the siddhis, i.e., the super-duper mantras, so I could hear other peoples' thoughts, become invisible, fly, and things like that. There definitely is something profound about meditating with a bunch of people, regardless of the other stuff. I even got my flying badge. It was more like hopping on your butt, to tell the truth.

After I'd left the house and got my little apartment, I reflected on life a lot. What the *hell* was I doing? I had a blow-up bed, a beanbag chair, and a phone. That's it. I was afraid everyone would hate me, I'd lose my job, and die a drunk in the gutter.

But I didn't *really* think that. I was witnessing what was happening in my life, as Avery advised during our regular chats. And the point of all this, is that Missy gave me some things to read. One was an article that said forgiveness was *the* most important thing. I didn't understand that. I didn't need to forgive her; I'd been the schmuck, right?

What I got into was, *A Course in Miracles*. She gave me the combined volume Second Edition, with the text of the *Course*, the *Workbook for Students*, and the *Manual for Teachers*. She also gave me another book, called *Gifts from A Course in Miracles*. I don't know if you're heard about the *Course*, but it has been very influential in the new spirituality movement. It was written in 1975; well, maybe not written. It was allegedly *received* by Helen Schucman, who was helped by her friend William Thetford, both professors of medical psychology at Columbia University. Yes, she *dictated what the Voice told her.* They weren't particularly religious, but had been discussing finding a better way to live than with angry and repressive feelings. I've read it was Jesus who dictated the *Course* to her, but I'm not believing that now.

In any event, it's a profound tome. The *Course* is about 700 pages, the *Workbook* has 360 lessons in about 485 pages, and the *Manual* is 100 pages. I didn't read it all, but I read a lot of it. It's challenging and intense. The *Course* is summarized as follows:

> *Nothing real can be threatened.*
> *Nothing unreal exists.*
> *Herein lies the peace of God.*

Hmm. Need a little more explanation.

> *You are a miracle, capable of creating in the likeness of your Creator. ... Miracles represent freedom from fear. Everything else is your own nightmare, and does not exist. Only the creations of light exist. ... You who want peace can find it only by complete forgiveness. ... A sense of separation is the only lack you need to correct. ... All aspects of fear are untrue because they do not exist at the creative level, and therefore do not exist at all. ... In sorting out the false from the true, the miracle proceeds along these lines:*
> *Perfect love chases out fear.*
> *If fear exists,*
> *Then there is not perfect love.*

But:

> *Only perfect love exists.*
> *If there is fear,*
> *It produces a state that does not exist.*

Well, in my little apartment, I wasn't feeling perfect love, just a lot of anxiety. I didn't understand about fear. I didn't think I had any fears. But there's that concept of separation and love again.

After the "hug" and before moving out, I'd been working on a novel. Missy took the kids out of town one weekend so I could work on it. I wrote for about 15 straight hours; it was great. During a break, I was talking on the phone with my friend Bob, who'd been going to AA for years. Somehow in the conversation about my anxiousness, Bob said, "Well, in AA, we ask what you think about sex, resentments, and fears." I said, "I don't think I have any fears." He said, "Everyone has fears."

"Well, I guess I'm afraid of being with a group of other lawyers and saying something dumb."

Bob said, "There you go, that's a fear."

I suppose he was right. Now, as I read the *Course*, it seemed to be saying that: if there's fear, there is not love, but there's no such thing as fear. So, if I had fear, it wouldn't be real, but I wouldn't have love.

Is that what the *Course* was teaching? Seriously, the language is so frickin' thick, it was hard to tell. The text includes many references to God, the Holy Spirit, and the Son, but I wasn't thinking in those terms. I was interested in the basic message that could help *me*. I was just trying to figure out what love was at the time. Did I love Mary, or not, and what did *that* mean?

The *Course* says, "Because of guilt, all special relationships have elements of fear in them. This is why they shift and change so frequently." I was experiencing that. The *Course* is basically saying you can't substitute a relationship for true love, but that seemed to cut against our social upbringing and finding perfect love in another person. "No one who has not yet experienced the lifting of the veil, and felt himself drawn irresistibly into the light behind it can have faith in love without fear. In the holy instant [of the lifting of the veil] God is remembered." So this was going to happen all at once, understanding what love is? What do I have to do?

"It is still up to you to choose to join with truth or illusion. But remember that to choose one is to let the other go."

OK, so I choose truth. There I did it; now what?

"But truth is real in its own right, and to believe in truth *you do not have to do anything*."

You mean I didn't have to do anything to get rid of this anxiety? I could know the truth just by contemplating my navel? I was getting lots of cryptic messages from the *Course*. Here are some more:

> *The past becomes the justification for entering into a continuing, unholy alliance with the ego against the present. For the present is forgiveness. ... If you were one with God and recognized this oneness, you would know his power is yours. But you will not remember this while you believe attack of any kind means anything. ... No one is strong who has an enemy, and no one can attack unless he thinks he has. Belief in enemies is therefore belief in weakness ... .*
> *Forgiveness is your peace.*

Here's that stuff about forgiveness again. The *present* is forgiveness, and forgiveness is peace? Sheesh.

Obviously, I can't explain this whole book to you, because I'm not sure I understand it. Maybe it's like the Bible, and not really comprehensible. But there's an underlying *knowingness* about it that pushes me to try and understand. Let me give you a few more tidbits from the actual *Course* lessons, each from a different day:

> *Nothing I see in this room means anything.*
>
> *I am never upset for the reason I think.*
>
> *A meaningless world engenders fear. (It is essential, therefore, that you learn to recognize the meaningless, and accept it without fear.)*
>
> *I have no neutral thoughts (… every thought you have brings either peace or war; either love or fear … .)*
>
> *It is with your thoughts, then, that we must work, if your perception of the world is to be changed.*
>
> *God is in everything I see. (The idea for today explains why you can see all purpose in everything. It explains why nothing is separate, by itself or in itself.)*
>
> *There is nothing to fear. (The idea for today simply states a fact. It is not a fact to those who believe in illusions, but illusions are not facts. In truth there is nothing to fear. It is very easy to recognize this.)*
>
> *The peace of God is shining in me now.*
>
> *The instant is the only time there is.*
>
> *Teach only love, for that is what you are.*

This was calming and comforting. I'm love. How 'bout that?

The *Workbook* answered that other question I had:

What is Forgiveness?

> *Forgiveness recognizes what you thought your brother did to you has not occurred. It does not pardon sins and make them real. It sees there was no sin. Forgiveness … is still, and quietly does nothing. … It merely looks, and waits, and judges not. He who would not forgive must judge, for he must justify his failure to forgive. But he who would forgive himself must learn to welcome truth exactly as it is.*

Finally, I grabbed this gem out of the *Manual for Teachers*:

> *How many teachers of God are needed to save the world?*
> *The answer to this question is – one.*

Ah, ha. I *can* save the world.

I never read all this entirely (was that important?), but it had a profound affect on me. It made me think deeply about many simple concepts, which didn't seem so simple anymore.

Was forgiveness really just not judging? Are there *no sins*? Could I become a teacher and really save the world – just me?

I was hoping I could find someone to help simplify the teachings of the *Course*. You'll never believe it, but Missy gave that book to me, too. I don't think she ever read the *Course*, but I know she read *Love is Letting Go of Fear*, by Gerald Jampolsky, MD, because her handwritten comments are on the copy I have.

This book was published in 1979, just a few years after the *Course*. Jampolsky apparently knew Schucman and Thetford, who developed the *Course*. Jampolsky was then a successful psychiatrist who had been going through a difficult divorce. He was drinking heavily, but somehow learned about the Course and wrote this book, in which he tried to explain the core concepts:

"The underlying truth for all of us is that the essence of our being is Love." That was nice.

"The *Course* states that there are only two emotions, love and fear. ... The course suggests that we can learn to let go of fear by practicing forgiveness and seeing everyone, including ourselves, as blameless and guiltless."

Did that mean Osama bin Laden was blameless?

"The world we see that seems so insane is the result of a belief system that is not working," says Jampolsky. True enough, I thought. The *Course* suggests "we can have the single goal of peace of mind and a single *function* of practicing forgiveness."

You know, that's what Avery said. "If you have peace of mind, you can live under a cardboard box in the woods." It was looking like this forgiveness concept was pretty important, after all.

The statement that there's only love and fear really intrigued me. I hadn't got that so clearly in trying to read the *Course*. But Gerry Jampolsky makes it simple.

> *Wouldn't our lives be more meaningful if we looked to what has no beginning and no ending in our reality? Only Love fits this definition of the eternal. Everything else is transitory and therefore meaningless. Fear always distorts our perception and confuses us as to what is going on. Love is the total absence of fear. Love asks no questions. Its natural state is one of extension and expansion, not comparison and measurement. Love, then, is really everything that is of value, and fear can offer us nothing because it is nothing. ... With Love as our only reality, health and wholeness can be viewed as inner peace, and healing can be seen as letting go of fear. Love, then, is letting go of fear.*

Wow.

> *We often believe that the fears of the past can successfully predict the fears of the future. The results of this type of thinking are that we spend most of our time worrying about both the past and future, creating a vicious circle of fear, which leaves little room for Love and joy in the present. ... We can choose to experience this instant as the only time there is, and live in a reality of now.*

So, if I let go of fear, then I'll have love in the present, right? It seemed to make sense, but at the same time I'd never heard this concept. Not at Mass and not in the *Baltimore Catechism*. If it was so easy, why didn't everyone know about it? Where did the concept of forgiveness fit in? I'd always understood forgiveness to mean being absolved of sins. We went to confession, said "Bless me father, yada yada," got a few Hail Mary's to say as penance, and *bye-bye sin*. Didn't someone have to do *wrong* to you to forgive them? Reconciliation, atonement, and all that? Jampolsky says:

> *Simply stated, to forgive is to let go. ... Inner peace can be reached only when we practice forgiveness. Forgiveness is letting go of the past, and is therefore the means for correcting our misperceptions. ... Forgiveness releases all thoughts that seem to separate us from each other. Without the belief in separation, we can accept our own healing and extend healing Love to all those around us.*

Learning about Love

You mean, even our enemies? I remember liking the "love thy neighbor" concept, and that we're supposed to extend love to our enemies. Really though, love Osama bin Laden?

> *When we perceive another person as attacking us, we usually feel defensive, and find a way, directly or indirectly, to attack back. Attacking always stems from fear and guilt. No one attacks unless he first feels threatened and believes that through attack he can demonstrate his own strength, at the expense of another's vulnerability.*
>
> *Attack is really a defense and, as with all defenses ... actually preserves the problem. ... In order to experience peace instead of conflict, it is necessary to shift our perception. Instead of seeing others as attacking us, we can see them as fearful. We are always expressing either Love or fear. Fear is really a call for help, and therefore a request for Love. It is apparent, then, that to experience peace we must recognize that we do have a choice in determining what we perceive.*
> *... **We all have the power to direct our minds to replace the feelings of being upset, depressed, and fearful with the feeling of inner peace**. (No emphasis added.)*

I think we do. *Of course, we do.*

Osama was fearful. Are all those militant extremists really just *afraid*? We attacked Iraq, so did we do that because *we* were afraid? These are things worth thinking about.

I have to say the lessons from Jampolsky have resonated with me more than anything so far. Choosing peace, and peace of mind, as a single goal seemed so – *peaceful*. Could I develop a personal allegory based on that? Could I demonstrate there's only love and fear in a way that would contribute to rectifying some of the world's problems?

The first step, I guess, would be to figure out what love is. The *Course* said, "Teach only love, for that is what you are." That didn't fit what I thought love was. In the olden days, you'd look up the definition in your dictionary; nowadays, of course, you go to dictionary.com:

<u>Love -- noun</u>

Heaven is Everywhere

1. a profoundly tender, passionate affection for another person.
2. a feeling of warm personal attachment or deep affection, as for a parent, child, or friend.
3. sexual passion or desire.
4. a person toward whom love is felt; beloved person; sweetheart.
5. (used in direct address as a term of endearment, affection, or the like): Would you like to see a movie, love?
6. a love affair; an intensely amorous incident; amour.
7. sexual intercourse; copulation.
8. (initial capital letter) a personification of sexual affection, as Eros or Cupid.
9. affectionate concern for the well-being of others: the love of one's neighbor.
10. strong predilection, enthusiasm, or liking for anything: her love of books.
11. the object or thing so liked: The theater was her great love.
12. the benevolent affection of God for His creatures, or the reverent affection due from them to God.
13. Chiefly Tennis. a score of zero; nothing.
14. a word formerly used in communications to represent the letter L.

Love –verb (used with object)

15. to have love or affection for: All her pupils love her.
16. to have a profoundly tender, passionate affection for (another person).
17. to have a strong liking for; take great pleasure in: to love music.
18. to need or require; benefit greatly from: Plants love sunlight.
19. to embrace and kiss (someone), as a lover.
20. to have sexual intercourse with.

–verb (used without object)

21. to have love or affection for another person; be in love.

Many of those definitions sounded like the love I knew from the movies. For example: *West Side Story.* "Maria; I just met a girl named Maria!" *The Sound of Music.* "Here you are standing there, loving me, whether or not you should."

True love, so sweet. And of course, the love God has for us, and that we *owe* to him.

But Scott Peck in *The Road Less Travelled* says that *love is a choice*. Was love only a choice; a *decision* to extend ourselves to encompass another person? Love couldn't be just a choice or decision, could it? Is crazy eyed, panting infatuation like I'd experienced a *choice*? That didn't seem right to me. Maybe deciding to get married was a choice, or to stay married. I don't think that was the type of love the *Course* had in mind.

As they say, ask and you shall receive. Whenever I look for an answer, I find it – eventually, but often in strange places. *Time* magazine ran a piece on January 26, 2009 about the death of Andrew Wyeth. He was perhaps most famous for his painting "Christina's World," where she's laying in a field like an invalid, looking toward a house on the prairie. The article recounted the story of some 240 works he painted of a mysterious woman name Helga, who'd been secretly posing for him for years.

In describing the paintings, Wyeth explained, "Painting is about love."

How could painting be about love? Maybe painting Helga had something to do with feelings of love Wyeth may have had, but was "Christina's World" about love? What about the "Mona Lisa," "Starry Night," "The Scream," or "The Garden of Earthly Delights?" The "Scream" didn't seem to be about love. Neither did the damned sinners in Bosch's triptych.

I then remembered a segment in my cousin Henry Glassie's book titled, *Turkish Traditional Art Today*. He wrote about what made a master artist. "Ease and perfection," said his Turkish friends, were the signs of a true artist. And that inner spirit of the artist, Henry said, was "love, devotion, passion. ... [W]hen artists love, when they give the wholeness of their being to their work, when passion governs their actions, the product is art. Art is the object that contains love."

Art contains love; holds love? Nothing in dictionary.com about that. If you follow the idea, though, it says to me that the *energy* of art is love. Oil on canvas, yes, but possessing some sort of intangible feeling or vibration. Could it be the same energy vibration we have? Thinking about it that way, wouldn't we be love?

Well, then, would the energy of other things be love, too? A sculpture, a painting, or a fine Turkish hand-made rug; wouldn't they contain love? Where do you draw the line? Would a fine winter coat be love? How about a puppy or a kitten? What do they contain? What gives them life or meaning? *Love* seems to be a pretty good answer, doesn't it?

Does my 1928 house contain love? What about my Toyota Highlander? Richard Carrier says a Dodge Caravan and a pencil are basically just energy. Is all the energy really just the same as love? Does the sun contain love? Is the earth made of love, as well?

The definitions of love from dictionary.com now seem hollow. I feel my heart swell with the thought that *everything is love*. Or, that I could at least *think* of everything as love! Wouldn't that be a good perspective?

Sure, love is an emotion, but even then it has a special energy. Something we seek, want, and try to emulate.

Jampolsky says, "The essence of our being is love, ... love is our natural state, ... we are – and all that we are – is pure love. ... Love is everything there is and the only thing that is real. ... [I]t is the total absence of fear and the recognition of complete union with all life. We love one another when we see that our interests are *not separate*." (Emphasis mine.)

That's what Freke and Gandy said (remember them?) in *The Laughing Jesus*. "We appear to be separate, but essentially we are all one. The awareness that is conscious in you is the same awareness that is conscious in everyone. And if you recognize this, you will find yourself in love with all. ... There is no 'us versus them.' There is only us. ... Love is what we feel when we know we are one."

We are one? I thought I was me, and you were you; and they were them, and *spiders certainly aren't like us*. But, maybe that's just how we *perceive* it?

Have you seen the movie, *What the BLEEP do we know?* It's pretty wild. The meaning is, we really don't know much about life. I'll talk more about it later, but there's one statement made by Dr. John Hagelin in the movie that's pretty profound on this score. He's a physicist. He says that modern physics now teaches that, at the deepest level of life – deeper than protons, neutrons, and quarks – *there's no separation between anything, at all!*

That would mean we're all connected, with no separation. All of us on this spaceship earth, a part of the same energy system. Maybe from way out in space, that's what we look like? One blue ball, of love.

Thinking of life that way, if love is the energy that's everything there is, and we're not separate, then we're all love and everything's love. *Whoa. That's* what the *Course* must mean; *that's* what Jampolsky means. Love is what you are, inside and out.

Well, maybe not quite literally. There's not a chemical on the periodic table of elements next to hydrogen or oxygen called love. It's more allegorical. It's a way of looking at things; it's a *perspective*. A perspective that love is *oneness*.

I could get into thinking of life as just being love, that it's all good.

Isn't that silly, though? Wouldn't everyone say that's some sort of pie-in-the-sky, rose-colored-glasses, new age, meaningless, mumbo jumbo? It's not *realistic*, right?

Yea, but is it realistic to have a life view based on non-existent mythical characters? Or a religion where there's nothing *but* separation from God and other people, and where anger is laudable?

Wait a minute. … If we assume there's only love and fear, and oneness is love, does that mean that the opposite, *separateness*, is fear? Well, it wouldn't just be separateness, because there's no such thing. If everything is one, nothing is separate, right? So, would fear just be the *perception of separation*?

If people think they're separate and not together with others in love, anything that makes them feel different, alienated, alone, or unlovable would be equivalent to fear; would cause them to be fearful. Their perspective would be they're not deserving of love or maybe they're separate from God (like that dude Josh McDowell believes). That doesn't sound like much fun, but a lot of people feel that way. So, how do we all get over the perception of separation to find love? I suppose we first have to understand fear.

Five | **Spitting Blood**

I was brushing my teeth, just before bed. I rinsed my mouth and spit out *blood*. Bright red blood. A lot of blood. All over the sink.

"Julie, look at this."

We stood staring at the bloody sink. This wasn't good. I looked at my mouth in the mirror, but couldn't tell where the blood was coming from. It was all over my tongue.

I checked the Internet. *Bleeding from the mouth is a symptom of oral cancer, including tongue cancer,* according to about.com. Oh, great. Drinking and smoking are risk factors. I like my beverages, and started smoking in college, though it's been a sporadic habit. My ex-wife and I would often go out on the porch and smoke two cigarettes after we put the kids to bed.

I called the doctor about spitting blood and made an appointment to see him in two days. The next day, my 94-year-old mother had a stroke.

Three days later, I saw a ginormous "floater" go across my eye. It was a brownish color. Checked the Internet again. *Seek immediate medical attention; your retina may be detaching and you could go blind.* Yikes; my grandfather had detached retinas and went blind! Off I went to the emergency room.

This was all the week after we spread the ashes of my wife's wonderful friend Eileen on the Potomac River near our house. She had ovarian cancer, fought it for a couple years, but lost the fight.

Sheeeet. Was I about to die? Was this the beginning of the end? I know I'm gonna die. We all are. But these things happening at the same time put death right up in my face.

Jefferson Glassie - Dead of mouth cancer at 57.

It's *scary*; death is scary. You don't know when you're going, what takes you, or what happens when you die. Now, being scared is *clearly* fear. I mean, by definition, if I say, "I'm scared of something," then I'm afraid of it and that's a fear. So, what was I afraid of? Was my fear really just the *perception of separation*?

Or was I afraid of dying because of the unknown? Was I just scared I'd get cancer, be in misery and pain the rest of my life, then die an awful death? Or maybe I was afraid I might die and go to hell? I used to feel confident I'd be going to heaven; up to that "cocktail party in the sky," as my mother calls it. But I don't think I believe in heaven anymore. Is that a bad thing? Not sure I believe in hell either, so maybe it balances out.

Darrell Ray said those infected with the religious virus are terrified of damnation, gnashing their teeth in the fires of hell and living till the end of days without the love of God. Is that the ultimate fear? Damnation.

After reading Jampolsky's book *Love is Letting Go of Fear*, I'd thought about fear a lot more. Remember my mid-life crisis? Well, during all that reading, learning, and introspecting, I decided to go somewhere and write down everything I thought was true. I had a business trip to Tucson and visited an old mission church south of town. I took a spiral notebook, sat in a pew about five hours, and wrote down all the truths I knew. I came home and wrote a book; I told you about that. That was in about 2003. My soul decided to call it *Peace and Forgiveness* (I'd wanted to change the title, but couldn't; damn book had a mind of its own). It has chapters on Us, Love, Fear, Forgiveness, and Peace. This is what I had to say about fear:

> *Our fear arises simply from the fact that we're human and are limited in our perceptions and understandings by what we can sense through these bodies we inhabit. We have fears based on our beliefs and expectations. Ultimately, we fear not having the peace and love of the Universe; we are afraid we will never have god's love. We are afraid of being alone. ... We are afraid of not being anymore. We are afraid to die. We are afraid of the future.*

That's a lot of things to be afraid of, but at the time, I thought the *ultimate* fear was not having love. Not having the love of God, if you want to look at it that way. That could be right, but some friends say that's not their ultimate fear. One neighbor says her ultimate fear is not being able to drink anymore.

Recently, after my bloody mouth incident, I've also been thinking about not being here. Not whether I go to heaven or hell, or have love. Just *not being here*. You know, like every other living thing eventually isn't here anymore. Remember Richard Carrier's point about there being nothing after life. It seems most plausible. As my friend David said, "I don't think there's shit on the other side."

What's so special about being here? Many folks seem to believe "Life sucks, and then you die." But I think it's pretty neat, being alive.

I go outside and look at the sky. Check out the trees, the clouds, and the sun. It's nice. There's no fear looking up at the sky. We take a walk with the dog and go down to the River; to a place we call Serenity Point. Julie and I call it that; not sure if anyone else does. That's where we let Eileen's ashes flow away. It's so peaceful. Saw some white-breasted mergansers out there recently.

But what happens when I'm not living any more? There will still be water flowing at Serenity Point. Hopefully, the mergansers will come back. Things will go on, until this planet stops rotating or we all get vanished by a nuclear bomb that some poor despairing stranger manages to explode downtown.

OK, so I guess that's a fear, too. The terrorists will kill us! Why do they wanna do that? It seems so nice outside; don't they go outside? Are they so stuck in dogma and religious rubrics that that don't see how *fabulous* it is outside?

Do you remember the underwear bomber? His name is Umar Farouk Abdulmatallab. He got caught trying to set off a bomb in his pants on a plane bound for Detroit on Christmas Day 2009. The *Washington Post* article on December 29, 2009, titled, "In Online Posts Apparently by Detroit Suspect, Religious Ideals Collide," by Philip Rucker and Julie Tate, said the "'terrorist' wrote 'I feel depressed and lonely; I do not know what to do.'"

The al-Qaeda group in the Arabian Peninsula that claimed credit for the attempt said it was a response to "U.S.-backed airstrikes against the group in Yemen." They were feeling attacked and defensive, so that's why they struck out by trying to exact murderous revenge. Hmm, if people feel attacked and defensive, then they must be afraid of something; maybe that the United States and the behemoth Western juggernaut would erase or destroy their culture - or perhaps there's a reason they're so backward technologically. Maybe they just fear they'll be ignorant and poor forever. Anyway I try to look at it, it seems like their attacks are based on their fear.

Poor Umar, though. According to the article, he wrote on his Facebook page about his disappointment at scoring only a 1200 on his SAT test. "It was a disaster." He must have thought he was stupid and wouldn't get into college. That's a fear.

His online persona at boarding school in Togo, *Farouk1986*, wrote: "I strive to live my daily live (sic) according to the quran and sunnah to the best of my ability. I do almost everything, sports, TV, books … (of course trying not to cross the deen.)"

The deen is the religious way of life, according to the *Post* article.

Wouldn't want to cross the deen, you know, violate the rules and be *afraid* of the repercussions. For example, *Farouk1986* also wrote on Facebook about the tension between his sexual desires and his religious duty of "lowering the gaze" when women were around. "The Prophet(S) advised young men to fast if they can't get married but it has not been helping me much and I seriously don't want to wait for years before I get married." According to the *Post*, he hadn't started looking for partners because of social norms, such as "a degree, a job, a house, etc. before getting married."

He wrote of his "dilemma between liberalism and extremism" as a Muslim. "The Prophet(S) said religion is easy and anyone who tries to overburden themselves will find it hard and will not be able to continue. … So anytime I relax, I deviate sometime and then when I strive hard, I get tired of what I am doing, i.e., memorizing the quran, etc. How should one put the balance right?"

So, we have here with this "terrorist" a good example of a kid not sure of what he's doing, ostensibly trying to stick with some arcane religious precepts, but really only trying to get laid. It doesn't strike me that Umar is evil; just trying to do the right thing, but *afraid* he isn't or can't. It's beginning to dawn on me the *source* of anger, attack, and murder really could be fear. I wonder what the experts have to say about why humans kill, for example, like those guys who murder people.

"We're still in the dark about where this comes from," said Jack Levin, a director of the Brudnick Center on Violence and Conflict at Northeaster University in Boston. This quote came from an article entitled, "Dark Matter: the Psychology of Mass Murder," written by Neely Tucker in the April 17, 2007 *Washington Post*, just after the Virginia Tech shootings. Tucker says, " Some of the research tells us the obvious: About 95 percent of mass killers are men, they tend to be loners, they feel alienated. They look normal on the outside and are really, really angry inside. … [T]hey're depressed, angry and humiliated. They tend to be rejected in some romantic relationship, or are sexually incompetent, are paranoid, and their resentment builds." Some event sets them off – "an argument, a small personal loss that magnifies a sense of catastrophic failure," and all the anger comes out.

Another article in the *Post* on the same day, "With Each Shooting, Common Threads," by Joe Holley, says the ones likely to commit mass killings "have the same common denominator. ... They're very, very self frustrated people who are so self-centered they feel the whole world is against them."

I get depressed reading all this; *how sad*. These men are angry about so many things; feeling rejected, incompetent, humiliated, and isolated. Basically, it seems to me they're *afraid they're no good*.

Hey, that's fear again; when they're afraid they just suck and no one loves them. Does their fear lead to anger? A big problem is, when they get angry, they can go out and buy high powered automatic weapons, at least here in the good old US of A. That's what happened at Columbine.

I'm sure you know about it. In 1999, Eric Harris and Dylan Klebold, two seniors at Columbine high school in Colorado killed 12 students, a teacher, and wounded 21 others. Then they killed themselves. It was an awful killing spree, one of the worst school incidents in U.S. history. There's been a lot written about Columbine, trying to understand *why they did it?*

They had a history of drinking and threatening other students. They loved violent video games and Harris took the antidepressant Zoloft. But what intrigued me was their *anger*. These were not happy dudes. A *Washington Post* article on July 6, 2006, "Newly Released Columbine Writing Reveal Killers' Mind Set," by T.R. Reid, talked about material written by the pair. "The documents depict intense levels of nihilism, anger and contempt for the boys' schoolmates." Here are some of the things they wrote, according to the article:

> *"I hate you people for leaving me out of so many things,"* high school senior Eric Harris wrote of his classmates. *"You had my phone #, and I asked you and all, but no, no no don't let that weird looking Eric kid come along. . . . I HATE PEOPLE and they better [bleeping] fear me."*
>
> *"Different is good,"* Harris wrote. *"I don't want to be like you or anyone."* Harris wrote a note asking a girl if she would like to go out with him. *"If you don't,"* he wrote, *"I'll understand, I'm used to it."*
>
> *"I know that i am different,"* Klebold echoed in his journal. *"As I look for love, i feel i can't find it, ever."*

Golly. They felt different, separate, and unloved. And they were very angry about it.

They got a slew of guns and murdered their classmates, because they were so desperately pissed off about being disrespected. It seems the *fear* of not being loved caused anger in Dylan and Eric. But wouldn't some people say they were just *evil*?

Well, what is evil? Is it a physical state? Is it the devil, Mr. Satan? Are some people actually evil? Or is it more aptly a description we simply hang on humans, things, or events that are truly awful and hideous when we don't know what else to call them? If a person does something dastardly, baby-killing bad, like the Nazis, it's handy to declare them evil. Plus, there's the extra benefit of being able to think, "We're not like them. They're evil; we're good."

But isn't there a perspective to this? Remember when George Bush in his State of the Union Address on January 29, 2002 called out Iran, Iraq, and North Korea as the "Axis of Evil." You know Bush believed Osama bin Laden was evil. Don't you think, however, that Osama believed George Bush was evil? If they each thought the other was evil, who was right?

Most people would say Adolf Hitler was evil, too, and as an incarnation of the devil he somehow converted the German people to evil. But what if they were all simply *afraid*? Germany had lost World War I; the country and its people were decimated. Their economy was in ruins. Fear and suspicion were rampant.

People don't like to hear that Hitler might *not* have been evil. Humans like to cling to their judgments, I think because it's easier to retain a feeling of superiority. Otherwise, we'd all be just the same and, therefore, could be awful like Hitler, too. For example, as reported in the *Washington Post* on December 27, 2007, Will Smith said in Scotland's *Daily Record* in 2007: "Even Hitler didn't wake up going, 'Let me do the most evil thing I can do today.' I think he woke up in the morning and, using twisted, backwards, logic, set out to do what he thought was 'good.'" I agree with that, Will. According to the article, "The quote was preceded by the observation that 'Will believes everyone is basically good,' prompting gossip Web sites to post stories alleging Smith believe that Hitler was a good person." But the Anti-Defamation League objected and Smith had to recant.

I'm not defending Hitler. He rallied the German people against others, including the countries that had humbled Germany and, of course, the Jews. What really could cause the holocaust, and other genocides like it in human history?

We could ask Ed Gernon. He was head of the Canadian production company Alliance Atlantis and was fired for an interview he gave to *TV Guide* regarding a CBS miniseries, *Hitler: The Rise of Evil*. You may recall that the United States invaded Iraq in March 2003, and there was a lot of fear in the country because of the attacks on September 11, 2001. Lisa de Moraes in "The TV Column" in the *Washington Post* reported on Friday, April 11, 2003: "Gernon stated his belief that fear fueled both the Bush administration's adoption of a preemptive-strike policy and the public's acceptance of it. ... Gernon said a similar fearfulness in a devastated post-World War I Germany was 'absolutely' behind that nation's acceptance of Hitler's extremism."

Gernon is quoted as saying of the miniseries, which tracks Hitler's rise to power in 1930s Germany:

> *It basically boils down to an entire nation gripped by fear, who ultimately chose to give up their civil rights and plunged the whole world into war. ... When an entire country becomes afraid for their sovereignty, for their safety, they will embrace ideas and strategies and positions that they might not embrace otherwise.*

Rather than attribute Germany's fanaticism in World War II to evil, Gernon said it was basically a *national fear* and compared it to the United States in 2003. That wasn't too politically correct, and CBS fired him rather than risk a backlash from those in the "home of the brave," who don't like to be told they're afraid. Another analogy here, based on these assumptions, might be that Germany struck out and attacked when it was afraid, as did the United States in Iraq. The Germans were afraid they weren't the master race, and we Americans were afraid we weren't the greatest country on earth or that we weren't powerful enough to protect against other attacks.

But, I digress. My point is we shouldn't rule out the ramifications of fear on human conduct. Let's look a little further into newspaper articles about other killers and see if we can discern a source for their actions consistent with the theories we're discussing.

OK, let's talk about Cho Seung Hui, the perp in the Virginia Tech killings on April 16, 2007. Remember him? He murdered 32 students and teachers with two illegally purchased handguns in the worst mass killing by a lone gunman in U.S. history. I wonder if he had some of these elements of fear?

According to the *Washington Post* article titled, "Killer's Parents Describe Attempts Over the Years to Help Isolated Son," by Brigid Schulte, on August 27, 2007, he had a social anxiety disorder (meaning *fear* of other people) so severe that the county school system developed "a detailed special education plan to help ease Cho's fears so he might be able to talk more openly." He wasn't just painfully shy, in class he was often "too paralyzed to speak." His fellow students would make fun of him, which "exacerbate[d] the anxiety."

He spoke with hardly anyone except his sister, but didn't tell her his feelings. If a visitor came to the family home, according to the *Washington Post* article on August 31, 2007, "Gunman Mailed Menacing Parcel During Lull in Slaughter," by Michael Ruane and Chris Jenkins, his "palms would become sweaty, he would freeze, would sometimes cry and was able only to nod yes or no." The article uses words like "isolated," "angry," "inadequate," and "violent."

After the murders, according to the *Post* article on April 19, 2007, police found an "angry" document in his room, along with a novel and some disturbing poems and plays. Here are some excerpts from a video that he sent to NBC News before the shootings:

> You have vandalized my heart, raped my soul and torched my conscience. You thought it was one pathetic boy's life you were extinguishing. Thanks to you, I die like Jesus Christ, to inspire generations of the weak and the defenseless people.
>
> Do you know what it feels like to be spit on your face and have trash shoved down your throat? Do you know what it feels like to dig your own grave?
>
> You have never felt a single ounce of pain your whole life. And you want to inject as much misery in our lives because you can, just because you can. Your Mercedes wasn't enough, you brats. Your golden necklaces weren't enough, you snobs. Your trust fund wasn't enough. Your vodka and cognac wasn't enough. ... You had everything.
>
> You had a hundred billion chances and ways to have avoided today. But you decided to spill my blood. You forced me into a corner and gave me only one option. ... I didn't have to do it. I could have left. I could have fled. But now I am no longer running.

Cho doesn't sound evil to me. He sounds fucking scared, literally, to death. He seems lonely, depressed, sad, and unloved. He believed all of the other people were separate and different from him; he was afraid he wasn't good enough. His fear led to an anger that killed 32 innocent people. Yep, I'd say, fear did it.

I know this is disturbing. But if we don't ever try and understand fear, how will we ever get to love? We have to look at more examples, to see if fear underlies other violent acts.

Robert Hawkins, 19, armed with an AK-47, opened fire in a shopping mall in Omaha, Nebraska killed 8 people and wounded 5 others. He was "a high school dropout who had recently lost his job at a fast food restaurant and broken up with his girlfriend," according the *Washington Post*, "Mall Killer Had History of Mental Woes," by Kari Lydersen and William Branigan, on December 7, 2007. In his suicide note, he "denigrated himself and said he would no longer be a burden to anyone, ... [He was] sorry he was a burden to everybody and his whole life was a piece of [expletive]."

Sounds to me like he didn't feel worthy or loved, *afraid* of being a worthless sack of [expletive].

In El Reno, Oklahoma, Joshua Steven Durcho, 25, was charged with five counts of first-degree murder after strangling to death his ex-girlfriend, who had dumped Durcho the week before, and her four children ages 3 to 7. Her mother told the press her daughter, "said it was over and it was done, and apparently that didn't go over too well." This was from an Associated Press story on January 14, 2009, "Man Charged with Strangling Ex-girlfriend, 4 kids."

Seems he was angry at being told she didn't want him anymore. He was *afraid* of rejection.

Twenty year-old Bernard Bellamy was charged with first-degree murder in the death of his 19 year-old girlfriend, who was pregnant with his child. His father said, "He did not want her to have the baby at all," according to the *Washington Post*, May 12, 2009, "Boyfriend Charged in Killing of District Heights Woman," by Matt Zapotsky.

There's almost an epidemic of men killing their pregnant girlfriends in this country; it happens way too often. I think they're *afraid* of being fathers with all those responsibilities and restrictions on their lives.

George Sodini seethed with anger and frustration toward women. He couldn't understand why they ignored him, despite his best efforts to look nice. He hadn't had a girlfriend since 1984, hadn't slept with a woman in 19 years. "Every evening I am alone and then go to bed alone," he wrote. "I see twenty something couples everywhere. I see a twenty something guy with a nice twentyish young woman. I think those years slipped right by me. Why should I continue for another 20+ years alone?" This was an Associated Press story on August 5, 2009, titled "Gunman at Pa. Health Club was Bitter Over Women," by Michael Rubinkam.

He went to an L.A. fitness Club in Pittsburgh and opened fire with three guns on a dance class filled with women, firing 36 bullets and killing 3 women and wounding 9 before committing suicide. The *fear* of being alone is quite strong.

Virginia Tech has had its share of tragedy recently, but this one was horrific. Xin Yang, 22, was a female grad student who had just started her graduate studies in accounting, according to another *Washington Post* article, "Inexplicable Violence Again Shakes Va. Tech," by Brigid Schulte and Theresa Vargas, on January 23, 2009. As she sat with her friend Haiyang Zhu, 25, drinking coffee at Au Bon Pain, he silently pulled out a kitchen knife and cut off her head. Turns out she had "rejected Zhu's romantic overtures," according to a later article from the Post dated December 22, 2009.

Hmm; his love overtures were rejected; he must have felt kinda bad about that, *afraid* of being rejected.

Jason Rodriquez, 40, "was a compilation of the front page of the entire [2009] year – unemployment, foreclosure, bankruptcy, divorce – all of the stresses. ... He has been declining in mental health. ... It looks like a classic case of stress overload," according to public defender Bob Wesley in a *Washington Post* article, "Orlando Shooting Suspect Mentally Ill, Lawyer Says," by Antonio Gonzales and Mike Schneider, on November 8, 2009.

Ah, stress; the *fear* of not being able to get it all done. Rodriquez was making less than thirty grand a year, but had debts of about ninety grand. He was accused of fatally shooting one employee and wounding 5 others at an engineering firm where he had been fired in Orlando, Florida.

Christopher Speight of Appomattox, Virginia, plunged into despair after his mother died in 2006. He then became close to his sister and her husband who moved from Georgia to be with him. They were trying to help him and make sure he could stay on the property his mother had left them. But he told another co-worker he thought the government might take his property and that his sister and brother-in-law were trying to cut him out of his inheritance. So, according to a *Washington Post* article, "Before Explosive Violence, No Hint of a Time Bomb," by Frederick Kunkle and Josh White, on January 25, 2010, he went on a 19-hour rampage shooting them, plus six other family and friends at the property.

If you were *afraid* your sister might take your house, would you kill her?

Christopher Wood, 34, left six notes describing his financial hardships, stresses, depression, and anxiety, and was at least $460,000 in debt on his family home in Middletown, Maryland, with another Florida house in foreclosure. To me, that situation doesn't justify shooting his wife and three young children and then slicing their necks to the point of near decapitation. But you know, according to the *Washington Post* article, "In Notes Left in Family's Killings, Md. Man Details Debts, Depression," by Matt Zapotosky, from April 22, 2009, "Several experts said slayings of entire families by fathers and husbands are often associated with economic hardship. Some men get to the point where it becomes impossible to tell family members that they're going to lose the house or that the kids can't go to college. ... If you've built your identity around that you're the breadwinner, you're the backbone, and that becomes unglued, it undermines your sense of self."

In other words, *fear* of dealing with difficult financial situations can destroy self-worth, which is really self-love.

The depressed economic situation in the United States claimed other lives. According to the Associated Press article, "Man Kills Wife, 5 Children, Self After Losing Job," by Thomas Watkins, on January 27, 2009, "A man who fatally shot his wife, five young children and himself ... had earlier faxed a note to a TV station claiming the couple had just been fired from their hospital jobs and together planned the killings as a final escape from the whole family. 'Why leave the children to a stranger?' Ervin Lupoe wrote." The violence stemming from the fears of the recession was rampant. The Lupoe killings were the, "fifth mass death of a Southern California family by murder or suicide in a year. Police urged those facing tough economic times to get help rather then resort to violence."

An article in the *Washington Post* on April 8, 2009, titled "Traumatic Events Fuel Killings," by Philip Rucker, noted that many experts have linked the economy to 57 deaths in eight mass murders in one month, as the "epidemic of layoffs, the meltdown of storied American corporations and the uncertainty of recovery have stoked **fear, anxiety and desperation** across society and unnerved its most vulnerable and dangerous." (Emphasis added.)

On September 12, 2010, I received an email from the managing partner of our law firm office. It was Sunday night, and I picked up the message on my blackberry. The subject line was "Our Clare Stoudt." I knew Clare, she was a tax lawyer, about three years out of law school. She was an attractive, pleasant young woman. Looked about twenty-five with curly red hair and a fabulous figure. She was very nice and had worked with me on several client projects. Turns out she was thirty-five and had *five children!* Two teenagers (apparently from when she was really young) and three (ages 7, 5, and 2) with another man she'd been living with.

Well, one never knows the whole story, but this much we know. She had decided to leave him, took the children, and wanted custody of the kids. They said he was a stay-at-home dad and now she was making pretty good dough. I guess that made him insecure or angry. Anyway, she went to his house and, while the three little ones played outside, he shot her to death. Then blew his brains out.

Why? Why would anyone feel so alone or angry that he'd kill the mother of his children? Yet, we see it *all the time*. http://murder-suicide.blogspot.com/p/murdersuicide-usa-2011.html

Yep, it just continues, guys shooting people up. There are too many incidents to mention, but one of the worst shootings was at Sandy Hook Elementary School in Newtown, Connecticut. Adam Lanza, 20 years young, shot his mother (four bullet wounds to the head as she lay in bed – imagine that) and then 20 children (8 boys and 12 girls) who were six and seven years old and 6 women who worked at the school. He shot 154 bullets with his mother's Bushmaster XM15-E2S rifle, and blasted each of his victims multiple times, including one 6 year old eleven times. The sad saga even has its own Wikipedia page: http://en.wikipedia.org/wiki/Sandy_Hook_Elementary_School_shooting.

What about Adam? He was smart, but shy. In fact, Richard Novia, the school district's head of security for years, said of Adam, "You had yourself a very scared young boy, who was very nervous around people … ." *Washington Post*, December 16, 2012 from the Associated Press, in an article titled "Reports: Shooter was 'Loner' Who Could Barely Feel Pain." The article continued, "His anxieties appeared to ease somewhat [after joining a tech club at school], but they never disappeared. When people approach him in the hallways, he would press himself against the wall or walk in a different direction, clutching tight to his black case." That sounds like he was *afraid*, doesn't it?

While women also kill people, it tends to be people they know, and hate. *Washington Post*, "Pattern in shooters: Most are white men," written by Paul Farhi, December 21, 2012. But white men often kill for no apparent reason. Randolph Roth, a history and sociology professor at Ohio State, says, "Men carry this kind of anger, … It's there." More from the article:

> *Similar violence has been observed in male chimps; after being defeated by a dominant rival, they will sulk for days – and then attack another chimp in a sudden, seemingly random outburst. [B]oth Roth and [criminologist James Alan] Fox suggest that many violent epi-*

sodes are linked by the killers' sense of having been denied something he thought was rightfully his - a job, a promotion, status of some kind. The profound sense of failure and resentment that accompanies such setbacks may be stronger in white men than others. Accused Aurora, Colo., shooter James Holmes had failed his PhD program; Oklahoma City bomber Timothy McVeigh had washed out of Ranger School.

"There's a feeling of entitlement that white men have that black men don't, says Fox, a professor at Northeaster University and co-author of "Extreme Killing." They often complain that their job was taken by blacks or Mexicans. They feel that a well-paid job is their birthright. It's a blow to their psyche when they lose that. If you're a member of a group that hasn't historically experienced unemployment, there's a far greater stigma to [losing a job] than those who have."

A perhaps telling statistic: About 60 percent of mass killers are older than 30, an indication, Fox says, that "it takes years to develop the level of frustration and anger" that expresses itself. Wow; anger, entitlement, failure, and resentment. Not consistent with love.

I understand this is depressing, and no one wants to read about or deal with this. However, it's *very* important. If fear causes anger, violence, and murder, we need to understand fear, fully. I don't think any of these truly unfortunate people were evil. I just don't believe it. But they do sound scared, frightened, and fearful, and that's exactly the point. Their symptoms are entirely consistent with my speculation that fear is the underlying cause of their actions. Whether it's a fear of dying or being without love or God, that doesn't matter. But if it's the case that fear is responsible, then we *must* address it.

My fear from spitting blood or having retinas detach seems to pale compared to the stories I've recounted about these murders. I have an inkling the ultimate source is exactly the same.

By the way, the doctors said I didn't have mouth cancer or detached retinas. Whew; I can live to die another day.

Six | **Post Dog Bite Disorder**

I didn't even know her name, but she terrified us. She scared the bejeezus out of me and my friends. I wasn't even sure she was a she. Might have been a he. Whatever, the dog had a bad reputation.

We only knew her as the "Morgan's dog." That's because the Morgans lived next door. They had a big piece of property, stretching all the way from Bradley Lane to Quincy Street, bordering Connecticut Avenue. The problem was, if a ball went over our fence, it went into the Morgan's yard.

I had a pretty good arm, and my best friend and I used to play catch with a baseball for hours back there. My strong arm, though, was directionally challenged. I would whip that ball here and there, and poor Johnny would have to chase it down and toss it back to me. He'd put it right into my mitt, and I'd zip it back to him over his head.

The fence was about six feet high, so we didn't throw it over often, but enough. The kid who threw it, had to go get it. It was usually me.

I'd scrawl up the fence and look around. If I didn't see the dog, I'd climb over and leap down into the garden. Then, wait; hoping the sound of my landing didn't attract the dog's attention. I was terrified that he (whatever) would come racing at me, pin me against the fence, and maul me to death.

Truthfully, the dog never once caught me in the yard chasing a ball. I'd crouch on the ground like an army man (having seen plenty of that on TV), run as fast as I could and grab the ball, chuck it back, and clamber over the fence. It was a heart attack every time.

It makes me a little nervous just thinking about it. The one time she did attack seems like a dream.

I was walking down the street past the Morgan's driveway. We always tensed up, hoping the dog wasn't out. I think I had my good clothes on for church or something. All of a sudden, there she was. A brown boxer, with those awful teeth. I saw her, and backed up against the bushes. She came right at me, growled, and bit me in the thigh. I screamed, but my mother was there, and *saved me*.

To this day, I'm wary of dogs. I take a walking stick around the neighborhood when we go on walks. I say it's to beat off the women if they attack me, but you never know when you might have to fend off a dog.

I asked my mother if she remembered when the Morgan's dog bit me. She didn't, but she was 98, after all. Mom seemed pretty brave back then. It was only recently I realized she's a worrier. Always afraid and worrying about stuff. It's part of her persona now.

She told me that, as a child, she wondered if her mother really loved her. My Mom was the last of nine children. She said her mother was pretty stressed with all those kids. My mother's father had been fairly well-to-do, owning a couple farms and a creamery in Wisconsin. But he went on a ride in one of the first cars back around 1900 or so, and had an accident. He flew out of the car and hit his head; never was right again, Mom said. There was no health insurance and no social security. He gradually had to sell everything and they ended up on a small farm with the nine kids. Lost his arm in a grain thrasher, too.

I don't know how that feels; not to be sure if your mother loved you. I never felt that, at all, even for a moment. But that fear of not being loved can take a toll. I think it lays a foundation that can remain under the surface your entire life, at least if you don't recognize it and let it go. I think that's one reason Mom is always worrying. She's just used to being afraid about things.

If phobias and fears like that can last a lifetime, I wonder what really acute fear can do? I find myself thinking about post-traumatic stress disorder – PTSD. The name says it all: a psychological (or maybe physiological) disorder that results from a seriously fearful situation. Not just a dog bite, but a life threatening incident. According to an article called, "The Peace Drug," by Tom Schroder, in the *Washington Post Magazine* of November 25, 2007:

> *PTSD is usually triggered by combat, rape, childhood abuse, a serious*

> *accident or natural disaster – any situation in which someone believes death is imminent, or in which a significant threat of serious injury is accompanied by an intense sense of helplessness or horror. ... The very symptoms – acute anxiety, heightened fear, diminished trust and inability to revisit the trauma – are a direct roadblock to healing.*

The article tells this story:

> *Fourteen years ago, Donna Kilgore was raped. When the stranger at the door asked if her husband was home, she hesitated. Not long, but long enough. That was her mistake.*
>
> *"That was it," Donna, 39 now, is saying. "He pushed in, I backed up and picked up a poker from the fireplace. I was screaming. He says, 'I've got a gun. If you cooperate, I won't kill you.' He unzipped his jacket and reached in. I thought, this is it. This is how I'm going to die. My life didn't flash before my eyes. I wasn't thinking about my daughter. Just that one cold, hard fact. I checked out. I could feel it, like hot molasses pouring all over my body. I went completely numb."*
>
> *She dropped the poker.*
>
> *Afterward, she stayed strong. ... And bottom line, she'd survived. She'd be fine, she told herself. She was wrong.*
>
> *"It was what it must feel like to have no soul," she says. She quit all her hobbies. A passion for tennis died. Devastating nightmares woke her in the dark, her heart racing and palms slick. She dreamed of explosions, tornadoes, bears eating people. ... She was often irritable, and felt an unaccountable anger, which sometimes morphed into a heavy-breathing, sweat streaming rage. Almost worse, she couldn't feel the love around her [from her family.] ... [F]or those five years: "I would put my finger on my arm, and it would be like touching a dead body."*

She finally sought treatment, after developing panic attacks, fainting spells, and migraines, telling her doctor, "Things don't feel real to me." He said that was *dissociation*, a prime symptom of PTSD. Wikipedia says, "Dissociation is a mental process that severs a connection to a person's thoughts, memories, feelings, actions, or sense of identity." Hmm, talk about fear being a *perception* of separation. It seems like PTSD may be actual separation from oneself, from love.

People often associate PTSD with soldiers at war, and there've been a lot of them. It used to be called 'combat fatigue,' 'shell shock,' or 'battle fatigue.' That makes it sound like the person suffering from it was a softie or just tired out. But PTSD, the residual effect of great fear, is profoundly strong. Like its effects on Donna in the story, who was totally stressed out for years after the rape.

According to that article from 2007, estimates were about 15 percent of those returning from Iraq and 10 percent of those from Afghanistan had PTSD. They figure the cost of treatment of Iraq vets over the next 50 years will be *$100 billion*. That's one pretty good reason *not* to start a war, huh?

Actually, other figures paint a worse picture. A *Washington Post* editorial titled, "Healing Warriors," on May 28, 2007, referenced a Pentagon study finding that 30 percent of soldiers experiencing heavy combat in Iraq had mental health problems. For those outside combat areas in Iraq, *only* 20 percent. A *Time* magazine article called, "The Hell of PTSD," by Tim McGirk, cited a Rand Corporation study finding that one of almost every 5 military personnel on combat tours have symptoms of PTSD or major depression. The article notes alcohol and drug abuse is common for those with PTSD, coupled with "an overall emotional numbness punctuated by outbursts of rage, severe depression and recurring nightmares," that can also lead to suicide or murder.

The article told the story of a former Navy SEAL, Mark Waddell, who had to seek help as a result of the effects of PTSD. His wife and kids noticed, when he returned from war, the noises of everyday life triggered his rages.

> "I'd come back from stepping over corpses with their entrails hanging out and my kids would be upset because their TiVo wasn't working," he recalls. Arriving home from one combat mission, Waddell insisted on sleeping with a gun under his pillow. Another night, he woke up from a nightmare with this fingers wrapped around his wife's throat, her face turning blue. [His wife] had to change the sheets every morning because of her husband's night sweats.

Another Iraq veteran described in the *Time* article, Sergeant Clint Hollibaugh, was the sole survivor of an ambush. He felt fine when he returned home. But one night he, "woke up outside his house; he had been patrolling the yard while sleepwalking. He kept a gun in every room of his house, one of them under the mattress. When his neighbor started firing off a shotgun, Hollibaugh instinctively leaped off the porch and began crawling through the grass while his wife looked on in horror and pity."

Michelle Turner's story was told in a *Washington Post* article from October 14, 2007, titled, "An Entire Family Struggles with PTSD." Her husband had "chronic and severe" PTSD after his Iraq duty. "It's like they took Troy and put him in a different person." His Army medical records say:

> He has nightmares frequently, two to three times a week, in which he sees himself back in Iraq ... and Baghdad. He sees himself fighting, sees dead bodies, parts of bodies, blood rushing from bodies. In the dreams he smells blood and burnt flesh and he hears bullets passing over his head. He is fearful and scared and wakes up in cold sweats. Flashbacks are also frequent, 2 or 3 times a week, triggered by helicopters passing over, burn flesh smell, barbecue, current Iraq news and sometimes seeing military vehicles brings flashbacks. ... He has a lot of guilt feelings that he could not save his sergeant.

Another article reports how difficult it is for wounded soldiers returning home and recovering at Fort Benning, Georgia, because there's a firing range across the street at the base. "Nearby Firing Range Complicates Soldiers' Recovery From Stress," by Ann Scott Tyson, *Washington Post,* June 3, 2008. One of the vets has what he calls "daymares – flashbacks caused by chronic PTSD that has left him paranoid. 'Anytime I see a U-Haul truck pull up, in my mind, I think it might be a car bomb,' he said." Another soldier reports living in *"near constant fear* of being shot or killed." (Emphasis added.)

Retired Air Force Tech Sgt. Scott Shore had multiple deployments, like a lot of those in the Army and Marines in Iraq, and his PTSD contributed to the breakup of his first marriage, as told in a *Washinton Post* article, "Troops at Odds With Ethics Standards," by Thomas Ricks and Ann Scott Tyson, on May 15, 2007. When he was home, he said, "I don't go into crowds, I don't like driving, I don't like doing a lot of things because I'm always on the lookout for the next ambush, the next IED." Meaning, improvised explosive device, which is how they killed a lot of our boys who were trying to kill them.

These traumatic stress situations lead to heavy physical and mental damage. There has been a lot more awareness of PTSD as the wars in Iraq and Afghanistan ended, and that's good. They're real life changing events, with sad consequences. But fear isn't really a *thing*; it's not a truck, lemonade, or a potpie. It's not a chemical or a mineral. It's a specific reaction to what a person feels or experiences. It's a *perception*, isn't it? Fear's simply an ethereal series of thoughts. How can it have such acute and chronic effects?

Well, frankly, I don't know. I suppose that's why they say no one knows what causes the serious symptoms of PTSD. But what seems true to me is that, *it's fear*; plain and simple, fear is the cause of PTSD. It's not easy to get rid of fear; it's like my Mom's worrying and anxiety. She can't really stop it. Her circuits are wired that way, even though she never went to war.

Once the soldiers experience those traumatic life-threatening events, it can send many of them over the edge. They're taught to tough it all out, which actually is pretty fearful, too. They need to learn to be numb to be able to go kill other people and deal with the fear of death all the time. Boot camp starts them off, according to a *Washington Post* article dated November 29, 2009, titled, "It's a job interview like no other," by Christian Davenport.

> *Nothing is ever fast enough. Nothing is good enough. The instructors find fault in everything, ... The yelling lasts all day, sergeant instructors, working in relentless shifts. The top of their lungs screaming wrecks their vocal chords and forces them to yell from their diaphragm, which produces an unnatural guttural growl that almost incomprehensible, but always intimidating.*

In January 2009, a number of suicides and attempts by West Point cadets lead to a formal investigation of the situation. "This is a stressful place. It's the United States Military Academy," said the Army's spokesperson in a January 30, 2009, *Washington Post* article, "Military Investigates West Point Suicides," by Ann Scott Tyson. Hazing contributes to the high stress level. One plebe who'd attempted suicide had been mercilessly teased after upper-classmen discovered he'd once worked for a private security contractor. "A lot of guys gave him a lot of crap. No one beat him up, but kids called him cruel names. That kind of mentality grows here; once someone gets ostracized, it snowballs."

Wait a second. I thought they were all on the same team up at West Point, you know, Band of Brothers. That's interesting, though. All these guys getting ready for war, and all of those suffering the awful effects of the war, were all in it to be fighting against some other guys. Why are we doing this to each other? The army men are stressing each other out, so they can go fight the other men and stress them out, and then everyone gets stressed out, angry, vengeful, hateful, mean, and nasty, and the whole cycle just keeps repeating itself.

Fear causes a lot of bad stuff.

Hmm. I wonder what causes terrorists to set those IEDs and do the suicide bomber thing? It seems the terrorists want to *cause* fear. That's almost by definition. They want other people to be afraid of them. Why? I mentioned before what Sam Harris points out – the Koran tells the Muslims to kill the non-believers. Does that mean, kill them and also make them afraid? There's an element of vengefulness, or maybe revenge, in some of the terrorist acts. They're not motivated by love. I wonder what's going on with them. Why do all those young guys do that?

In "The Homegrown Young Radicals of Next-Gen Jihad," by Marc Sageman in the *Washington Post* on June 8, 2008, he says:

> *What makes next-gen terrorists tick? … Here's a recipe: having a sense of moral outrage, seeing this anger as part of a 'war on Islam,' believing that this view is consistent with one's everyday grievances, and mobilizing through networks."* Interesting, this concept of a war against Islam. Did we declare a war against Islam, I mean, before 9/11? I know it was called a 'war on terror,' but the Muslims must view it as a war on their religion. Sageman says that, while Muslim Americans don't feel the attacks on Islam so much, in Europe it is different: *"The children of unskilled Muslim immigrants there face discrimination across the continent, resulting in striking unemployment rates. … Anti-immigrant sentiment … only reinforces the message of rejection – and produces grist for the terrorists mill.*

Dinesh D'Souza, writing in *USA Today* (you can find good information pretty much anywhere) on November 16, 2009, argues in the title of his piece "Don't Blame God for Terrorism." He says, rather than having a spiritual goal of 72 virgins in paradise, terrorists act out of secular goals. "The predominate theme … is that 'Islam is under attack'" from the forces of global atheism and immorality, and that Muslims should fight back to protect their religion, their values and their way of life." Seems over-reactive and kinda defensive to me, if that's their reason for murdering innocent people.

Jessica Stern writing in the *Washington Post* on January 10, 2010, "Five Myths on Who Becomes a Terrorist," would agree that terrorists act primarily for secular reasons. "Of the 25,000 insurgents and terrorism suspects detained by U.S. forces in Iraq as of 2007 [I had no idea we'd taken that many people into custody], nearly all were previously underemployed. ... [R]ank and file terrorists who claim to be motivated by religious ideology often turn out to be ignorant of Islam, ... the majority did not have much formal religious instruction and had only a limited understanding of Islam."

While some may practice a more pure or radical version, she says, "... it may be ignorance of Islam that renders youths vulnerable to al-Qaeda's violent ideology. ... Many speak, in particular about being motivated by a feeling of humiliation. A Kashmiri militant founded his group because, he said, 'Muslims have been overpowered by the West. Our ego hurts ... we are not able to live up to our own standards for ourselves.'" She also agrees with Sageman's point about the networks; knowing someone else in al-Qaeda was more important in joining than religious ideology. In other words, the need for *community* is a powerful factor driving terrorism, and other events in our world, too. Wanting to belong, like gangs or even soldiers, can serve as a remedy to loneliness or isolation.

"You have high rates of young guys unemployed [in Somalia]. You have a high rate of dropouts. They're difficult to integrate and work into the mainstream," says Omar Jamal, executive director of the Somali Justice Advocacy Center in Minnesota in another *Washington Post* article on March 11, 2009, titled, "Somali Americans Recruited by Extremists," by Spencer Hsu and Carrie Johnson. The religious extremists worked with the Somali youth and "gave them hope in their lives – and then indoctrinated them into this violent, radical ideology."

In the *Washington Post* on February 26, 2010, in an article titled, "Taliban Defectors Accept U.S. Approach But Wait for Promises to be Kept,' by Joshua Partlow, about the war in Afghanistan, it's reported that, "Several ex-fighters said they joined the Taliban not out of religious zealotry but for far more mundane reasons: anger at the government in Kabul, revenge for losing a government job, pressure from family or tribe members [interesting they are still in tribes] – or simply because they were broke."

There's also a real defensiveness with the terrorists, at least in what I read. Remember the terrorist attacks in Mumbai, India? Know why they did it? One of the Muslim attackers, Imran Babar, rattled off a litany of things he was mad about to a TV reporter, according to the *Washington Post* article, titled, "Cell Phone Calls Offer Clues to Motives of Mumbai Attackers," by Emily Wax, from December 16, 2008:

> The 2002 riots in Gujarat state during which more than 1,000 people, mostly Muslims, were killed; the 1992 demolition of the centuries old Babri mosque by Hindu mobs; and India's control over part of the disputed Himalayan region of Kashmir. Are you aware of how many people have been killed in Kashmir? Are you aware how your army has killed Muslims? We die everyday. ... It's better to win one day as a lion than die this way," Imran said.

So, there are a number of things going on here. These reports indicate many of the terrorists are not well-educated, un- or underemployed, humiliated, and embarrassed dudes who just want to belong. Looking at it in terms of the love-fear meme, they're afraid they aren't smart or good enough, don't have enough money, feel their culture is less successful (and may be jealous of the West, with its movies, music, fast cars, and sexy women), and want some love and recognition in their lives. It doesn't sound to me like they're evil; they're acting out of fear.

Something else I've wondered about; do the civilians in war torn countries also get PTSD from having so much fear of death all around? I read about terrible car bombings, with limbs, bodies, and shoes all over the place. If that was happening routinely at the nearby CVS, we'd be totally freaked out. What if neighbors and relatives were just disappearing before our eyes?

A March 19, 2007 report in the *Washington Post*, "Poll: Iraqis Gripped by Fear and Anger," by Will Lester, said that, "about three-fourths of Iraqis report feelings of anger, depression and difficulty concentrating; more than half of Iraqis have curtailed activities like going out of their homes, going to markets or other crowded places and traveling through police checkpoints." And we think we have something to be wigged out about here in the United States.

"I don't consider this post-traumatic, I consider it continuous traumatic, because the trauma they have is ongoing," said Abdul Razak, who ran a clinic in Sadr City in Iraq, according to a *Washington Post* article, "Iraq's Crisis of Scarred Pysches," by Jonathan Finer and Omar Fekeiki, from March 6, 2006. "In a survey of just over 1,000 randomly selected people across five Baghdad neighborhoods, completed ... by psychiatrists at Baghdad's Mustansariyah University, about 890 reported having experienced a violent incident first-hand, including all 27 children under 12 in the sample." "Iraq's conflict is exacting an immense and largely unnoticed psychological toll on children and youth that will have long-term consequences," according to another *Washington Post* article, "Iraqi Youth Face Lasting Scars of War," by Sudarsan Raghavan, June 26, 2007. The article continued about the epidemic of war and trauma-induced mental illness in Iraq:

> *On this morning, 4-year-old Muhammad Amar had a blank look on his soft, round face framed with curls of black hair. When mortar shells pummeled his street even months ago, he was too terrified to cry. 'He remained still, in shock. He froze,' said his father, Amar Jabur. ... Muhammad is showing signs of epilepsy and had a mild seizure the night before. Abdul Muhsin said he believes there could be a link between the explosions and the seizure .... Jabur cast a glance at his silent son. 'It is quite possibly because of the fear,' he said."*

I guess 'shock and awe' worked, huh?

As bad as this is, can you conceive of a situation where *women* would blow up other people? Well, you may know that many women in Iraq did just that. And what would cause such an act? The July 5, 2008 *New York Times* article entitled, "Despair Drives Suicide Attacks by Iraqi Women," by Alissa Rubin, explains the reason:

> *The women who become suicide bombers often have lost close male relatives – a husband, a brother, a son – in fighting, because they became suicide bombers or because they were detained by American or Iraqi security forces. ... "Although she is bombing herself and aiming to kill people, I feel these women are really victims of terrorism," said Mrs. Qaduri, who is Shiite and whose husband was kidnapped two years ago and has not been heard from since. "Only women in despair, in desperate situations, would do this. ... It seems in many of these cases the women have had their husband killed or sent to prison and she feels she has no choice, she is very depressed."*
>
> *In [the Iraqi] countryside, most women cannot imagine the world beyond the date palms they see on the horizon. It might be an hour-*

> long walk to the next village, there are no telephones, and cell phones often do not work. Most of the women cannot read.

I'm not going to say all suicide bombers do so for secular reasons. We've talked about the 'kill the infidels' theme in Islam, and there's also the 'you can go to paradise' theme. A *Time* magazine article on May 14, 2007 called, "Moms and Martyrs," by Tim McGirk, looked at the phenomenon of female suicide bombers. The article said:

> At least some of the captured suicide bombers ... say their motivation is the promise of paradise. Terrorist recruiters often tell male martyrs that a bevy of 72 virgins awaits them in heaven. But some women suicide bombers believe that in paradise they will become queens, while others are told by recruiters that no matter how old or grotesque they may be in this life, they will become the fairest of the 72 virgins that await each jihad warrior.

I tell you, they're pretty gullible. Is the religious reward theory really that convincing? We can look behind some of the stories from the *Time* article.

> In October 2003, a glamorous, well-to-do 29-year-old lawyer named Hanadi Jaradat calmly walked into a restaurant in Haifa and blew herself up, killing 21 Israelis and wounding 48 others. In her case, revenge was the motive; Israeli soldiers had raided her home, killing her brother and fiance, both militants, as she helplessly watched.

I read into that anger and hopelessness, i.e., fear of the future. Another female suicider, 22-year-old Reem Riyashi was honored as a hero after she blew herself up in January 2004 and took with her a number of Israeli soldiers at a Gaza border crossing. What caused her to do that, when she had two young children, was that her husband, "discovered that she was having an affair with a senior Hamas commander. Among conservative Palestinians, as in other parts of the Islamic world, an adulterous woman is often punished with death. Riyashi was given a second option: she could become a martyr."

Quite a price to restore her family's honor, I'd say. Wafa Samir al-Biss faced a similar choice. She was also only 22-years-old, and a burn victim from Gaza. She received free medical treatment for her burns from an Israeli hospital. "Militants convinced her and her family that since she was disfigured she would never get married and was better off becoming a martyr," according to the *Time* article. Her attempt to detonate her suicide belt at the hospital where she'd been receiving treatment failed when the bomb didn't explode. Later, in tears, she told the press, "Maybe I have been used." Do ya think? More from the article:

> *Some see becoming a suicide bomber as preferable to an arranged marriage, common in the Arab world. One teenager volunteered for suicide duty because her father refused to let her marry a boyfriend. 'I'd rather spend my life in an Israeli prison than trapped with a husband that I didn't love.'*

What if the effects of all those restrictions in Islamic society created the same type of fear that triggers PTSD? Traumatic life-threatening situations cause physical and physiological reactions that can stick around a long time. I wonder if a chronically oppressive society itself can cause the same type of reactions? It might be called "Perpetual Chronic Stress Disorder." Living under the sweltering restrictions of a repressive theocracy could give rise to irrational responses to social and political situations. When you combine the chronic stress of anger, hate, and random killings with stifling limitations on social interaction, the effects could be the same – though perhaps on a different level - as PTSD.

Jessica Stern surmises something similar in an article in the June 20, 2010 *Washington Post*. She is an author of a couple of books on terrorism, including *Terror in the Name of God; Why Religious Militants Kill*, and is a recognized expert on terrorism and PTSD. She and her sister as teenagers were victims of a serial rapist, which she writes about in the book, *Denial: A Memoir of Terror*. In the *Post* article, she tells about feeling less and less as she got older. In fact, in stressful situations, she would often be entirely calm and peaceful. But she also had "heightened sensitivity to sudden movements, scents, sounds and light." A therapist finally suggested to her that her symptoms were consistent with PTSD. She tackled her feelings head on, and worked with police to determine that the man who raped her had actually raped or attempted at least 44 girls between 1971 and 1973. He was convicted, went to prison, and eventually committed suicide. What's interesting is that she's interviewed many terrorists in the course of her work.

> It was only after I began research into my own rapist, whom the police and I discovered had probably been molested by a priest, that I thought more about the connection between the terrorized and the terrorists. I realized the possible importance of the frequency of rape in students at the radical madrassas I studied in Pakistan. I have felt, in my interviews of terrorists, that there was an element of sexual humiliation at work, but it was rarely more than an intuition on my part. Could sexual traumas contribute to contemporary terrorism?
> ... Is there a link between trauma and alienation and vulnerability to terrorist recruitment? Could terrorism sometimes reflect a kind of perverse post-traumatic evolution?

Another traumatically stressful situation leading to crazed behavior has been endemic in Africa. I'm talking about the rampant murder of civilians bordering on genocide in many African countries. What happens to those victimized by such violence? How do they make out?

Have you heard of the young rapper Emmanuel Jal? I read about him in *USAToday* on June 27, 2008, "Rapper Jal, living to tell about it," by Edna Gundersen. He was one of the 'lost boys of Sudan.'

> His ordeal began at age 7, when his father sent him, along with thousands of children, to the Ethiopian bush to train with the Sudan People's Liberation Army, in combat with government forces since 1983. "A seed was planted in me," he says. "I lived in war. Our village got burned, and I thought the world was ending with these loud bombs and houses burning and people screaming. My grandmother and uncle were beaten almost to death. When I saw my mom beaten and my auntie raped, a feeling of hatred was developing inside me."
>
> On the treacherous walk to the training ground, several children fell prey to wild animals or drowned crossing rivers. Boot camp conditions weren't much better. "I was beaten until I was dizzy. ... We were made hard. [Hmm, just like the U.S. Army recruits?] My desire was to kill as many Arabs as possible."
>
> After a year of training and five years of fighting, an exhausted Jal and about 400 "lost boys" deserted the army, fleeing across Sudan's

> *parched landscape with a meager supply of maize. "When the food got finished, we depended on snails, snakes and rats," he says. "Soldiers too weak to hunt were forced to eat bodies." Most succumbed to enemy attacks, disease, starvation, and suicide. Only 16 survived.*
>
> *Jal was lucky; he survived and was adopted. "I started going to church and singing in the choir. Love let me heal. Letting go of my hatred for Muslims helped the most. And music was my therapy."*

Jal now has a rap CD and a book. Another young boy who lived a similar life was Ishmael Beah, author of the book, *A Long Way Gone*. I was deeply moved by his story, but it wasn't pleasant. He lived in a small town in Sierra Leone. When the book begins, he's 12 years old, and with a few friends goes to a talent show in a nearby town called Mattru Jong. He and his brother and friends had started a rap and dance group when he was 8, having first seen rap music on the television. When they were gone, the rebels attacked his hometown, and from there, the young boys spent a long time running through the jungles trying to find their families.

He describes total chaos everywhere they went. Lots of screaming fleeing mothers and children, guns and bombs, and many shot up bodies. At various times, the rebels chased them through the bush. He finally was forced to join the army and became a boy soldier. After all of the horrible things he experienced, he became a killing machine. His family was all killed, his friends killed, and he had nothing to live for.

> *In the daytime, instead of playing soccer in the village square, I took turns at the guarding posts around the village, smoking marijuana and sniffing brown-brown, cocaine mixed with gunpowder, which was always spread out on the table, and of course taking more of the white capsules, as I had become addicted to them. They gave me a lot of energy. ... I walked around the village aimlessly, as I felt restless because I simultaneously felt a tremendous rush of energy and numbness. ... We watched movies at night. War movies, Rambo: First Blood, Rambo II, Commando, and so on, with the aid of a generator and sometimes a car battery. We all wanted to be like Rambo; we couldn't wait to implement his techniques.*
>
> *When we ran out of food, drugs, ammunition, and gasoline to watch war films, we raided rebel camps, in towns, villages, and*

Post Dog Bite Disorder

forests. We also attacked civilian villages to capture recruits and whatever else we could find. ... I ... felt special because I was part of something that took me seriously and I was not running from anyone anymore. [Community?]

A lot of things were done with no reason or explanation. Sometimes we were asked to leave for war in the middle of a movie. We would come back hours later after killing many people and continue the movie as if we had just returned from intermission. ... When we conversed with each other, we talked only about the war movies and how impressed we were with the way either the lieutenant, the corporal, or one of us had killed someone. It was as if nothing else existed outside our reality.

[W]e proceeded to practice killing the prisoners the way the lieutenant had done it. [The corporal] picked Kanei, three other boys, and me for the killing exhibition. The five men were lined up in front of us on the training ground with their hands tied. We were supposed to slice their throats at the corporal's command. The person whose prisoner died quickest would win the contest. ... I had already begun staring at my prisoner. His face was swollen from the beatings he had received. ... I didn't feel a thing for him, didn't think that much about what I was doing. I just waited for the corporal's order. The prisoner was simply another rebel who was responsible for the death of my family, as I had come to truly believe. The corporal gave the order with a pistol shot and I grabbed the man's head and slit his throat in one fluid motion.

Ishmael was eventually rescued by UNESCO and taken to New York. He was rehabilitated, and spoke at the United Nations. You can see the picture of this handsome young man at http://www.alongwaygone.com/. A 12-year-old boy, learned to kill with no feeling whatsoever. His real life terror led to murder and mayhem. Was he, or is he, evil? What could cause that kind of behavior? Is he any different than a terrorist, or one of the mass murderers, or me, or even my mother?

It seems these individual fears also can be played out collectively, in nations, for example. Here in the United States, we're trying to kill the terrorists, who apparently are afraid of us and want to kill us, and that's the reason we're afraid of them. I'm not saying it's wrong to protect oneself, but certainly our good old US of A is pretty fearful on that score. It's like we collectively have *Post-9/11 Disorder*. The fear of that attack totally changed our lives. Just drive around in my hometown, Washington, DC, and see all the police, army men, and barricades, and realize that the security lines at airports are also all about being *afraid* of an attack. I'm not saying those precautions aren't justified. But looking at whether actions are based on fear can be instructive in trying to understand what's really going on with us humans.

"The 'war on terror' has created a culture of fear in America … and has had a pernicious impact on American democracy, on America's psyche and on U.S. standing in the world." This according to Zbigniew Brzinski in the Outlook section of the *Washington Post* on March 25, 2007 in his article titled, "Terrorized by 'War on Terror.'" "Fear obscures reason, intensifies emotions and makes it easier for demagogic politicians to mobilize the public." So, what are the effects of this traumatic stress the United States suffered on 9/11? Zib says:

> *The culture of fear is like a genie that has been let out of its bottle. It acquires a life of its own – and can become demoralizing. … We are now divided, uncertain and potentially very susceptible to panic in the event of another terrorist act in the United States itself. … That America has become insecure and more paranoid is hardly debatable.*

That sounds exactly like PTSD, *for the whole country*, doesn't it? Can the psychological mindset of a nation be created by paranoia? Maybe, like Germany in the 1930s? If PTSD can affect a person, why not major fearful situations leading to national psychoses? Frankly, I think post 9/11 Disorder is responsible in many respects for the Tea Party movement that began in 2010; they are angry and afraid about so many things, the face of which was Barack Obama - socialist. This fear stuff is really wild when you start thinking about it.

What *else* does fear do?

Seven | Random Paranoias

I love to get up high. I like to go up on the roof of the house or to a high floor in a hotel and look out. I love to gaze out of airplane windows at the expanse below. Such awe; such perspective.

But if I'm on a balcony really high up, I freak out if I get too close to the edge. I'm not afraid to fall off; *I'm afraid I'll want to jump off*. I have to hold on tight to the railing to make sure I stay put. Where does *that* come from?

It's an odd juxtaposition. Love and fear at the same time.

There's such a thing as a building *too high*. I remember being on the top floor of one of the Twin Towers in New York maybe twenty years ago. I like looking out windows, but that swaying building was just too friggin' high. Scary. I'd be afraid of working there. Just unnatural.

Here's another one. I was quite shy as a little kid. I'd be afraid to speak up almost anywhere outside my home. I remember my friend Johnny's sister saying one time - I was maybe nine years old - "You don't need to ask Jeff anything, because he'll never say anything." That's right; I was afraid to open my mouth. I'm usually not shy now, but of course have moments when I don't want to speak up.

Is that nature or nurture? Was I born shy? Maybe it was good old BS. Fifty kids in a class and Sister Joel threatening to rap your wrists with a ruler. Or it might have been that big-time fear, you know, *of God*. Almighty, all-knowing, all-good God, compared to poor little me. We had that *Baltimore Catechism* stuff going into our fertile brains. I wonder how much of my personality now could be attributed to my Catholic upbringing? All that stress, being worried if you're good enough. It's one thing to worry about God, but another to worry about money.

> *Children raised in poverty suffer many ill effects: they often have health problems and tend to struggle in school, which can create a cycle of poverty across generations. Now, research is providing what could be crucial clues to explain how childhood poverty translates into dimmer chances of success: Chronic stress from growing up poor appears to have a direct impact on the brain, leaving children with impairment in at least one key area – working memory. 'There's been lots of evidence that low-income families are under tremendous amounts of stress, and we know that stress has many implications,' said Gary W. Evans, a professor of human ecology at Cornell University in Ithaca, N.Y., who led the research. 'What this data shows is the possibility that it's also related to cognitive development.' Washington Post,* "Study; Poor Kids Stress Affects Brain," by Rob Stein, April 6, 2009.

That makes a lot of sense. Stress can affect development of the brain. Wowzers. All the fears of growing up not being able to pay the bills, moving from place to place, not having a secure childhood - all that uncertainty - can affect kids' brains. The study found that, the longer the kids lived in poverty, the worse their scores on working-memory tests.

Lots of shades and colors of fear. Fear of heights. Shyness, the fear of other people, I guess you might call it. Fear of not having money and the things it brings. A whole range of perceptions *not* of oneness.

"I'm 27 years old, newly married, happily employed – and, for two months this fall, I was petrified to leave my house. I have panic disorder." So said Kara Baskin in the Health section of the January 9, 2007, *Washington Post* in an article titled, "Not Just Any Old Butterflies." Here's what else the article said:

> *According to statistics from the National Institute of Mental Health (NIMH), panic disorder afflicts roughly 6 million Americans, and women are twice as likely to suffer from it as men are. The attacks usually begin in one's 20's. Trembling, sweating, heart palpitations, shortness of breath, chest tightness and nausea are a few of the symptoms that come on like a lightning storm, out of the blue.*

Sometimes fear can get all out of hand. Then they call it *obsessive-compulsive disorder* (OCD). "OCD takes many forms, including excessive handwashing, fear of creating hazards for others, a need for order, anxiety over germs or contamination, repetitive checking of irons, ovens and door locks, and fear of harming others with knives or similar sharp objects," recounts a *Time* magazine article, "When Worry Hijacks the Brain," by Jeffrey Kluger, on August 13, 2007. These can have fancy names, like "scrupulosity – an intolerance of disorder or asymmetry, this is a fastidiousness that goes way beyond mere tidiness."

The article said the root of the obsession is anxiety, argued to be a good thing – you know, watching out for tigers and things that might hurt you – but these disorders are problems when they interfere with your life. Ask Michael Gowland:

> *A gravel-voiced fire department captain, Michael Gowland says he had never been a big crier. ... Now, sometimes, he cries two or three hours at a stretch. Other times, his temper has exploded, prompting him one day to pick up a crescent wrench and chase an auto mechanic around a garage. Even more perplexing to him, the once devout Roman Catholic now wonders "if there's anything out there."*
>
> *More than two years after the storm, it is not Hurricane Katrina itself but the persistent frustrations of the delayed recovery that are exacting a high psychological toll on people who never before had such troubles, psychiatrists and a major study say. ... [W]hile signs of depression and other ills doubled after the hurricane, two years later, those levels have not subsided, they have risen.*
>
> *Depression is often discussed in terms of chemical causes, but interviews with psychiatrists and patients here ascribed its appearance in post-Katrina New Orleans to the stresses of rebuilding.* From "Hurricane Katrina Exacts Another Toll: Enduring Depresssion," by Peter Whoriskey in the *Washington Post*, September 23, 2007.

You'd expect 9/11 to have created problems for people who were in New York, but "Americans who said they became anxious and stressed after the September 11, 2001 attacks – some just from watching the collapse of the World Trade Center on television – reported higher rates of heart disease up to three years later." A news service story in the *Washington Post*, January 8, 2008.

Remember Tim Russert? That lovable NBC anchor of "Meet the Press." There was a lot of discussion when he died at age 58 of a heart attack. It's always shocking when young, enthusiastic people die too soon. Why? Everyone asks how could this happen?

The experts at the time talked a lot about heart disease, plaque buildup in the arteries, high blood pressure, obesity, age, etc. "Another common concern is stress. Douglas Zipes, emeritus director of the cardiology division at Indiana University School of Medicine, said traumatic events such as terrorist attacks, natural disasters or a spouse's death can prompt a rise in sudden cardiac death, the abrupt stoppage of the heart," according to the article "Doctors Renew Heart Advice," by Brittany Johnson, in the June 24, 2008 *Washington Post*.

That basically means stress can kill you. But we knew that, didn't we? We all have stress, so does that kill us? By the way, what really is stress? Here's what the American Institute of Stress says:

> *If you were to ask a dozen people to define stress, or explain what causes stress for them, or how stress affects them, you would likely get 12 different answers to each of these requests. The reason for this is that there is no definition of stress that everyone agrees on, what is stressful for one person may be pleasurable or have little effect on others and we all react to stress differently. The term "stress," as it is currently used was coined by Hans Selye in 1936, who defined it as "the non-specific response of the body to any demand for change."*
>
> *Selye had noted in numerous experiments that laboratory animals subjected to acute but different noxious physical and emotional stimuli (blaring light, deafening noise, extremes of heat or cold, perpetual frustration) all exhibited the same pathologic changes of stomach ulcerations, shrinkage of lymphoid tissue and enlargement of the adrenals. He later demonstrated that persistent stress could cause these animals to develop various diseases similar to those seen in humans, such as heart attacks, stroke, kidney disease and rheumatoid arthritis. At the time, it was believed that most diseases were caused by specific but different pathogens. Tuberculosis was due to the tubercle bacillus, anthrax by the anthrax bacillus, syphilis by a spirochete,*

> etc. What Selye proposed was just the opposite, namely that many different insults could cause the same disease, not only in animals, but in humans as well.
>
> Stress was generally considered as being synonymous with distress and dictionaries defined it as "physical, mental, or emotional strain or tension" or "a condition or feeling experienced when a person perceives that demands exceed the personal and social resources the individual is able to mobilize." Thus, stress was put in a negative light and its positive effects ignored. However, stress can be helpful and good when it motivates people to accomplish more. Any definition of stress should therefore also include good stress, or what Selye called eustress. For example, winning a race or election can be just as stressful as losing, or more so. A passionate kiss and contemplating what might follow is stressful, but hardly the same as having a root canal procedure.
>
> Many times we create our own stress because of faulty perceptions you can learn to correct. [A]s Eleanor Roosevelt noted, nobody can make you feel inferior without your consent. While everyone can't agree on a definition of stress, all of our experimental and clinical research confirms that the sense of having little or no control is always distressful – and that's what stress is all about.

I get several things from this little exposition. Stress is not bad or good, or right or wrong. (Maybe there isn't right or wrong?) Clearly, however, different "insults" - from within or without – are considered stressful and can have serious physical or physiological impacts on us, and on animals (plants, too?). In his experiments, Selye subjected his animal subjects (poor little dudes) to stress in the form of "blaring light, deafening noise, extremes of heat or cold, [and] perpetual frustration." Let's think about those.

"Blaring light" – I didn't know light 'blared,' but whatever. Seems like that would be really bright light, the kind that was so bright you'd be afraid it wouldn't turn off and would blind you. Huh? *Afraid* it wouldn't stop and you'd go blind. OK, next.

"Deafening noise." Really loud, motorcycle, jet engine, explosion type noise, where you'd be afraid it would burst your eardrums, you'd go crazy, or it'd never stop. *Afraid* again, indeed. Next.

"Extremes of heat or cold." Fucking freezing, or hot as hell; and you'd be *afraid* you'd flat out expire.

Finally, "perpetual frustration." That'd be, to me, being *afraid* I'd never get it done. Constantly frustrated would mean you'd be afraid that you could never get off that track.

Each of these comes down to being *afraid*, and that always means having fear. Anytime you're afraid of something that's fear. *It's not wrong.* The American Institute of Stress, that bastion of thinking about stress, says stress can be good or bad, or even pleasurable. The Institute says many times people create their own stress through "faulty perceptions." Maybe like, perceptions of separation?

The Institute also says, "having little or no control is always stressful." Um, that would be, *afraid* of not having control, right? And also implies that you can't be stressed without your consent. Jeepers. You have to *decide* you want to be stressed, I think is what they're saying.

So, when the *Course in Miracles,* Gerry Jampolsky, and others say there's only love and fear, *maybe this is what they're getting at?* All this stuff we've been talking about – stress, murder, depression, anger, hate, and random paranoias – comes from fear. It's not a mathematical equation; not a physical law of the Universe. But, an allegorical theorem, maybe, to help form a perspective on things.

Here's something that can stress me: billable hours. I may have mentioned I'm a lawyer. It's a 'good' kind of law; I represent nonprofit organizations, mainly trade and professional membership associations. I like it. Associations do good things for society; I believe that. But I've always been in big law firms, until my most recent firm. Started out my career in the biggest law firm in the world. (I know, another scary thought.)

Anyway, for many lawyers, it's no longer a gentlemen's profession; it's all about the *money*. And for these firms to have enough money, you need to charge the clients (a lot) for what you do. The current method of determining how much to charge the client is by adding up the number of hours you spend and multiplying that by the hourly rate. For these gargantuan firms to make tons of dough, everyone needs to have lots of billable hours. Not just time I spend doing good work for the firm. I can spend time writing articles, giving legal speeches, helping younger associates learn the law, or even providing *pro bono* services (literally, 'for the good'), but if I can't charge, the firm doesn't get money. Fair enough, but here's the thing.

In order to keep my job at my old firm, I was supposed to bill enough hours so the firm is happy with me. I had a fairly low rate for being a partner in the firm – my time was billed at $550 per hour in 2010. My old firm wanted me to bill almost 1900 hours each year so I could be a good do-bee, advance, and make more money.

Well, though I do work pretty hard, I usually only billed about 1600 hours a year. I think that's a lot of hours, but I couldn't get ahead unless I spend an extra few hundred hours a year out of my life (which I'm *afraid* is my only life), so we can make the clients pay more and we can all be rich and happy.

It stressed the *bejesus* out of me. Everyday - *bill, baby, bill.* Now don't get me wrong, I've got a good gig. Nice house, beautiful wife and kids, cool stuff in my life. Gotta work for a living, I know, but billable hours can *stress me out*. I was afraid I'd work billing all those hours for years and not get to spend time outside watching the clouds and enjoying life, then just kick the fookin' bucket. That stress of billable hours sure seems based on fear.

I am happy to report, though, in May of 2011 AD, I took another job with a different firm, and it's much better. It's not as big a firm, so they're not as focused on hours for partners. I'm really happy about that.

Here's another of my fears. That I won't be able to pay off the mortgage by the time I want to retire. Seems like I paid alimony forever; gave the first wife the house, and bought our current home at the age of 50. Couldn't afford the 15-year mortgage at the time, so went with a 30 year. *What was I thinking?*

But we did get a good deal on the house, and we love it. It's a 1928 stucco house, a couple hundred feet from the canal, and not much farther to the Potomac River. I don't want to ever leave it. Gonna die right there in the upstairs room off the back of the house; the one that faces the River. I just have to pay off the mortgage first. So, trying to advance the payments and pay it off early. I find myself worried I won't be able to do that.

Of course, from all my reading about fear and stress, I know that debt stress can make you sick. Maybe that's what caused my bleeding mouth and floaters? Who knows? An article titled "AP-AOL Poll: Debt Stress Tears at Your Body, Too," by Jeannine Aversa, on June 9, 2008, said that:

> *The stress from deepening debt is becoming a major pain in the neck – and the back and the head and the stomach – for millions of Americans. When people are dealing with mountains of debt, they're much more likely to report health problems, too, according to an Associated Press-AOL Health poll. And not just little stuff, this means ulcers,*

> *severe depression, even heart attacks. ... That finding is supported by medical research that has linked chronic stress to a wide range of ailments. ... Among those reporting high debt stress in the new poll:*
>
> *27 percent had ulcers or digestive tract problems, compared with 8 percent of those with low levels of debt stress.*
>
> *44 percent had migraines or other headaches, compared with 15 percent.*
>
> *29 percent had severe depression, compared with 4 percent.*
>
> *6 percent reported heart attacks, double the rate for those with low debt stress.*
>
> *More than half, 51 percent, had muscle tension, including pain in the lower back. That compared with 31 percent of those with low levels of debt stress.*
>
> *People who reported high stress also were much more likely to have trouble concentrating and sleeping and were more prone to getting upset for no good reason.*

So, if I'm stressed about the mortgage, odds are I'll get (another) ulcer or something. Stress can do some mean stuff. None of it's right or wrong, though. It's my decision, just like we all make decisions. I could stop being a lawyer, sell the house, and get a job as a park ranger or something. I'd just have to live with the consequences. Maybe I make these choices because I mainly like the money?

Well, you know, that's a question; what's money? I think of money as a medium of exchange. Wikipedia bears me out. But today, it's really part of a global economic system. According to the International Society for Complexity, Information, and Design, "The concept of *scarcity* is essential to the field of economics. A resource is considered scarce when its availability is not enough to meet its demand. ... Most goods and services can be definable as scarce since individuals desire more of them than they already possess (scarcity is maintained by demand)." http://www.iscid.org/encyclopedia/Economic_Scarcity

I remember that from my college economics course. If there's a scarcity of supply compared to demand, the price goes up. The quote above says most goods and services are scarce; I guess that means there's not an infinite supply of things. So, the whole concept about money and our economic system, is that it's based on *the perception of there not being enough*. People are afraid they won't be able to get enough of what they want, so they desire more money than they need to be sure they can get what they want in the future.

I think this also plays out between countries as trading partners. One nation's people need more of one commodity and if several countries are *afraid* they won't have enough at a low price, everyone gets all flustered, upset, and angry. Think *oil*.

So, is our economic system based on the fear of not having enough? It makes me have to work harder, so may be a good thing for the firm; is it ultimately for me, though? I don't know, but it seems relevant here to inquire as to what effects the desire for money may have. A study in *Spirituality & Health* magazine in the November-December 2008 issue, discusses "How Money Hardens the Heart." Psychologist researchers Kathleen D. Vohs of the University of Minnesota and others at Florida State University and the University of British Columbia undertook the study, and reported, "In every condition, all participants who were reminded of money demonstrated behaviors consistent with decreased interpersonal skills and increased personal performance." (Kinda sounds like people in the United States, doesn't it?)

> *Specifically the participants who were exposed to money spent less time helping a person who needed it, sat farther way from another person, and preferred solitary activities. In addition, the money-motivated people showed preferences for working alone and asked for help less frequently.*

Well, if the economic system were imbued with fear, you'd think my trusty *Washington Post* shoulda had some articles about this during the economic recession. Let me look in my stack of papers; here are some random article headlines:

> *Fear Again Takes Over the Markets – October 16, 2008*
> *Stocks Sink as Gloom Seizes Wall St. – October 16, 2008*
> *As Markets Slump, U.S. Tries to Halt Cycle of Fear – March 4, 2009*
> *Fears On Global Economy Sink U.S. Stocks – November 17, 2010*
> *Global Markets Plunge on Economic Fears – September 23, 2011*
> *As Mortgage Rates Rise, so do Fears about Economy – June 19, 2013*
> *Around the World, Fears Rise and Markets Fall, June 21, 2013*

"The stock market is nothing if not a psychological barometer. The present signal is unmistakable: fear. ... The wild stock swings confirm the palpable fear and uncertainty." That was Robert J. Samuelson writing in the November 25, 2008 *Washington Post*, in an article titled, "A 'Wealth Effect' in Reverse." He had written on September 17, 2008 that, "Greed and fear, which routinely govern financial markets, have seeded this global crisis." Later, he wrote, on February 9, 2009, that, "banks and investors have become terrified of almost any risk." On October 5, 2009, "[P]anic, driven by the acute fear of the unknown, feeds on itself and disarms the stabilizing tendencies. In this situation, only government can protect the economy as a whole, because most individuals and companies are involved in the self-defeating behavior of self-protection." And, on May 24, 2010, he wrote about the Greek financial debacle that, "Panics thrive on fear and ignorance."

Wow, that's interesting. A respected economist identifying fear as the driving force in cataclysmic financial events, and also saying government – with its more detached perspective – is necessary to stem the tide. And what does all that fear lead to unabated?

"We know that fear and anger are linked emotions," said Matt Wallaert, a behavioral psychologist. "It's natural that at a time like this when everything seems uncertain, fear is going to produce anger." *Washington Post*, "Recession Has Changed Lifestyles, Poll Shows," by Nancy Trejos and Jon Cohen, April 29, 2009.

It's true. Back then, the country was awash in uncertainty and fear. Right before Barack Obama took office and during the time of his Inauguration, *everyone* was scared *shitless* we were going down the tubes - fast. That fear helped lead to the Tea Party anger. Wild thing about it, anger seems downright infectious.

My good buddy, (Not!) George Will, admitted as much in his article in the *Washington Post* on March 25, 2007, titled, "Anger is all the Rage." He says, "Americans are infatuated with anger." Of course, he was caustically implying that liberals and Democrats are the angry ones. I would say the shoe is dramatically on the other foot, but hear him out: "Today's anger is a coping device for everyday life. It also is the defining attribute of an increasingly common personality type: the person who, unless he is angry, feels he is nothing at all. That type, infatuated with anger, uses it to express identity. Anger as an expression of selfhood is its own vindication."

In fact, fear and anger *are* contagious. "When people have limited information about something important – a potential crisis in a building several stories high, a fire alarm goes off in a crowded theater or a sudden drop in the stock market – they use other people as guides to their own behavior. This can be smart if the people on whom you are modeling your behavior know more than you do. When you follow people who don't know what they are doing, and other people follow you, the resulting feedback allows small events to trigger huge and irrational changes in group behavior." The article titled, "Bad Ideas Can Be Contagious" from the *Washington Post* on December 17, 2007, by Shankar Vedantam, also cites studies saying that contagion clearly played a role in recent stock market crashes. The herd mentality, if you know what I mean. I'm thinking *group fear* might be a bad idea.

Greed might be a "virtue" that's not helpful, too. I always liked reading Steven Pearlstein in the *Washington Post*. He had a number of articles about the great recession of 2008 that focused on the impact of greed. "[M]ost definitions of greed refer to an excessive desire for wealth that is beyond what anyone really needs or deserves." He said that on October 3, 2008, just before the election that year, in an article titled, "Greed is Fine. It's Stupidity That Hurts." "Wall Street is nothing if not an organized system of greed, a high-stakes game in which the object is to take advantage of customers and counterparties by buying pieces of paper from them at less than they are really worth and selling them to others for more than they are worth."

Is this behavior based on a feeling of oneness and love for one another? I don't think so.

In his article, "Wall Street's Know-it-alls Can't Tell Right From Wrong," in the *Washington Post* on April 23, 2010, Pearlstein condemns, "Wall Street's complete and utter amorality. There, concepts like truth, justice, fairness, trustworthiness, duty of care, right and wrong are now totally without meaning." It just *feels* to me those concepts are similar to love, and greed is absolutely not. *Greed is simply a fear of not having enough.* I mean, if we trusted life and that it'd provide for us - meaning, you wouldn't starve to death – maybe we could all actually enjoy life and wouldn't need to fight so much for money, or power.

I thought our country was above that. But now, I'm thinking maybe not. In a November 22, 2010, *Time* magazine review of the book, *The Monster: How a Gang of Predatory Lenders and Wall Street Bankers Fleeced America – and Spawned a Global Crisis*, by Michael Hudson, the author quotes an Ameriquest manager's email, which said, "We are all here to make as much f_____ money as possible. Bottom line. Nothing else matters."

Hey, what about love?

There are so many books and articles on the 2008 Recession and its aftermath, that I don't need to go on more here. But there's another aspect to this money and greed thing. Alpha Males. That's the implication of an article titled, "Alphas in Their Bunkers," from the October 30, 2008 *Washington Post*, by David Ignatius:

> *The hedge fund industry coined a term several years ago for the idea that special people (i.e., hedge fund managers) could achieve above-average returns without taking commensurate risk. They called this investment nirvana "alpha," to distinguish it from the "beta" of average market returns … . Average compensation for the top 225 [hedge fund] managers last year was a jaw-dropping $892 million, up from $532 million in 2006. Five managers "earned" (if that's the right word for it) more than $1 billion each.*
>
> *The idea that you could use financial engineering to achieve high returns on capital with low risk became contagious. That make-believe world began to crash in August 2007. Suddenly, there was no market for the paper assets that had been created out of pools of mortgages – because of a falling market, nobody knew what they were worth. All the smarter-than-average people who had been chasing better-than-average returns began to get frightened. And over the past year, that fear became toxic. It sucked the trust and confidence out of the market. …*
>
> *And then the panic: That has been the most unattractive part of this story. The greed side of the alpha world was bad enough, with its $100 million homes and private art galleries. But the fear side has been more destructive. What's driving the severe financial downturn is the quest for "alpha security" among the richest and most powerful. Having made their loot, the very rich are desperate to protect it.*

No more desperately rich than Goldman Sachs. Not sure how I came across this article in *The Sunday Times* from the United Kingdom, but it had a fascinating look at the firm in November 2009. "I'm Doing 'God's Work.' Meet Mr. Goldman Sachs."

> *Goldman Sachs isn't nicknamed "Goldmine Sachs" for nothing. There's so much of the stuff sloshing around that in an average year a good investment banking partner will make $3.5m, a good trading partner $7-10m and a management committee member $15-25m.*

> *Some 953 employees got bonuses of at least $1m in 2008. [Lloyd] Blankfein [Goldman's Chair and CEO] may insist he is still a blue-collar guy, but he manages to have a $30m apartment on Central Park West and a 6,500-square-foot home in the Hamptons, the summer playground of New York's elite. One former Goldman banker describes the culture as "completely money-obsessed. I was like a donkey driven forward by the biggest, juiciest carrot I could imagine. Money is the way you define your success. There's always room — need — for more. If you are not getting a bigger house or a bigger boat, you're falling behind. It's an addiction."*
>
> *But there's another powerful motivator: doubt. There may be arrogance at 85 Broad Street — behind closed doors, Blankfein likes to joke (but not really) that he has "attained perfection" — but behind the bravado, Goldmanites, curiously, question their ability. "There is a deep and constant paranoia about everything we do," says [Michael] Sherwood [co-boss of Europe]. Insecurity is hard-wired into the system.*

You know, one could write a ton about Goldman Sachs, but I don't really care about them. It'd be nice to have the money, but I don't want my life dominated by that sort of insecurity; worrying about money all the time. Somehow, the people most flipped about money end up being the people in charge of the system. I do wonder what makes them tick.

Pearlstein in another article titled, "Failure In Need of a Theory," in the January 8, 2008 *Washington Post,* talks about how economics is standing on its head, where nothing seems to make sense any more and market failures are common. He points out the ridiculously high compensation paid to chief execs, "who would be willing to take a job for half of what they are being paid." Or the absurdity that first year law associates at big firms are paid more than sitting federal judges. Believe me, I know about that, and it makes no sense. Pearlstein talks about a new economic theory where people make really counter-productive decisions - "the goal is not so much to maximize profits, income or welfare, as economic models assume, but to beat the competition." He continues:

> *Perhaps nobody has done more to expand our understanding of rela-*

> tive competition than Robert H. Frank of Cornell University. Frank's particular focus has been on the importance of status in consumer choices. His point is that the desire for ever-bigger homes, ever-fancier gas grilles, ever-more powerful SUVs is based not on some absolute notion of what is good or sufficient, but rather on the relative basis of what everyone else has.
>
> It is this compulsion to keep up with the Joneses, Frank argues, which leads us to over-spend on status goods that, in the end make us no happier. Meanwhile, we wind up under-investing in leisure time or "public goods," such as better schools and parks, that would give us more satisfaction. ...
>
> [W]hat explains the herd behavior by Wall Street's money managers is the fear they might not do as well as their peers. By that logic, it's okay to lose money as long as everyone else does. ...
>
> Part of this motivation has to do with the human desire to be respected and avoid embarrassment, ... [b]ut just as important is the fear by money managers that they will be priced out of the market for the things they value at the margin – a nicer co-op on the Upper East Side, a fancier school for their kids ... .

This is *not* love and oneness, people. This is a perception that what one has – how much money, in particular – actually distinguishes and separates the person from others. Money differentiates, some think. Because, if I don't have more money or better things than *them*, then I certainly can't be any better, sheesh, I might even be worse. *Afraid* that I'd be no good or – heaven forbid – unlovable.

Yea, it's hard when you've got too much money. But what if you lost it? An American Public Media report from the Marketplace program on National Public Radio on August 24, 2009, talked about that. Kai Ryssdal started off the show by saying: "This isn't something a lot of people like to admit, but a good part of the way we see ourselves, how we identify ourselves, is shaped by how much we have. ... Ashley Milne-Tyte reports that, even for the super rich, their identity takes a beating when their portfolios do."

She described one man, who had lost $100 million in the recession (still had another $100 million), and didn't feel invincible any longer. "Impotence was the first problem, erratic behavior soon followed and drinking picked up afterwards."

Another very successful businessman who had lots of money had been traveling by private jet for thirty years. He lost tens of millions and had to fly commercial. He had all the instructions for how to fly on a regular plane, but got confused, couldn't find the plane, and it left without him. He was so unnerved he avoids planes now.

Another guy named Josh Harris made $80 million in the 1990's in the dot.com era. He spent all his money on some start up companies and lost everything, living now in a friend's pool house in Los Angeles. He wants to get married, but feels he needs to have money. How much? Twenty Million. Ashley says, "Hang on, are you saying you need to have $20 million before you can, quote, find a wife?" That's what he was saying.

I wonder if the guy was *lonely* without all that money, and no wife? I read that loneliness was a major factor in the rising crime rate among the elderly in Japan. Usually, it's the young people committing more crimes. In Japan in recent years, old people in some areas have been arrested more than the young.

"The elderly in Japan are committing crimes – nearly all of them nonviolent offenses, mostly petty theft – because of loneliness, social isolation and poverty, according to a [Japanese] Justice Ministry white paper .... 'When people feel lonely, there is an impulse to commit a crime so they will somehow connect with people,'" said a police officer reported in a *Washington Post* article on November 30, 2008. The police have found "a consistent pattern of isolation and anxiety among elderly who commit crimes. The police officer said, "They are not in touch with their children and have no connection with their brothers and sisters. ... They want somebody to talk to, ... If they get caught, they can talk to the police. They are very easy to catch."

A site called Web of Loneliness says, "a human being's existence is a lonely existence. At the end of the day, we are all alone. ... There is a painful reality that ultimately we are alone, by ourselves, and ultimately lonely." http://www.webofloneliness.com.

So, loneliness is about *not feeling connected*. In spite of how intimately everything is intertwined in the Universe, the lonely person feels alone. *Truly, a perception of separation.* The stress of loneliness also can make you sick, or kill you. According to the website Live Science, about one in five Americans experience loneliness, and it can be as harmful to your health as smoking. http://www.livescience.com/health/090218-lonely-brain.html. Lonely people have been found to have less blood flow throughout the body, poorer immune systems, and increased levels of depression. Some studies have said they die sooner. *Washington Post*, September 14, 2007, in an article, "Loneliness is Detected at the Genetic Level." Is loneliness sorta like money?

According to the article on Live Science, in the book *Loneliness: Human Nature and the Need for Social Connection* (W.W. Norton, 2008) ,written by John Cacioppo, a professor of psychology at the University of Chicago, and William Patrick, former science editor at Harvard University Press, the authors, "argue that loneliness creates a feedback loop that reinforces social anxiety, fear and other negative feelings. Getting out of the loop requires first recognizing it and overcoming the fear related with connecting with others. 'The process begins in rediscovering those positive, physiological sensations that come during the simplest moments of human contact,' Patrick said. 'But that means overcoming the fear and reaching out.' 'Lonely people feel a hunger,' Cacioppo added. 'The key is to realize that the solution lies not in being fed, but in cooking for and enjoying a meal with others.'"

These professors are saying loneliness is based on fear and the cure is to overcome that fear and go out for dinner and drinks with some folks. I like that idea; makes a ton of sense. Loneliness is also contagious, like fear and anger. Cacioppo led another study reported in the *Washington Post* on December 1, 2009, in an article titled, "Feeling Lonely? Chances are you're not alone," by Rob Stein, which concluded loneliness *can* be transmitted.

> *Although it may sound counter-intuitive, loneliness can spread from one person to another ... [and that] underscores the power of one person's emotions to affect friends family and neighbors.*
>
> *"Let's say for whatever reason – the loss of a spouse, a divorce – you get lonely. You then interact with other people in a more negative fashion. That puts them in a negative mood and makes them more likely to interact with other people in a negative fashion and them minimize their social ties and become lonely," said Cacioppo.*
>
> *[T]he new analysis, involving 4,793 people ... showed that having a social connection to a lonely person increased the chances of developing feelings of loneliness. ... The effect was most powerful for a friend, followed by a neighbor, and was much weaker on spouses and siblings, the researchers found. Loneliness spread more easily among women than men, perhaps because women were more likely to articulate emotions, Cacioppo said.*

Maybe fear does act sort of like a virus, as Darrel Ray argues about religion. I haven't been lonely much at all in my life, that I can remember. Even after I moved out of the house and had my little apartment with the blow-up bed, I wasn't lonely. I remember thinking one night – probably a Friday night – as I sat around my apartment, "I'll bet not even one person in the whole world is thinking of me right now."

That was probably true. I was keeping a low profile, not going to parties, and just trying to see the kids as much as I could. I was also working on a novel at the time I called *The Point* (though I wasn't sure what the point was; it was cathartic), reading spiritual and self-help books, and exercising. But I didn't get lonely.

So, here's another thing. Stress (remember, we were talking about stress) also can cause obesity. There apparently is some kind of biological switch that caused mice who were subjected to chronic stress and eating the equivalent of a junk-food diet to gain significantly more weight than other mice not being stressed. The kind of fat the stressed mice gained is also a problem. They put on belly fat, which is worse than other fat.

What's more peculiar, though maybe not unexpected, is that it's the same kind of visceral fat that people with depression get. "Older people who are depressed are much more likely to develop a dangerous type of internal body fat – the kind that can lead to diabetes and heart disease – than people who are not depressed. … People with depression were twice as likely as others to gain visceral fat – the kind that surrounds internal organs and often shows up as belly fat." I found the article, "Depression in Elderly Tied to Heart Disease," at http://www.msnbc.msn.com/id/28002128/ns/health-mental_health/t/depression-elderly-tied-heart-disease/#.TrHyDnGPZGA.

Depression is sort of a sadness, right? I haven't had any serious depression, but it seems to me it's also ultimately fear-based. All these random paranoias we are talking about have their roots in fear. We've seen how powerful fear can be when looking into PTSD. If fear can cause depression, which leads to physiological changes, then it has the power to affect humans, animals, and probably plants and other things, too.

It's commonly understood that depression is caused by chemical imbalances in the brain. If you go to www.depression.com, that's what it says there (of course, the site is funded and developed by GlaxoSmithKline). But here is what Google health says:

> *The exact cause of depression is not known. Many researchers believe it is caused by chemical imbalances in the brain, which may be hereditary or caused by events in a person's life.*
>
> *Some types of depression seem to run in families, but depression can*

> *also occur in people who have no family history of the illness. Stressful life changes or events can trigger depression in some people. Usually, a combination of factors is involved.*
>
> *Men and women of all ages, races, and economic levels can have depression. It occurs more often in women.* https://health.google.com/health/ref/Major+depression

It's not clear what causes depression, *but what if it's fear?* An article titled, "Depressed? Or Just Reacting Normally?" by Shankar Vedantam in the *Washington Post* on April 3, 2007, said, "Up to 25 percent of people in whom psychiatrists would currently diagnose depression may only be reacting normally to stressful events such as a divorce or losing a job. ... The new study ... found that extended periods of depression-like symptoms are common in people who have been through other life stresses such as divorce or a natural disaster and that they do not necessarily constitute illness." Jerome Wakefield, a New York University researcher who led the study, and another researcher Allan Horwitz from Rutgers University, "pointed out that sadness has increasingly come to be seen as pathological in the United States." They have written a book called: *The Loss of Sadness; How Psychiatry Transformed Normal Sorrow Into Depressive Disorder.* Something to think about.

OK, so another random paranoia has to do with guns. I was brought up around guns. My father loved to hunt ducks, geese, and deer. He used to go hunting a lot. Dad liked to get out with the guys in nature and shoot animals. He first took me hunting when I was about eight years old. We went down to the Island to shoot ducks. I was bundled up warm in wool sweaters, thick socks, and big boots (my feet have never been as cold as when we used to go duck hunting in those outdoor blinds).

I'd wait quietly with Dad, peering out over the water to see if any ducks would fly by. Actually, when I first started hunting, I wasn't tall enough to see over the camouflaged wall of the duck blind. I used to stand on a wooden whisky box. The anticipation of the hunt is quiet an experience. We'd get all excited if we saw a duck from far over there start to fly toward us. We'd crouch low, not making a move or a sound. Slowly, we'd reach a hand over to the gun, pull it up in position, and take the safety switch off if they came close.

It never occurred to me that maybe we didn't need to kill the little critters; it was just what I was brought up doing. I learned gun safety and to respect guns. My father even had a rack in our kitchen *with guns on it*. He had another gun display out in the

TV room. I distinctly recall he had an old World War II carbine. It was a nice small rifle. I'd take it out and show my friends when they came over. I knew never to point a gun at anyone, whether or not it was loaded. There was ammunition in a locked cabinet, but I knew how to get into it without the key.

Guns were for hunting food and, at times of war, for killing the enemy, whoever it happened to be at the time. Had to be afraid of those other armies. Hard to imagine what it's like. I watched that HBO series *Pacific* about World War II. It was awful, really horrible, what war is about. The Japs were pretty fired up soldiers. But now the Japanese people are one of our nation's close allies and they're our friends. Funny how that works.

I never had a gun for protection. I never felt so afraid that I needed to have a gun.

Wait a second; *I lied*. That Island I've been talking about has been an integral part of my life. My father was a member of the Jefferson Island Club that owned the Island in the Potomac River. We spent weekends and even weeks there all my life; still go down as much as I can. Anyway, after college I was hired as the manager and moved to the Island; the watermen who'd run the place for decades had retired. I frankly didn't know what I was doing. A friend of mine, Terry, came along and we worked and lived on the Island. I stayed there about a year and a half. Terry was with me the first summer, then my brother John came the second summer.

Point being, we had a gun; a shotgun that was mine from duck hunting days. The reason we had a gun was, before we moved to the Island, there'd been some vandals who'd come on the Island and totally ransacked the Clubhouse. They broke furniture, busted up the lockers, drank and stole the liquor, and just smashed everything up. One of them took a dump right on the floor. So, we were *afraid* they might come back and figured we'd better have some protection.

Of course, I'd never shot at a person before, so who knows what would have happened if I had to use it. I had seen a blowed-apart rabbit that a friend of my father's shot at the Island one time, so I knew what a shotgun could do.

I had two guns, but I recently gave them to my youngest son. They have some sentimental value from the old days with my Dad, but I don't have any shells and couldn't use them. I don't feel any need to have a gun now. I'm not afraid in my daily life. The key here is that the only time I ever had a gun for protection was when I was *afraid* of some other people. So, right or wrong, all the people who have guns for protection *do so out of fear*. Seriously, why would anyone have a handgun or some kind of automatic weapon if he or she wasn't afraid? And they almost seem *proud* to be so afraid.

I see those pictures in the newspaper of the terrorists all bandoliered-up showing how strong and powerful they are. Another way to look at it is to understand how fearful they are. Afraid of losing their religion, their values, or whatever. And always with big scarves masking who they are, *afraid* that someone will recognize them. They got a *lotta fear* going on.

If you agree that the terrorists have a lot of fear, then you probably think Americans are very brave. Well, the brave folks of the United States of America have more guns than the people of most nations. The Federal Bureau of Investigation estimates there are about 200 million privately owned guns in the country. With guns for police, military, etc., there could be one gun for each American. According to Wikipedia, there are more guns proportionate to population in the U.S. of A. than any other country. http://en.wikipedia.org/wiki/List_of_countries_by_gun_ownership. We have way more than the next country on the list, Yemen, about nine times more than Pakistan, and twenty times more than the United Kingdom. Aren't we cool?

Owning a gun is pretty much a religious-type obsession for a lot of people here in the U.S. I understand the Second Amendment to the Constitution gives us these rights, but are people so frightened of living in America that they need guns for protection? I thought we were the land of the free and the brave; how come everyone is so terrified of one another? Is our country that scary a place to live?

We talked in Chapter 5 about the mass murders here in the United States; most of them having been committed with guns. The rest of the world thinks we're nuts for having so many guns.

> *In Britain, there was shock at the scale of the [Virginia Tech] killings, but many people were not surprised, seeing the United States as a nation obsesses with guns, where firearms are easy to obtain.*
>
> *"I think the reason it happens in America is there's access to weapons – you can go into a supermarket and get powerful automatic weapons," Keith Ashcroft, a psychologist, told the Press Association. Ashcroft said he believed such access, along with a culture that makes gun ownership seem normal, increases the likelihood of such attacks in the United States. ...*
>
> *In France, news of the shootings dominated the Web pages of every major French newspaper. Bloggers responding to the reports*

> *overwhelmingly blamed the tragedy on what they called lax American rules on gun ownership.*
>
> *"In France, it is incomprehensible to understand what could prompt someone to own a handgun," a blogger identified as Aliosha wrote on the Web site of the daily newspaper Liberation, adding that it is "the right (almost the duty) for each American to be able to obtain a weapon without much trouble.* Washington Post, *"Shock, Sympathy and Denunciation of U.S. Gun Laws," by Kevin Sullivan, April 17, 2007.*

What about the gun violence on the streets of the inner city? It doesn't seem that hard to figure out now, does it? In a book review in the *Washington Post* on January 24, 2010, Colbert King wrote an article titled "Why Does a Young Man Buy a Gun?" reviewing the book, *Wrong Place, Wrong Time – Trauma and Violence in the Lives of Young Black Men*. King says of the book, "We learn, too, what it means to feel physically, psychologically and socially unsafe, and how that translates into getting a weapon for self-defense or retaliation. ... Pervasive fear and the instinct for physical and emotional survival, not economic necessity, turned them to violence."

Feeling unsafe, of course, stems from fear. Does the fear that drives a person to get a gun play out in other areas? Like the basic fear of life itself and all its troubles?

> *Suicide Rates Higher in States with the Most Gun Owners - Nearly twice as many people commit suicide in the 15 U.S. states with the highest rates of gun ownership than in the six states with the lowest rates of gun ownership, although the population of the two groups is about the same, researchers said last week. ...*
>
> *States with higher levels of gun ownership consistently have higher levels of suicide, and that is not because of differences in poverty, unemployment, drug addiction or mental illness, according to [the] study.* Washington Post, *April 16, 2007.*

Does that mean fear of others is coincident with fear of life? The author of that article, Shankar Vedantam, also wrote on the subject in the July 7, 2008, *Washington Post*, in an article titled, "Packing Protection or Packing Suicide risk?"

> There are many ways to read the Second Amendment to the United States Constitution, but all the versions point to one core idea: Americans have the right to own guns to protect themselves against outside threats, whether the danger comes from a school shooter, a vicious mugger, a robber breaking into a house, a lawless neighborhood – even the government itself.
>
> What the authors of the Second Amendment did not foresee, however, is that when people own a gun, they unwittingly raise their risk of getting hurt and killed – because the odds that they will one day use their gun to commit suicide are much larger than the odds they will use their gun to defend themselves against intruders, muggers or killers. ...
>
> [S]uicide – and not assaults, break-ins, muggings, school shootings and other fatal attacks by sinister strangers – ... dwarfs homicide as a killer in the United States. There were 32,637 suicides in the country in 2005 ... . That year, the collective homicidal mayhem caused by domestic abusers, violent criminals, gang fights, drug wars, break-ins, shootouts with cops, accidental gun discharges and cold, premeditated murder produces 18,538 deaths. Even the risk of terrorism doesn't begin to come close to the risk of suicide.

This is stunning to me. I can't comprehend suicide. *Life is really cool.* Life goes down as one of my *all-time greatest experiences.* But a lot of people apparently have such hopelessness and dread of living they end it all. If one commits suicide, isn't that a fear of living? So, self-hatred is more lethal than murder by someone else? Lots of Americans are so afraid of their fellow citizens they need a gun, but statistics show that their own fear is worse for society than killing by other people. What has gotten into people? *Where's the love?*

Maybe it has to do with misperception? Our buddy Shankar Vedantam opines it is. His article, "In Judging Risk, Our Fears are Often Misplaced," appearing in the *Washington Post* on September 24, 2007, basically says that people worry about the wrong things and that fear clouds their perceptions. All these random paranoias – and many, many more; believe me, I have lots more articles about fear-based behavior – are just *illusions.* They aren't real.

In the study by Jennifer Lerner at the Center for Public Leadership at Harvard's

Kennedy School discussed in that article, the conclusion is that "when people are asked to make judgments about risk in uncertain situations, they fall back on mental rules of thumb that regularly turn out to be preposterously wrong. ... Research going back three decades shows that people are more likely to worry about unusual risks and less likely to worry about everyday dangers." Yea, even an article in *Parade* confirmed that. "What Should You Worry About?" was in the October 18, 2009 magazine and said:

> *Humans are good at many things – typing, inventing stuff – but we're quite bad at assessing risk. Day after day, we get bent out of shape over things we shouldn't worry about so much, like airplane crashes and lightning strikes, instead of things we should, like heart disease and the flu. ... Fear sometimes distorts our thinking to the point where we become convinced that certain threats are unstoppable. ... We humans tend to respond to uncertainty with more emotion – fear, blame, paralysis – than advisable. Uncertainty has a nasty way of making us conjure the very worst possibilities.*

That means to me our perceptions of risk are often misplaced. The Tea Party folks were very angry; the anger was caused by plain old fear. It might be fear of taxes, big government, or minorities, but fear bred their anger, even though it could be completely misplaced as to the *source* of their discomfort.

Ocupy Wall Street also was based on a fear – that the elite didn't care and the rest of the folks would be screwed. There's been such a pomeling of the poor and middle class. But OWS versus the Tea Party was based on concern for everyone as opposed to distress about one's own situation.

As I read and think about this love – fear stuff, it makes more and more sense. I know it may seem simplistic, but fear-based emotions have significant consequences. I had no idea when I began researching all this. But there's one aspect of life that illustrates for me how really powerful fear is – sex.

Eight | **Who's Afraid of Daddy?**

It was cold in those duck blinds. That's why I was wearing an old wool facemask. Yea, *wool*. It was 1965, close as I can reckon. I was about twelve. Dad and I were in the blind, and not many ducks were flying. We were cold, *ergo* the mask. It was dark blue on the outside, with eyes, nose, and mouth cut out. *Not ergonomically designed.*

Dad couldn't see my face, or me smiling, when he gave me the old 'birds and the bees' talk. It was something like, "You see, son, the man puts his penis in the woman's vagina, and the sperm goes in there and makes a baby."

Little did Dad know I'd already heard this fantastic tale from a friend. He'd sworn the man *pees* into the woman's hole. That sounded somehow wrong, but somehow right at the same time. Funny how we often have a sense about things. Dad's explanation seemed more accurate, but the basic physical act was the same. I guess Dad should know.

It didn't take much longer for me to discover various aspects of this. I remember the first time lying in bed at night absent-mindedly fiddling with myself. Then – *Holy smokes! What's going on? I'm about to wet myself!* - Oooooo. … I turned on the light and looked down at a little white pearl. Oh, *that's* it. Now I can make a baby.

Girls changed around that time, too. They grew breasts and started wearing bras. I was intrigued. Went to dances and stood around hoping they'd play "My Girl." You know, a slow dance. The next decade or more was about trying to get horizontal with girls, or vertical, or even perpendicular. Ya know, women are *so nice*, I mean, *sweethearts*. And mothers are the greatest things on earth. I totally love women. I enjoy working with women; they're often more interesting than men. They're cuter by far. Sometimes a little wacky, but much cuter.

My Mom says, "The Lord made the sex urge a little too strong." It sure is strong, though I guess I can only speak for men. It's almost like you want it because you're afraid that, if you don't, you'll miss some opportunity to go down in history. Hmm, could some aspects of sex be fear-based?

My favorite Pulitzer Prize winning anthropologist author, Jared Diamond, has written a book entitled, *Why is Sex Fun?* Funny thing, he doesn't really explain why it's fun, but it did entice me to buy the book. In the fascinating manner of some of his other books, like *Guns, Germs and Steel* and *Collapse*, Diamond explains human sexuality from the perspective of evolutionary biology.

> *Attitude surveys in a wide variety of human societies around the world have shown that men tend to be more interested than women in sexual variety, including casual sex and brief relationships. That attitude is readily understandable because it tends to maximize transmission of the genes of a man but not of a woman. In contrast, the motivation of a woman participating in extra-marital sex is more often self-reported as marital dissatisfaction. Such a woman tends to be searching for a new lasting relationship; either a new marriage or a lengthy extramarital relationship with a man better able than her husband to provide goods or resources.*

Men and women play out their fears in different ways. I don't have all the answers on that, but it seems right. I wrote earlier that studies show women have more issues with depression than men. Different aspects of depression, too. For example, according to the website www.helpguide.org:

- *Role strain - Women often suffer from role strain over conflicting and overwhelming responsibilities in their life. The more roles a woman is expected to play (mother, wife, working woman), the more vulnerable she is to role strain and subsequent stress and depression. Depression is more common in women who receive little help with housework and childcare. Single mothers are particularly at risk. ...*
- *Unequal power and status - Women's relative lack of power and*

status in our society may lead to feelings of helplessness. This sense of helplessness puts women at greater risk for depression. ...

- *Sexual and physical abuse - Sexual and physical abuse may play a role in depression in women. Girls are much more likely to be sexually abused than boys, and researchers have found that sexual abuse in childhood puts one at increased risk for depression in adulthood. Higher rates of depression are also found among victims of rape, a crime almost exclusively committed against women. Other common forms of abuse, including physical abuse and sexual harassment, may also contribute to depression.*
- *Relationship dissatisfaction - While rates of depression are lower for the married than for the single and divorced, the benefits of marriage and its general contribution to well-being are greater for men than for women. Furthermore, the benefits disappear entirely for women whose marital satisfaction is low. Lack of intimacy and marital strife are linked to depression in women.*
- *Poverty - Poverty is more common among women than men. Single mothers have the highest rates of poverty across all demographic groups. Poverty is a severe, chronic stressor than can lead to depression.*
- *Coping mechanisms - Women are more likely to ruminate when they are depressed. This includes crying to relieve emotional tension, trying to figure out why you're depressed, and talking to your friends about your depression. However, rumination has been found to maintain depression and even make it worse. Men, on the other hand, tend to distract themselves when they are depressed. Unlike rumination, distraction can reduce depression.*
- *Stress response - Some studies show that women are more likely than men to develop depression under lower levels of stress. Furthermore, the female physiological response to stress is different. Women produce more stress hormones than men do, and the female sex hormone progesterone prevents the stress hormone system from turning itself off as it does in men.*
- *Puberty and body image - The gender difference in depression begins in adolescence. The emergence of sex differences during*

> *puberty likely plays a role. Some researchers point to body dissatisfaction, which increases in girls during the sexual development of puberty. Body image is closely linked to self-esteem in women, and low self-esteem is a risk factor for depression.* http://helpguide.org/mental/depression_women.htm

Interesting; a lot of these points involve relationships with men. I also don't mean to say men don't get depressed. Men go through lots of changes and often monetary problems. Men are the "providers." Or at least, we fashion ourselves that way. However, stresses, "including severe economic dislocations, can cause a man to lose faith in his abilities and prospects." "Economic Crisis Hits Men Harder," interview of psychotherapist Jed Diamond, by Sindya Bhanoo, *Washington Post*, February 17, 2009.

But even becoming a mother is so stressful that many women develop long-term depression. According to Sandra Boodman writing in the *Washington Post* on September 19, 2006, in an article titled, "Legacy in Blues," "unrealistic expectations about motherhood may be increasing the risk of depression in women who feel they can't measure up." And being afraid you can't measure up is a *fear*, just like low self-esteem is being afraid you're not good enough.

Women get angry about different things than men, and react differently. There's that saying, "Hell hath no fury like a woman scorned." My father was divorced from his first wife back in the 1940's. She got all his money and eventually took their daughter to California; she and Dad didn't see much of one another after that. For that matter, I rarely see her and she's my sister. Of course, I had my own divorce. You learn a lot in a divorce. I heard that, for my ex-wife, *rejection* was the big sadness. I didn't feel I'd rejected her; my soul decided to move on, but I honestly didn't intend to reject her. Truly.

A divorced friend of mine, who's been pretty much shut out of his daughters' lives, found out he'd become a grandfather when my wife told him she'd seen pictures of the baby on *Facebook!* He had no idea. It's sort of hard to imagine someone can feel so full of vengeance that they wouldn't tell the ex about his own grandson.

Another (female) friend of mine says women trade sex for security, and men trade independence for sex. That makes some sense, too. They give it up to reduce their fears of being alone or not being safe. We agree to stick around, 'cause we're afraid otherwise not to get as much as we'd like. Likely never as often as a guy'd want, but not *none*. Maybe the rejection my-ex wife felt was more deeply related to losing security, which is a *fear* of being alone or vulnerable.

Returning to the different fears of men and women, my wife Julie is afraid of running by herself on the canal. She's a big time runner. Ran in several marathons, including the friggin' *Siberian* marathon! She's not a fearful type. She'll go down for a walk on the canal with the dog and loves to run with me on the canal, but won't go running there by herself. Reason? She was attacked once by a man while running. She's jogging along and the dude jumps out of the bushes at her. She screams and manages to take off, so avoided getting raped or beaten. But it sticks with her.

Rightly so, because most often men are the perps. A random review of the *Washington Post* weekly crime report for Montgomery County, Maryland where I live on June 3, 2010 listed 95 incidents, such as break-ins, thefts, assaults, etc. Thirty-seven identify men as the culprits, two women, and all of the others didn't indicate the sex of the person committing the crime – but I bet they were men.

I wonder if the reason women may be more prone to depression in some respects is because of a *fear of men*? When you look at most of those murderers and terrorists we talked about, they're men. Women and girls have been raped and murdered by men for eons. Not being as strong physically may have something to do with it, but that's only a factor. I've never been afraid of being assaulted and killed by a woman. I've been afraid of being mugged by men, walking down a dark alley, but haven't been afraid of women.

Might women actually have an intrinsic, *evolutionary* fear of men? And rather than women simply being born with more tendency for depression, for example, maybe men somehow *cause* that in women? From the book *Zero Limits*, by Joe Vitale and Ihaleakala Hew Len, PhD (a book I will return to later), Dr. Lew states it even more bluntly, "One of the most insistent programs in the world is women's hatred of me, … . There is a deep-seated hatred of men on the part of women."

We're talking serious stuff here. If there're only two sexes in the human race on this planet, and one of them's afraid of or hates the other, that wouldn't be cool. Not saying it's so, but I don't think we can shy away from facing ideas that are difficult if tackling them might help us.

OK, let's do it. Listen to Jim Hoagland on April 17, 2009 in the *Washington Post*, in an article titled, "The War Within Islam."

> *"Leave me for the moment – you can beat me again later,"* a 17-year-

old girl begs between sobs in a video airing on Pakistan's private television networks and circulating on the Internet. But the local Taliban commander continues to flog her without mercy as a group of village men watch in silence.

Hoagland writes that we've read of the "brutal subjugation" of poor women in Pakistan and Afghanistan, but don't really understand it. We think they're fighting us over there because of their "fierce nationalism, … ancient warrior culture," and terrorism. But he says this bootleg video reminds us of another reason they fight us so tenaciously:

The desire of Pakistani and Afghan men to be left in peace to deal with their womenfolk as they see fit. There may be no more important recruiting tool for the Taliban and other Islamic extremist organizations. … All religions are absorbing the shock of globalization. But none has felt more besieged than Islam … . And none has produced as violent of a backlash from some of its adherents.

Again, we see threats to a way of life causing fear-based behavior. Why do the men keep the women down? Let me ask that another way; *why would anyone keep someone else down unless they were afraid of them?* I don't think the people of any particular religion are any stupider than the global population as a whole. But doesn't the basic concept of the story in Hoagland's article sound much like the underlying belief system in the following statement by Hajatoleslam Kazem Sdighi, an Iranian Muslim cleric quoted in the May 3, 2010, *Time* magazine as saying: "Many women who do not dress modestly lead young men astray, corrupt their chastity and spread adultery in society, *which increases earthquakes.*" (Yea, I did add that emphasis.)

What we have here are males afraid of women dressing immodestly and corrupting fine young men. (Why? Because they totally freak out and start masturbating at the sight of a women's neck or elbow?) As a result, they're *forced* to keep women down to prevent such debauchery. I was saying earlier that maybe men cause women to be stressed, and it might be that the reason men want to keep the bitches down is because the *men are afraid of the women!* Maybe the men just might find out they're not the stronger sex; that we're not so cool as we think?

I know I have a desire to be loved and admired. Maybe that's why I'm writing this book. I want to be recognized as a wonderful and smart person. That's not a bad thing. But if my desire to be special leads me to put women down or kill them, that's working out *my fear of not being loved* in a way that isn't helpful to women or society. It's not just a third world situation though.

Here in the land of the free and the home of the brave, "We're suffering under the mass delusion that women in America have achieved equality," Jessica Valenti writes in the February 10, 2010, *Washington Post*, "For Women in America, Equality is still an Ilusion," that:

> *Because despite the indisputable gains over the years, women are still being raped, trafficked, violated and discriminated against – not just in the rest of the world, but here in the United States. ... It's time to stop fooling ourselves. For all our "empowered" rhetoric, women in this country aren't doing nearly as well as we'd like to think. After all, women are being shot dead in the streets here, too.*
>
> *And it's not just strangers who are killing women; more than 1,000 women were killed by their partners in 2005, and of all the women murdered in the United States, about a third are killed by a husband or boyfriend. A leading cause of death for a pregnant woman? Death by a partner. ... In Iraq, women serving in the military are more likely to be raped by a fellow soldier than killed by enemy fire. ... The actual number of U.S. women raped in 2008 was more than 1 million.*

Again from the *Washington Post* on December 15, 2011, in an article entitled "Survey: Nearly fifth of women are victims of sex assault," by Rob Stein, the following:

> *Nearly one-fifth of U.S. women have been the victim of a sexual assault at some time in their lives, ... In addition, one in four has been the victim of severe physical violence by a boyfriend or husband, according to the National Intimate Partner and Sexual Violence Survey, which was conducted by the Federal Center of Diseases Control and Prevention. ... "The prevalence of sexual and intimate-partner violence is staggering," said Esta Soler of Futures Without Violence, a San Francisco-based group.*

And a June 20, 2013 report of the World Health Organization (WHO) titled, "WHO report highlights violence against women as a 'global health problem of epidemic proportions,'" said one-third of women around the planet have been victims of sexual or physical violence. Thirty-eight percent of women who were murdered were actually killed by their intimate partners. Here, check it out: http://www.who.int/mediacentre/news/releases/2013/violence_against_women_20130620/en/.

I mean, holy shit! Her point is that sexism and misogyny are rampant not just in Muslim countries, and not just in the United States, but all around the globe. Golly, with all that anger targeted at you, no wonder women would be afraid of men. They can trust Christian leaders though, right?

> *One in every 33 women who attend worship services regularly has been the target of sexual advances by a religious leader, a survey … says. The study, by Baylor University researchers, found that the problem is so pervasive that it almost certainly involves a wide range of denominations, religious traditions and leaders. … It found that two-thirds of the offenders were married to someone else at the time of the advance.*
>
> *Carolyn Waterstradt, 42, a graduate student who lives in the Midwest, said she was coerced into a sexual relationship with a married minister in the Evangelical Lutheran Church in America for 18 months. He had been her pastor for a decade, she said, and told her the relationship was ordained by God. "I believed him because I was looking for direction and for help," said Waterstradt, who ended the relationship years ago and entered therapy. Washington Post, "Many Women Targeted by Faith Leaders," by Jacqueline Salmon, September 10, 2009.*

There's, like, no place for women to hide.

> *When dealing with a "disobedient wife," a Muslim man has a number of options. First, he should remind her of "the importance of following the instructions of the husband in Islam." If that doesn't work, he can "leave the wife's bed." Finally, he may "beat" her, though it must be*

without "hurting, breaking a bone, leaving blue or black marks on the body and avoiding hitting the face at any cost."

That quote from a *Washington Post* article, "Clothes Aren't the Issue," on October 22, 2006 was written by Asra Q. Nomani, referencing a book by a Saudi scholar interpreting the 34th verse of the 4th chapter of the Koran. She says the idea of using physical punishment as "disciplinary action ... for controlling or mastering women" is common in the Muslim world. This is what the clerics are *preaching* to the men.

In Africa, too, the men have implemented a vile strategy of frightening women. An article in *Parade* magazine on March 22, 2009, titled, "We Must Stop the Rape and Terror," talks about the campaign to rape and terrorize women in Zimbabwe. According to Paula Donovan, an AIDS worker, "The bad guys, from the Congo to Zimbabwe to everywhere, have figured out that if you can't afford guns and bullets, and you have a political or military objective, the most effective, efficient strategy to employ is sexual violence."

In our intense world of sports here in the United States, athletes have developed the same reputation. Sally Jenkins asks in the *Washington Post* on May 8, 2010, "If women should fear athletes?" She contends there's a cone of silence where menacing athletes are given a free ride. We know about the situations involving many pro sports figures and sexual assaults, from Kobe Bryant to Ben Roethlisberger to Lawrence Taylor. But the trigger for her article was the tragic murder of 22-year-old Yeardley Love, a University of Virginia lacrosse star, at the hands of her boyfriend and men's lacrosse player George Huguely. He'd threatened her before, and his teammates knew it, but no one said anything. He beat her and left her to die in a pool of blood in her bed. Jenkins calls out athletes, but she might as well be targeting *all males* for the millennial conspiracy against women.

> *The truth is, women can't do anything about this problem. Men are the only ones who can change it – by taking responsibility for their locker room culture, and the behavior and language of their teammates. Nothing will change until the biggest stars in the clubhouse are mortally offended, until their grief and remorse over an assault trumps their solidarity.*

That's right. **Men have to change.**

But in the interest of comity, it seems women can do something to help. Maybe society needs to change, too. We'll have to think about what that might be.

Really, you may ask, how bad is it? How awful are men to women? Well, my recent awareness came from a conversation with a neighbor. In the way everything comes around, I was discussing this book at a neighborhood arts group. I mentioned that things around the world, though often bleak through a certain lens, were better than they used to be; I gave as an example the transition from slavery to our multiracial society. Janet said, there's more slavery now than ever before. She gave me a book to read, *Half the Sky, Turning Oppression into Opportunity for Women Worldwide*, by Nicholas Kristoff and Sheryl WuDunn. It's a story of how our global society has discriminated against and held women down in many different ways, from neglect of young girls to forced prostitution. I'm going to quote several passages from their book:

> *The global statistics on the abuse of girls are numbing. It appears that more girls have been killed in the last fifty years, precisely because they were girls, than men were killed in all the battles of the twentieth century. More girls were killed in this routine "gendercide" in any one decade than people were slaughtered in all the genocides of the twentieth century.*
>
> *Paradoxically, it is the countries with the most straight-laced and sexually conservative societies, such as India, Pakistan, and Iran, that have disproportionately large numbers of forced prostitutes. Since young men in those societies rarely sleep with their girlfriends, it has become acceptable for them to relieve their sexual frustrations with prostitutes. The implicit social contract is that upper-class girls will keep their virtue, while young men will find satisfaction in the brothels. And the brothels will be staffed from slave girls trafficked from Nepal, Bangladesh or poor Indian villages. ...*
>
> *Far more women and girls are shipped to brothels each year in the early twenty-first century than African slaves were shipped into slave plantations in the eighteenth or nineteenth centuries ... .*

Kristoff and WuDunn tell the individual heart-wrenching stories of girls captured or sold into sex slavery; many very young. And it's often a death sentence, since they don't use condoms.

> [I]t's not hyperbole to say that millions of women and girls are actually enslaved today. (The biggest difference from nineteenth-century slavery is that many die of AIDS by their late twenties.) The term that is usually used for this phenomenon, "sex trafficking," is a misnomer. The problem isn't sex, nor is it prostitution as such. In many countries – China, Brazil, and most of sub-Saharan Africa – prostitution is widespread, but mostly voluntary (in the sense that it is driven by economic pressure rather than physical compulsion). ... The problem of sex trafficking can more properly be labeled slavery.

The numbers of such modern slaves vary from the estimate of the U.N.'s International Labour Organization of over 12 million persons in forced labor of all kinds, including sex services, with one million children in Asia alone. The numbers of prostituted children each year is in the range of one million up to ten million. My question; *who are all those guys doing it?* Hey, we talked about the little-too-strong-sex-urge, but are human males so weak that they need to get their kicks from kids? Maybe God is telling them to do this stuff? After all, the Bible issues this commandment:

> If a man takes a wife and, after lying with her, dislikes her and slanders her and gives her a bad name, saying, "I married this woman, but when I approached her, I did not find proof of her virginity," then the girl's father and mother ... shall display the cloth [that the couple slept on] before the elders of the town. ... If, however, the charge is true and no proof of the girl's virginity can be found, she shall be brought to the door of her father's house and there the men of her town shall stone her to death. Deuteronomy 22:13-21.

I agree with Kristoff and WuDunn that: "Of all the things that people do in the name of God, killing a girl because she doesn't bleed on her wedding night is the most cruel." They go on to say that, "this emphasis on sexual honor is today a major reason for violence against women."

> Of the countries where women are held back and subjected to systematic abuses such as honor killings and genital cutting, a very large proportion are predominately Muslim. Most Muslims worldwide

> don't believe in such practices, and some Christians do – but the fact remains that the countries where girls are cut, killed for honor, or kept out of school or the workplace typically have large Muslim populations. Hinduism has similar problems, not to mention vicious burnings of brides by their new families, but Hindu women in India are more autonomous and more likely to be educated than their Muslim women neighbors.
>
> [O]ver the centuries Christianity has mostly moved beyond ... , [but] conservative Islam has barely budged. It is still frozen in the worldview of seventh-century Arabia, amid attitudes that were progressive for the time but are a millennium out of date. When a girls' junior high school caught fire in Saudi Arabia in 2002, the religious police allegedly forced the teenage girls back into the burning building rather than allow them to escape without head coverings and long black cloaks. Fourteen girls were reportedly burned to death.

Those girls sure had a reason to fear men. Given our discussion about the powerful effects of fear, I'm wondering what women's fear of men may have led to over time. We'll return to Jared Diamond's book to set the scene for a startling theory. He's talking about how humans don't know when a female is ovulating.

> Neither a male nor female baboon needs a hormonal test kit to detect the female's ovulation, the sole time when her ovary releases an egg and when she can be fertilized. Instead, the skin around the female's vagina swells and turns a bright red color visible at a distance. She also gives off a distinctive smell. In case a dumb male still misses the point, she crouches in front of him and presents her hindquarters. Most other female animals are equally aware of their own ovulation and advertize it to males with equally bold visual signals, odors, or behaviors. We consider female baboons with bright red hindquarters bizarre. In fact, we humans are the ones whose scarcely detectable ovulations make us members of a small minority of the animal world.
> ...
> We're also bizarre in our nearly continuous practice of sex, a behavior that is a direct consequence of our concealed ovulations.

> *Most other animal species confine sex to a brief estrous period around the advertised time of ovulation. ... We humans, though, practice sex on any day of the estrus cycle. Women solicit it on any day, and men perform without being choosy about whether their partner is fertile or ovulating.*

Why are concealed ovulation and constant sex important?

> *Since we humans are exceptional in our concealed ovulations, unceasing receptivity, and recreational sex, it can only be because we evolved that way. It's especially paradoxical that in Homo Sapiens, the species unique in its self-consciousness, females should be unconscious of their own ovulation, when female animals as dumb as cows are aware of it. Something special was required to conceal ovulation from a female as smart and aware as a woman.*

Diamond goes on to say that this constant human sex (as opposed to only doing it at ovulation time) would generally *not* be a good idea. Sex takes extra energy and time, and fornicating couples could be attacked by predators when they weren't watching. He discusses the two leading viewpoints as to these evolutionary developments; the 'daddy-at-home' and the 'many-fathers' theories. The first opines that concealed ovulation promotes monogamy so the man would stay home, which encourages parental care by the male. That one sounds nice. The second is that it allowed the women to have many sexual partners so the man wouldn't be certain whether her children were his. Hmm.

The 'many-fathers' theory is based on the historical fact of infanticide in many animal species over the centuries. I've heard from TV shows and movies that a male lion that becomes head of a pride will often kill the offspring of the females he takes on. Chimpanzees, gorillas, and some dogs do the same. Something like one third of infant gorilla deaths are due to this baby killing. It's to make sure that some other guy's progeny don't survive. Infanticide is especially likely to be committed by adult males against infants of females with whom they have never copulated – for example, when intruding males supplant resident males and acquire their harem of females. The usurper 'knows' the infants are not his own and kills them.

From an evolutionary basis, the female whose infants are murdered will re-start her cycle and be ready for sexual activity sooner. The male then has a better chance to pass his own genetic seeds along.

If ovulation is apparent, the male will know which infants are his, but the other males will also know, so they don't have any problem killing them if they get the chance. If the female's ovulation is *not* apparent – concealed from the male – and she's ready almost anytime, none of the males will know for sure if the infants are his or not. Since they don't know, they won't risk killing their own offspring and may even help in the care of the young-uns. Diamond explains that this evolutionary development helped females ward off murderous male suitors.

So, why is all this important or relevant to our discussion? According to Diamond, **"Anthropologists have long recognized that infanticide used to be common in many traditional human societies!"** Whoa, holy smokes! You mean that human males in the past murdered the infants of human females just like lions? **And the many-fathers theory would say that, because of women's fear of men murdering their children, they evolved so that no one would know when they were ovulating.**

You know, I'm sorry, but this is *profound*. Human men used to kill women's children, so the women's fear caused them to modify their own bodies to protect their children. This to me is the most dramatic indication of what fear can do. It makes evolutionary modifications to our bodies.

That's not the end of the story, of course. Diamond examines the historical mating systems in primates; promiscuity, harems, and monogamy. These were all prevalent at times in various species of us two legged animals. He concludes, "promiscuity or harems, not monogamy, is the mating system that leads to concealed ovulation. This is the conclusion predicted by the many-fathers theory. It doesn't agree with daddy-at-home theory."

Diamond goes on to explain that the daddy-at-home theory may have been at work, too. The male doesn't know when his woman is fertile because of concealed ovulation, so he stays at home to make sure he nails her at the right time. "[W]omen's concealed ovulation and constant receptivity evolved in order to promote monogamy, paternal care, and father's confidence in their paternity. ... *Then*, with concealed ovulation already present, humans switched to monogamy."

Or, to put it another way, Diamond says:

> *What it boils down to is that concealed ovulation has repeatedly changed, and actually reversed, its function during primate evolu-*

tionary history. It arose at a time when our ancestors were still promiscuous or living in harems. At such times, concealed ovulation let the ancestral ape-woman distribute her sexual favors to many males, none of which could swear that he was the father of her child but each of which knew that he might be. As a result, none of those potentially murderous males wanted to harm the ape-woman's baby, and some may actually have protected or helped feed it.

Once the ape-woman had evolved concealed ovulation for that purpose, she then used it to pick a good caveman, to entice him or force him to stay at home with her, and to get him to provide lots of protection or help for her baby – secure in the knowledge that it was his baby too.

The other thing this tells me is that, while fear may have triggered the evolutionary response in women to protect their children, it seems to have worked. The love they had for their children overcame the fear and they snookered the men out thumping their chests. The females' fear wasn't right or wrong, just a response to the situation. It would have been better for the men not to kill the infants in a selfish attempt to protect their DNA, but the monogamy that evolved has become a societal norm that works for some people.

I've attempted so far to point out what fear is and the powerful effects it can have. The fact that human females evolved so significantly in response to fear of males astounds me. Many physical changes in animals have likely been caused by fears. How many other profound actions are caused by fear that affects our species here on earth? Wars, murder, rape - all are fear-based. I think we can do better, especially if we can recognize fear and its effects to help us toward a better society. So, with this background, or perhaps at least suspended belief on your part, it's now time to turn our attention to love and oneness.

Nine | **A Dog's Life**

I am born. I suppose everyone starts off that way. All I can really remember is that one day I was looking at a lot of feet. Wherever I went I was looking at feet. And chairs. And stairs. Now, I've grown so that I'm eye level to knees. Which is better, but I'm so tired of looking up I could scream. Do you know how hungry you can get just looking up? And especially when every time you look up, it seems like somebody is eating. Really chowing-down on something good. Not Gaines-burgers. Well, they're not as bad as some of the other garbage that they try to give you. Ugh. That stuff is for the birds.

I do remember that for along time I lived in a big house with a fairly good sized back yard. But the family never wanted me to come inside any farther than the kitchen. What a drag. Can you imagine not being able to sleep in the TV room or see who is at the door? Whenever I can, I run through the house and check everything out. There are more neat things in houses. I get so excited that I can't control myself (of course, I am house-broken). I run all over the place and jump up on everything I can. I love to jump into walls and then bounce off and go running back and do it again and again. That's more fun than anything, except being petted. I love to be petted. I can just sit there and let my mind wander and it feels so good. Whenever somebody is petting me and they stop, I put my nose right under their hand and shake my head a little. The people usually laugh and keep petting. Oh, it's a dog's life!

Well, last year one of the boys in the family, Jeff, took me to this Island. It's fifty acres of the wide-openest country I every saw. I run around all the time, whenever I want and don't have to worry about cars. Those things scare me to death. You'll be running around, across the hard ground, and all of a sudden this huge car will be coming right toward you. If that isn't enough, then this loud sound goes off and I just about die!

Heaven is Everywhere

One of the other fun things about the island is the boats. You actually get out on the water and go faster than I could ever run, faster than deer even. It is just great. The first time I got into a boat I put my two front feet on the side of the boat and it started moving away! Jimmy Christmas, I couldn't move either way and, the first thing you know, I was in the water! Now I know not to do that.

After we go in the boat, we usually get into a car and drive off. I must admit that cars are okay to ride in. I sit in the back seat and watch everything go by. I think cars go faster than boats and it always makes me so excited I breathe really fast and hard. Jeff yells at me to sit down or lie down, and I do for a while, but I'd much rather stick my head out and lick the windows.

Sometimes we go home and other times, we go to this girl's apartment, which is fun because I like everybody there. Except Pete. Pete is another dog who made love to me once (it was really about ten times) at the Island. He was too short and no matter what anybody says, I am not pregnant. And I refuse to be until I meet some nice tall, brown and white, handsome dog. But Pete was the only one around, so what do you expect?

You know, there are lots of things to smell. People never seem to smell things much. I love to smell anything. And roll in dead fish! My goodness it's heaven. People always tell me that I smell. I think they smell! Some perfume is downright unnatural.

One thing that bugs me is that people are always waking me up. I'll be sleeping in a nice warm place somewhere and then they'll yell, "Hey, Muffin, whatcha doin?" Well, of course, I've been asleep, but I get up. Then, of course, I want to be petted. Whenever somebody wakes you up, they have to pet you. I wonder if people like to be petted as much as I do. Or whether they just disguise it. I'm glad I don't have to worry about what people think.

Have you ever heard of stereos? Well, I guess just about everybody likes music. I really enjoy a good tune every now and then. But why so loud? I have gotten used to it, but it is really enough to drive you to distraction. You can't sleep with Bruce Springsteen playing in your ear.

While I'm on the subject, what are those little white things that burn and people put in their mouths? It seems ridiculous to be chewing on something that is so hot. Whenever somebody puts some hot food in my dish, I wait until it cools off. I'm no fool.

You know what? It seems to me that I really couldn't be happier. Sure, there are a lot of things that upset me. But that's just part of it. When you come right down to it, there's nothing better. Well, enough for canine philosophy. I have many things yet to say, but I guess I'll have to wait. So long for now.

That was my dog Muffin. She lived with me on the Island for those 15 months when I was about 22 years old. A brown and white Springer Spaniel (pure bred), she was my constant companion. I wrote my first book that winter on the Island and called it *Phantasmagoria*. It was an eclectic compilation of musings, poems, and stories like the autobiographical one about Muffin.

Having spent so much time with her, we developed a real relationship. We loved each other. At least I think we did. Can a dog love? I'm sure many of you have had the same type of relationship with a dog, or maybe a cat, parrot, fish, or – well – almost any kind of animal. I mean, we know they're alive. We talk with them, and they communicate in various ways with us. I read something the other day saying that animals are a lot better at understanding our language than we are at having a clue about theirs (I forgot to cut out the article). They certainly can do many things much better than we can.

Did you know dogs can smell a thousand times better than humans? Little old Muffin could sniff out all sorts of birds, mice, and other animals around the Island. She also would look where I pointed. Ya think that's easy? According to a *Time* magazine article on September 21, 2009, entitled "The Secrets Inside Your Dog's Mind," by Carl Zimmer, "while humans and canines can [understand a pointed finger], *no other known species in the animal kingdom can*. Consider too all the mental work that goes into figuring out what a pointed finer means; paying close attention to a person, recognizing that a gesture reflects a thought, that another animal can even have a thought."

The article said that dogs are the most social animals, and that leads to the close relationships they have with humans. It may even mean nonvisual or non-auditory communication with animals. Here was a letter to the 'Animal Doctor' in the *Washington Post* on April 22, 2010:

> *Dear Dr. Fox:*
>
> *For quite some time, we've noticed that Trevor, our 6-year old Shar-Pei and retriever mix, was able to pick up on mind pictures. When my husband thinks, "I'll take the dogs to the field for a run," Trevor immediately starts jumping around and doing his little spread leg stance. All my husband has to do is think, "I'll take the dogs out," and Trevor starts his little excited dance. … We have been cognizant of any body language we might be portraying, such as picking up my*

> *purse or car keys or putting on a coat that our smart mutt might pick up on that we are going somewhere. We try to make sure that we don't give a hint, and yet he still somehow knows.*
> M.A.S.
> Arlington, Tex.

I don't remember if Muffin knew what I was thinking, but I can't say she didn't. I mentioned before that, at the time writing this, we had a 16 year old dog named Bella, who is a black and white Springer and Border Collie mix. She's a good girl, and very sweet. She always barked when someone came to the door (before she lost her hearing). But oddly enough, when my wife and I came to the door, she didn't even move. Sorta laid there and lifted her head when we came in, as if to say, "Hi, I was sleeping." How did she know it was us, and not bark? The other day, before I pulled into the drive-way from work, my wife said Bella went to the door and wagged her tail waiting for me. How does she know I'm almost home?

There must be some ethereal communication going on. In fact, sixty-seven percent of pet owners say they understand their animals' woofs, meows, or other sounds, including 18 percent who say they comprehend completely, according to an Associated Press-Petside.com poll, reported on December 17, 2008. Interesting also that more women than men say they understand their animals and more older and lower-income folks claim they do.

Are these imaginary feelings? Can anyone tell for sure that animals understand their thoughts? I know dogs can be trained to obey, but this is different. It's not what we have been taught to accept. My father thought Muffin was a dumb old smelly dog. In some ways, he was right. But I think we'd agree there's something else going on. If you don't think dogs can be happy, watch this: http://www.youtube.com/watch?v=ysKAVyXi0J4&feature=share.

What do dogs think about things? Is my story about Muffin accurate; is that what she might have thought? First of all, of course, they don't think in words or language, at least ours. There's a philosophical concept called linguistic determinism that says our words *limit* how much we can understand. According to Wikipedia, this theory holds that, "the words we possess determine the things that we can know. If we have an experience, we are confined not just in our communication of it, but also in our *knowledge* of it, by the words we possess." http://en.wikipedia.org/wiki/Linguistic_determinism

So, because Bella wasn't limited in her understanding of the world around her by specific words that have certain meanings, she may be better able to appreciate the subtle meaning of smells, sounds, or thoughts that we would have absolutely no concept about. We humans think we're so smart, but what does the smell of another animal's urine on a leaf on the path mean? Can Bella tell what that animal was doing, or what it was looking for or eating? Lord knows she likes to smell leaves and bushes for a *long* time! Doesn't it have to be the case that her mind is considering what she's smelling, and might it not be more than we could comprehend?

Point being, we and man's best friends likely have radical different concepts about life. We have different perspectives. Our brains may be bigger and we might be able to demonstrate our smarts in different ways, but how can we say that what we understand or appreciate about the cosmos is superior? It is the case, though, that dogs do have similar feelings to ours.

Before Juie and I got married, she had an old dog named Ponchik. You saw that right, *Ponchik*. Friends of ours would ask: "Pawn-chick?" - "Porno-chip?" Julie would politely respond, "Ponchik, it means donut in Russian. He was a little round puff of brown and white and looked like a sugar donut when he was a puppy." Julie had been in Moscow at the time and that's how he got his name.

Well, the little dude and I didn't see eye to eye at all. He was the most spoiled dog I'd ever met. Got up on the couch and would scrape his paws on the fabric to make his bed, and eventually *destroyed* Julie's couch. He didn't "come" or "sit" or even know how to eat steak bones! He didn't like to walk on grass, and insisted on pooping in the middle of the street! (I wouldn't let him; I was *not* gonna get run over picking up his crap.) Plus, he smelled so bad that, if the windows were closed, you couldn't breathe in Julie's apartment. I just didn't like the rascal. I couldn't help myself. I know I'm the peace and love guy, but I just wasn't into Ponchik.

Of course, he didn't like me either. He would avoid me like the plague. I would try to discipline him like the Dog Whisperer says to do, and he just paid no attention. For that matter, he paid no attention to Julie's efforts to train or discipline him either. But she loved him like crazy and treated him just like her baby, so he loved her, too. He fawned over her. She'd spend all this time petting him and walking him, and he was the *#1 man* in her universe. I mean, I wasn't jealous, I swear.

I wonder how he felt? Probably something like, "Ha, you douche bag, she loves *me!*" Truthfully, just hearing this much of the story, don't you think that's how he felt? Ponchik could *not* have felt *nothing* about it, right? He knew me, knew who I was, knew what I was doing (including in the bed with Julie, until I kicked him out of the room because his licking himself would wake me up at night). So the point is that the animals have got to have thoughts at least sort of like ours. In fact, ...

> [N]ew research is adding to the growing evidence that man's best friend thinks a lot more than many humans have believed. [A] provocative new experiment indicated that dogs can do something that previously only humans, including infants, have been shown capable of doing; decide how to imitate a behavior based on the specific circumstances in which the action takes place.
>
> The findings come amid a flurry of research that is revealing surprisingly complex abilities among dogs, chimps, birds and many other animals long dismissed as having little intellectual or emotional life. ... "Every day, we're discovering surprises about animals and finding out animals are far more intelligent and far more emotional than we previously thought," said Marc Bekoff, an animal behaviorist … . Washington Post, "What Were They Thinking? More Than We Knew," by Rob Stein, June 4, 2007.

The experiment showed that, while most dogs would move a bar with their teeth to get food, when a dog saw another dog use its paw to do that, she'd use her paw. I know, it doesn't sound that profound, but here's what the scientists said about it (and we know scientists are smarter than most humans):

> The findings stunned many researchers. "What's surprising and shocking about this is that we thought this sort of imitation was very sophisticated, something seen only in humans," said Brian Hare, who studies dogs at the Max Planck Institute for Evolutionary Anthropology in Germany. "Once again, it ends up dogs are smarter than scientists thought."
>
> The experiment suggests that dogs can put themselves inside the head of another dog – and perhaps people – to make relatively complex decisions. "This suggests they can actually think about your intention – they can look for explanations of your behavior and make inferences about what you are thinking," Hare said. Others go even further, suggesting that dogs have a sense of awareness.

> *"It really shows a higher level of consciousness," said Stanley Coren at the University of British Columbia, who studies how dogs think. "This takes a real degree of consciousness."*

Of course, dogs have a sense of awareness! You can't tell me that neither Muffin, Ponchik, nor Bella didn't have a sense of awareness. They always knew what they were doing. They can act downright human.

For example, Julie and I visited her sister's home just across the state line into New York. They had recently renovated the house big time, and the path from the front door to the kitchen had been rerouted; if you walked straight in the way it used to be, you'd stumble right into the bathroom. Bella had been going up to that house for years, so with the renovation, she kept walking into the new bathroom trying to get to the kitchen (for food, of course). Her understanding of the situation was not that different than mine. Dogs might not have it wrapped in the same thought packages we do, but isn't that the only real difference between anything; just the same energy in different packages?

Scientists also believe that individual animals have personalities. *Washington Post*, "Why Behavior Varies in Species," by Christopher Lee, June 11, 2007. That doesn't seem like rocket science to me; I know for sure that Ponchik, Muffin, and Bella had different personalities. When Julie walked Bella around the neighborhood, she knew the names and personalities of the other dogs more so than the humans attached to the other end of the leash. And the dogs seem to recognize other dogs, and shy away from the mean ones.

Can a dog be jealous? I know Ponchik wasn't jealous of me; he had *his* woman. But do they have feelings like ours?

> *Do dogs feel envy? A provocative new study indicates that they do, making man's best friend the first species after primates to appear to chafe at being treated differently. [There is a] growing body of literature suggesting that animals, including dogs, have much richer emotional lives and more sophisticated behavior than humans have traditionally believed.*
>
> *"The more we study animals and the more we learn about them, the more we are realizing that maybe humans are somewhat different but not really all that different," said Friederike Range of the Univer-*

> sity of Vienna in Austria, ... "They have these kinds of feelings, or at least the precursors of these feelings, that we thought were uniquely human." Washington Post, "Dog Feel Envy – or at Least Grasp Inequity When It Comes to Treats," by Rob Stein, December 15, 2008.

The study involved a bunch of border collies, shepherds, retrievers, mutts, and other types of dogs and a protocol by which dogs would get rewarded for a task. Sometimes the researchers would give a treat to some of the dogs that completed the task, but not others. What would you expect?

> [T]he deprived dogs would start acting frustrated, scratching themselves, licking their mouths and yawning. "They would refuse to look at you, start looking at their owners or at the other dog chewing, and eventually refuse to cooperate. ... It's not just, 'Oh, shoot. I'm not getting rewarded, so I stop working,'" Range said, "If both are not rewarded, that is not a big problem. But if you rewarded one and not the other, that's where you saw the problem."
>
> "As time goes on, we're collecting an astounding amount of information about the social skills that animals have for negotiating their social world," said Marc Bekoff of the University of Colorado, author of the forthcoming book "Wild Justice," about animal morality. Bekoff noted, for example, that mice have shown signs of empathy for other mice in danger, that elephants will go out of their way to care for other injured elephants, and that coyotes will ostracize other coyotes who play unfairly. ... "From an evolutionary point of view, there's very little that we have that other animals don't have."

I remember when my oldest son Jay was a toddler. His first word was "ball." Then he learned another word; I don't remember what it was. Then, another, and another, and I recall at one point he knew about eight words. After that, no counting. He seemed to learn a thousand words overnight. No way to keep track. How many words do you think a dog can learn?

I know Bella knew her name, but I'm not sure how many words she knew. A six-year-old border collie named Betsy knows a lot of words, as reported in the March

2008 issue of *National Geographic*, "Minds of their Own; Animals are Smarter Than You Think," by Virginia Morell:

> Betsy can put names to objects faster than a great ape, and her vocabulary is at 340 words and counting. Her smarts showed up early: At ten weeks she would sit on command and was picking up on names of items and rushing to retrieve them – ball, rope, paper, box, keys, and dozens more. She now knows at least 15 people by name, and in scientific tests she's proved skills at linking photographs with the objects they represent. ...
>
> [A] border collie named Rico appeared on a German TV game show in 2001. Rico knew the names of some 200 toys and acquired the names of new ones with ease. Researchers at the Max Plank Institute for Evolutionary Anthropology in Leipzig heard about Rico ... that led to a scientific report revealing Rico's uncanny language ability: He could learn and remember words as quickly as a toddler.

And I recently read about a dog whose vocabulary is in the six figures! "Dog Breaks Vocabularly Record: Chaser the Border Collie Understands Over 1,000 Words," by Joanna Zelman, in the *Huffington Post* on January 11, 2011. Watch the video here: http://www.huffingtonpost.com/2011/01/11/dog-breaks-vocabulary-rec_n_804728.html

Where did dogs even come from? Weren't they just wolves? Well, yes, probably so. And it musta taken a long time for them to transform into the some 160 breeds now on earth. Well, uh, maybe not.

There are about 25 breeds of working dogs, 29 of sporting dogs, 23 of herding dogs, 26 of our hound friends, 23 toys, 28 terriers, and some others. The current theory about how they became our BFFs is quite amazing. It didn't take long at all to domesticate dogs, and it happened pretty quickly around the globe, maybe 10,000 years ago. The thinking is that dogs became part of our families when humans started living in villages.

What's the first thing that happened when humans stopped being nomadic and gathered in more permanent living places? Trash. We (really them, but still us) started taking our leftover food and junk to a dump area, just like we do now. There'd be wolves around, and it was pretty enticing for them to see all this food they could munch on without having to hunt and kill it. They could go hang out at the dump, wait for the humans to leave, and pounce on the bits of food.

Some of the wolves had a long "flight distance" according to biologist Raymond Coppinger. That means they stayed pretty far away so they could vamoose if they wanted. Others weren't so *afraid* of humans and would come closer, realizing they'd get more food. Those dogs gradually became domesticated. Meaning, the nicer ones became our friends and moved in.

How fast were they domesticated? Well, a study conducted in Russia in the 1950's provided a hint. The goal was to breed the wild foxes that could be domesticated.
http://thoughtfulanimal.wordpress.com/2010/02/23/russian-fox-study/

Russian scientist Dmitri Belyaev had lost his job at the Department of Fur Animal Breeding at the Central Research Laboratory of Fur Breeding in Moscow, but wanted to conduct his own study. They had a bunch of foxes in cages and determined which ones to use for breeding based on the simple formula of sticking a hand with a glove on it into the cage, and only breeding the foxes that didn't bite. Within just a couple of generations, not long at all in fox time, the foxes were not only completely tame, but also had started displaying different features of fur color, etc.

In other words, evolution plucked out the nice foxes, essentially transformed them into dogs overnight, and changed their physical attributes. Voila, dogs! I wonder if we bred humans that way, could we evolve out the sickos who want to kill other people?

How much are they really like us? Do dogs experience the effects of fear like we do?

> *Gina was a playful 2-year-old German shepherd when she went to Iraq as a highly trained bomb-sniffing dog with the military, conducting door-to-door searches and witnessing all sorts of noisy explosions. She returned home to Colorado cowering and fearful. When her handlers tried to take her into a building, she would stiffen her legs and resist. Once inside, she would tuck her tail beneath her body and slink along the floor. She would hide under furniture or in a corner to avoid people.*
>
> *A military veterinarian diagnosed her with post-traumatic stress disorder — a condition that some experts say can afflict dogs just like it does humans.*
>
> *"She showed all the symptoms and she had all the signs," said Master Sgt. Eric Haynes, the kennel master at Peterson Air Force Base. "She was terrified of everybody and it was obviously a condition*

that led her down that road." "Military Dog Comes Home from Iraq Traumatized," by Dan Elliott, Associated Press, August 3, 2010.

So great, we take our dogs to war, and they get PTSD just like the humans. I don't think war is good for man or beast. Do they feel grief too? From "Animal Doctor," in the *Washington Post*, August 12, 2010,

> Dear Dr. Fox:
>
> Your recent column about how a "dog's devotion to master can lead to the grave" is similar to what occurred with our golden retriever more than 20 years ago, when my husband died at 46 after a four-year battle with cancer. During my husband's illness, Friday lay beside his bed, provided support when my husband walked and never left his side. Friday obviously knew that something was wrong. He was devoted to his master.
>
> Before my husband became ill, he was a senior sports-and-news cameraman for a major TV station. Because of the nature of his assignments, my husband's work hours were unpredictable. Regardless of the hour, Friday always knew when my husband was headed home. He ran to the front door, wagging his tail, and he sat patiently until my husband's car pulled into the driveway.
>
> After my husband's death, which took place in a hospital, Friday sat at the front door all day, every day, whining and waiting for my husband to return. He stopped eating and wouldn't leave the front hallway. He refused to play with our children, whom he loved, because "guard duty" was his only purpose. He left his post only when he needed to be walked. My heart was breaking for this dog.
>
> After one week of watching Friday's vigil, I decided to help him understand what had happened. Hesitantly, Friday left his post and got into the car with me. His car behavior was unusual: he paced from window to window, looking everywhere for my husband. I drove to the cemetery, and we walked together toward my husband's grave. As we got closer, Friday pulled away from me and ran directly to the grave. He lay down atop it, closed his eyes and just stayed there, quietly. I didn't try to talk to Friday or disturb him. He needed to grieve,

too. After an hour, Friday got up and walked over to me, using his mouth to hand me his leash. He was ready to go home.

On the way back home, Friday lay quietly in the back seat. After we arrived home, he kept kissing my hands as if to say "thank you," and he never again sat by the front door waiting for my husband to return home. He now understood. Although obviously sad, his behavior returned to normal around the children and he began eating again. In time, he healed, as did we.

L.B.J.

Lake Worth, Fla.

Ten | Elephants, Slime Mold, and Worms - Oh, My!

Can a dog really grieve?

What *is* grief?

The only time I really grieved, so far, was when my father died. It was 1991. He'd had a stroke in 1985, the last year I was in law school. The mountain of a man I revered my entire life was chopped off at the knees. Slowly, he declined. His grandchildren grew up around him, but they didn't know the man I knew.

I walked outside after he passed, and I wasn't all there. I felt like my arms, or legs, or maybe heart, had been cut out. My spirit was drained.

The love I felt was palpable. I'd never known the love for my father that I had when he was gone. The strongest feelings I'd experienced touched my core.

Maybe also because it was the closest I'd ever been to death.

All of sudden, I was *mortal*. I wasn't going to wake up every morning forever. All of a sudden, I was afraid I was gonna die, too. That fear of death thing again.

The deepest love and fear together; at the same time; right then.

That's what I've learned grief to be; extreme fear and love at the same time. I don't think we celebrate grief enough. We somehow seem to shy away from it, because of the feeling it evokes - that we're going to die.

So, could an animal experience grief? Could her dog Friday truly know grief for her husband? Well, *why* the hell not?

> It's well established that elephants appear to mourn their dead, lingering over a herd mate's body with what looks like sorrow. They show similar interest – even what appears to be respect – when they

> encounter elephant bones, gently examining them, paying special attention to the skull and tusks. "Inside the Minds of Animals," by Jeffrey Kruger, Time, August 16, 2010.

When Dad passed, we had the option of choosing a nice mahogany urn, ceramic container, or even a coffee can. But we decided to make our own. My brothers and I went to his workshop and we built a heart-of-pine box – with inlaid handles – to put his ashes in. We buried him near his parents, where he wanted to be. One might say we paid special attention to his skull and tusks.

> *Among the different emotions that animals display clearly and unambiguously is grief. Many animals display profound grief at the loss or absence of a close friend or loved one. Nobel laureate ethologist Konrad Lorenz writes: "A greylag goose that has lost its partner shows all the symptoms that [developmental psychologist] John Bowlby has described in young human children in his famous book Infant Grief . . . the eyes sink deep into their sockets, and the individual has an overall drooping experience, literally letting the head hang ... ." Sea lion mothers, watching their babies being eaten by killer whales, wail pitifully, anguishing their loss. Dolphins have been seen struggling to save a dead infant and mourn afterward.*
>
> *Wild animals also grieve. Among the best examples are grieving rituals of elephants in the wild observed by such renowned researchers as Iain Douglas-Hamilton, Cynthia Moss and Joyce Poole. Captive elephants also grieve. To quote Joyce Poole: "As I watched Tonie´s vigil over her dead newborn, I got my first very strong feeling that elephants grieve. I will never forget the expression on her face, her eyes, her mouth, the way she carried her ears, her head, and her body. Every part of her spelled grief." Young elephants who saw their mothers being killed often wake up screaming.* "Grief in Animals; It's Arrogant to Think We're the Only Ones Who Grieve," October 29, 2009, by Marc Bekoff, in *Animal Emotions*.

It is actually becoming quite accepted that many animals do grieve. "Scientists are finding new evidence that beasts, honor, mourn, and even hold wakes for their dead," said an article in *Time* magazine, by Jeffrey Kluger, April 15, 2013, titled "The Mystery of Animal Grief." The article told about crows – up to a hundred of them - basically holding memorial services for their dead!

And this floored me. Ever heard of the "Elephant Whisperer?" I know everyone watches the Dog Whisperer on television, but I hadn't heard about the elephant guy. Anyway, his name was Lawrence Anthony. He had saved and lived with elephants in South Africa. He had a real connection with them. When he died March 7, 2012, within three days *two herds of elephants he knew showed up to pay their respects;* no shit. You can watch a story about it here: http://delightmakers.com/news/wild-elephants-gather-inexplicably-mourn-death-of-elephant-whisperer/. How did they even know he had died?

Elephants are pretty smart, too. They can play drums and soccer, paint and use tools, and communicate many kilometers through infrasound (below the level of human hearing) vocalization. Some say they can actually talk. "Kosick, Talking Elephant, Attracts Researchers and Tourists in South Korea," (Video), *Huffington Post*, October 10, 2010. http://www.huffingtonpost.com/2010/10/11/kosik-talking-elephant-at_n_758335.html.

Female elephants are thought to call out when they're in heat, and the males show up quickly. They apparently can "hear" through the ground. The soft pads at the bottom of their feet pick up vibrations and sounds, including their infrasound vocalizations, and may even be able to detect seismic events (you know, like an earthquake). Researcher Joyce Poole has found that elephants use more than 70 kinds of vocal sounds and 160 different visual and tactile signals, expressions, and gestures in their day-to-day interactions. From *Nature*, online: http://www.pbs.org/wnet/nature/unforgettable/communication.html

Elephants also are self-aware, as shown in a study demonstrating that they "can recognize themselves in a mirror and use their reflections to explore hidden parts of themselves, a measure of subjective self-awareness that until now has been shown definitely only in humans and apes," according to an article titled, "Who's That Pretty Pachyderm?" by Rick Weiss, in the *Washington Post,* on October 31, 2006. Apparently, when most animals look at themselves in a mirror, they think their image is another animal. Elephants looking in a mirror will try and remove a smudge they see on their own face. They also have elaborate social systems and make friends, who they remember years later.

> *One of the strongest cases for elephant intelligence comes from the idea of elephants sharing a taught, collective culture. Scientists have studied and compared the behavior of wild elephants that have been hunted by humans and wild elephants that have not been hunted. The elephant herds that had never historically been hunted display no fear of humans, and react to human presence in a friendly and curious manner. Elephant herds that had been hunted displayed fear of humans.*
>
> *One particularly interesting study deals with an elephant herd that had been almost poached to death by a hired hunter. The few remaining elephants retreated to the forest, adopted a nocturnal lifestyle, and shied away from all human contact. They have not been hunted since. Three generations of elephants later and with none of the original herd remaining, the descendants of this particular herd exhibit the same fearful, nocturnal behavior. Researchers believe the mother elephants are teaching survival tactics to their calves that don't align with any other elephant herd's behavioral patterns based on the herd's shared history.* "Elephant Intelligence," by Esther November, Associated Content from Yahoo!, June 5, 2008. http://www.associatedcontent.com/article/805527/elephant_intelligence_why_elephants_pg2.html?cat=7.

So, elephants show the effects of fear if they've been hunted. But if they haven't been hunted, they don't. A friend of mine who visited the Galapagos Islands said that you can walk right up to animals there and they just look at you. They're not afraid of humans, unlike most non-domestic animals around here, e.g., squirrels, deer, geese, ducks, etc. I wonder how that might apply to humans? Maybe if we could learn to stop fearing one another, we wouldn't have war?

There're a lot of other interesting things about animals that one wouldn't necessarily know. For example, I bet you didn't know that rats *laugh*? True. And mice *sing*. Humans can't hear mouse sounds, because they are too high a frequency. (It's probably a good thing, because they we'd know when they were running around in the cabinets.) Scientists have known mice make sounds for some time, but with advances in technology, they now can tell they're songs and not just babbling. "It [] became … apparent that these vocalizations were not random twitterings but songs," said researcher Timothy Holy from the Washington University School of Medicine. "There was a pattern to them. They sounded a lot like bird songs." "Scientists Discover Singing Mice," by Walter Butler, at Buzzle.com. http://www.buzzle.com/editorials/11-5-2005-80691.asp. In fact, male mice often start singing when they picked up the scent of female mice! Oh, baby!

It seems pretty clear that bird songs aren't just random chirps either. Birds use songs for a number of purposes, including checking out birds of the other sex, attracting a mate, establishing territory, announcing changes in duties in caring for their young, holding a flock together, and intimidating enemies. Mockingbirds can learn 200 calls, and they learn new ones constantly. "Urban Jungle," by Patterson Clark, in the *Washington Post* on October 23, 2012. Nightingales can learn 60 different songs after only hearing them a few times, according to the article "Some animals look alike, but they may not talk alike," by Brian Palmer, in the *Washington Post*, June 26, 2012. There also is evidence they *practice* particularly hard songs to get them right. http://www.ornithology.com/songscalls.html. Sounds an awful lot like what we do! And an article in the *Washington Post,* on November 8, 2011, by Randolf Schmid, "Talk About Tweeting! Wrens Sing Duets," says the females use their songs as a test to pick a male who can keep up with the changing parts! Ya think they know love songs? Do they *love*?

Well, some scientists believe animals actually do feel love. I found an article in *Time* magazine by Carolyn Sayre, on January 8, 2008, titled, "Wildly in Love; Humans Aren't Alone. Romance Appears to Roam Among Animals Too." "They don't write each other love letters," says Helen Fisher, an anthropologist at Rutgers University and author of *Why We Love*. "But animals can definitely feel romantic love."

- *Burying beetles – so named because they bury the remains of small animals to use as food – may mate for life. They cooperate to raise their young, looking out for them even after they've hatched.*
- *Known for their elaborate displays of public affection, [white fronted parrots] are often spotted engaging in open-beak kissing. The Mexican and Central American natives form a life long bond with their mate and share the responsibility of caring for younglings. Like bereft widowed humans, some parrots will die shortly after their partner does.*

- *You won't find these rodents out on the prowl after a breakup. As one of the few mammals to display what scientists call social monogamy, [prarie voles] typically refuse to find another companion after a partner dies or otherwise goes missing.*
- *Male [elephants] can become so lovesick during courtship that they simply stop eating. Females are more levelheaded, at least when it comes to sex. They choose lovers wisely, typically waiting four years to mate.*
- *Commitment doesn't scare [titi monkeys.] Unlike most of their fellow primates, nearly 99% of these South American natives establish lifelong bonds with a partner.*
- *[Chimpanzees] would much rather find love in short term bonds than take the plunge. Lovers groom each other, kiss, make up after fights and can even take mini-vacations together. What's more, bonobos, close cousins of chimps, mate face to face.*

How 'bout them bonobos? They're *very* special, according to the *Time* magazine article, "Eden for the Peaceful Apes," by Alex Perry, published April 21, 2008. They "appear to be largely animals of peace. They live communally, enjoy gender equality and when disputes occur, resolve their differences through sex – straight sex, gay sex and sometimes, when different bonobo troops cross paths, group sex." Huh, monkey make-up sex? Animal group sex?

The following quote is from a website created by Courtney Laird, as a student web site for Biology 323, Animal Behavior, at Davidson College. The website is no longer active, so you'll have to trust me on this one:

Bonobos are a highly sexual species. ... Bonobos engage in many different sexual positions with many different sexual partners. They have been known to french kiss, ... participate in ventro-ventral copulation (front to front), which is made possible because the placement of the female bonobo's vaginal opening and clitoris, engage in ... female-female genital rubbing, ... and male-male penis fencing. Gibbons found that less than one-third of the sexual contact is between adults of the opposite sex. Bonobos use sexual contact for other purposes besides procreating. Bonobos use sexual contact for calming infants, resolving conflicts among adults of the same sex, food distribution, and social bonding/community organization. Bonobos use sex to release tension, form social bonds, and smooth over agonistic encounters. ...

Promiscuity serves as a "mechanism for socialization." It allows for affection, common goals, identification with others, and reconciliation. There are a couple of reasons that bonobos have a hierarchically promiscuous mating system. The first reason is that female bonobos prefer to mate with dominant males with good genes in order to insure the survival of their offspring. In order for the female bonobos to determine which male has the best genes there must be a male bonobo hierarchy. Secondly, female bonobos are found in large groups which are too big for one male to control. Therefore, there are no harem groups in bonobo society. Thirdly, since groups are composed of both sexes, multiple males associate with multiple females everyday. This makes the females readily accessible to the males, which leads to an increase in competition between the males. This competition aids in the establishment of the male bonobo hierarchy.

Bonobos probably developed into a promiscuous mating system in order to guard against infanticide. Females have sexual swellings almost all of the time, so males do not know when they are in estrus. Therefore the males do not know which copulations result in offspring. Females also have multiple partners, therefore males do not know which offspring are theirs. Since the males do not know which offspring are theirs, they are not going to risk killing their own kin, if they kill the infants.

Well, that sounds like my man, Jared Diamond! It supports the theory we were discussing a couple of chapters ago. Now it doesn't seem so kooky as when first hearing that a fear of infanticide may have led to evolutionary changes in female homo sapiens.

In addition, bonobos have a very complex societal organization, in which the females develop a very strong "artificial sisterhood." They leave their own families to avoid incest (with all that mating they do). The mothers are the most important figures in their social order, and the males typically end up having a strong bond with their mother (in part because they don't know their fathers as a result of the promiscuity).

Frans B. M. de Waal, in an article originally published in the March 1995 issue of *Scientific American,* believes that the bonobos do engage in sex to avoid and resolve disputes; he calls it *reconciliation.* He said in the article that his "study yielded the first solid evidence for sexual behavior as a mechanism to overcome aggression." Many are now looking to the bonobos as models for conflict resolution, such as therapist Susan Block in California, who according to that article in *Time* magazine on April 21, 2008, says "You can't very well fight a war when you're having an orgasm."

Anderson Cooper did a TV show on bonobos, focusing on Dr. Sue Savage-Rumbaugh at the Great Ape Trust teaching bonobos how to talk. Funny segment has Cooper dressing up in a bunny suit because the bonobos asked him to. Seriously. http://ac360.blogs.cnn.com/2010/11/17/ac360%C2%B0-preview-anderson-cooper-meets-bonobo-chimps/. In that same show, they looked at lemurs and how they can count, http://ac360.blogs.cnn.com/2010/11/16/ac360%C2%B0-preview-scientists-study-lemurs-for-insights-into-development-of-human-intelligence/?iref=obinsite, and how dolphins apparently can also recognize themselves in a mirror.

I also learned that bonobos don't have many natural predators, but they are endangered because it turns out *humans* are the most effective killers of bonobos. Poachers kill them for their meat and sell the young as pets. Humans also have been reducing the bonobos' natural habitat, which is only in the forests of the Congo. Way to go, guys, we're eliminating a peaceful species that's just like us and walks upright, having over 98% of the same DNA, and prefers sex to war. Great.

I never knew any bonobos or monkeys, but Jane Goodall does. At the age of 23, she went to Tanzania and began studying chimpanzees. They were afraid of her at first, so she could only observe them from afar. But within a few months, she observed chimps using tools. She watched a couple chimps use long pieces of plants they'd stripped of leaves to dip into a termite nest and fish out the tasty little insects. No one really thought animals could use tools before that.

Jane Goodall says, "Biologically, they're so close to us. Their brain is almost identical. We have fascinating similarities in social behavior: kissing, embracing, holding hands, and shaking the fist. These things are done in the same context that we do them and clearly mean the same thing." *Time,* "10 Questions," September 21, 2009. She also says monkeys do not make good pets; they are like six or seven times stronger than us. No pet monkeys!

Recently, chimpanzees living in West Africa also were "observed fashioning deadly spears from sticks and using the tools to hunt small animals." "For First

Time, Chimps Seen Making Weapons for Hunting," by Rick Weiss, *Washington Post*, February 23, 2007. They would sharpen sticks and use them as spears to stab bush babies (small nocturnal primates) out of tree hollows. Poor little babies. But another indication of animals using tools to kill others, just like humans do.

Another thing those researches concluded was that the female Senegalese ape-like creature were the main innovators and problem solvers in their society. Although chimps and baboons are male dominated, the babes also run bonobo societies. Frans B. M. de Waal wrote, "In both baboons and chimpanzees, males are conspicuously dominant over females; they reign supremely and often brutally. It is highly unusual for a fully-grown male chimpanzee to be dominated by any female. Enter the bonobo. Although female bonobos are much smaller than the males, they seem to rule." Funny how that a primate society run by females has more sex than the ones run by men? Maybe we should reconsider some things.

There's a lot of information out there about other things monkeys can do. "Researchers at Harvard University have evidence that monkeys can understand the difference between prefixes and suffixes," meaning they can understand principles of grammar. "Monkeys Recognize Grammar Principles, *Washington Post*, July 20, 2009. Moreover, chimps actually can outperform humans in some tests. The December 10, 2007 *Washington Post*, in a report titled, "Chimps Outdo Humans in Test," by Christopher Lee, says researchers at Kyoto University found that chimps could outperform students in quickly remembering the order of numerals. According to one of the study's authors, Tetsuro Matsuzawa, "Here we show for the first time that young chimpanzees have an extraordinary working-memory capability for numerical recollection – better than that of human adults tested in the same apparatus, following the same procedures."

We've just recently been learning a lot of these new things about animals, and about the world. Maybe some of our old beliefs should be examined to make sure what people *think* they know actually comports with what scientists tell us about our existence. I like scientists. They know a ton more than me about so many things.

A *Washington Post* article by David Brown on March 10, 2009 reported that chimps have now been observed planning. You know, like planning to go to the grocery store. "Arsenal Confirms Chimp's Ability to Plan" is the title of the article, and it talked about a 30-year-old captive chimp named Santino at a zoo north of Stockholm. He apparently gets really annoyed when visitors gather near his enclosure, where he's the dominant male. "He shows his displeasure by flinging stones or bits of concrete at the human intruders, but finding a suitable weapon on the spur of the moment perhaps isn't so easy."

So, in the morning to prepare, he'll wander the area, finding stones and pieces of concrete that he breaks up and puts in piles. Then, he throws them at visitors when they arrive, up to 20 in a row to demonstrate his unhappiness. "Many animals plan. But this is planning for a future psychological state. That is what is so advanced," said Mathias Osvath, director of the primate research station at the Lund University and author of the study published in the journal *Current Biology*. The *Post* article reported on other instances of planning by chimps, such as gathering particular plants and saving them in a specific area. Of course, for that matter, most of the squirrels that live around my house seem to do that, too.

Another article in the *Washington Post*, by Rob Stein on September 7, 2009 (there are a lot of good articles in the *Post*), titled, "It's Music to These Monkeys' Ears, and Also Their Hearts," recounted how a scientist studying a small type of monkey called a tamarin was able to show that music affects them emotionally. No, he didn't use our kind of music, but, well, you know, tamarin sounds. Charles T. Snowden is a psychologist at the University of Wisconsin in Madison, who teamed up with David Teie of the National Symphony Orchestra. Snowden had been studying the tamarin for years, and had a large recorded collection of their complex vocalizations.

He and Teie worked together to compile similar types of their sounds, that is, recordings of both hostile and alarming sounds the tamarins made and also of sounds produced when they were safe and calm. They arranged the sounds into songs, Teie recorded them with him playing the cello (sped up to a tamarin beat), and they watched the little monkeys carefully. They observed that the tamarins became agitated at the hostile music and calm during the soothing ballads. This was one of the first controlled studies accurately showing how animals respond to music. Of course, the tamarins didn't respond to human music. I wouldn't probably respond to tamarin tunes either, but hopefully you catch the drift.

I'd like to go back to the types of little insects the chimps were eating; well, not termites, but ants. Do you think an ant would be a good teacher? If you think yes, then you win. Because an article in the *Washington Post* by Shankar Vedandtam on January 16, 2006, proclaimed, "Ants are First Non-Humans to Teach, Study Says." How about that? These little dumb creatures we can squish with a finger or shoe are good teachers. What do they teach? Well, they can show other ants the trail to get food. Like if an ant finds some jelly on your counter, they can go back to the colony and bring others along. It's not just showing them a map, but actually the leaders take their time to show the followers how to learn the correct path to the jelly.

Nigel Franks, professor of animal behavior and ecology, whose paper recounted the results of the study, said, "Within the field of animal behavior, we would say an animal is a teacher if it modifies behavior in the presence of another, at cost to itself, so another individual can learn more quickly." Many times, the followers would take different routes back home, some quicker than the original route, indicating that the teaching was done in a way to increase knowledge of the environment and not just provide basic information. Honeybees tell others from their hive about food sources through ritual-like dances that use the "sun's position as a point of reference," as explained in an article titled, "Honeybee Genome May Shed Light on Social Evolution," by David Brown in the *Washington Post,* October 30, 2006.

Another thing I didn't know about ants is that, like honeybees, they have "astonishingly complex and interesting societies," according to Adrian Higgins in his article on gardening titled, "The Circle of Life, Squarely Underfoot," in the *Washington Post* on September 9, 2010. Apparently, ants like to work the soil, just like gardeners. Higgins quotes from a book titled, *Bees, Wasps, and Ants* written by Eric Grissell, who says "ants fulfill a major role in the environment by aerating and mixing the soil, enhancing water infiltration, recycling and incorporating dead and organic matter and nutrients." That's actually pretty cool, and a lot more than I could claim to do for the soil.

According to Grissell, "ants actively disperse the seeds of more than 3,000 species of herbaceous plants, and maybe many more. Trilliums, violas, corydalis and many other plants produce seeds coated with nutrients. Ants haul the seed home, feed the goodies to their young, and the seed then germinates." I wonder if the plants grow up right in their dining rooms?

Here's a scary thought though. There's a single mega-colony of ants that has colonized much of the world, across Europe, the United States, Japan, and South America. They started out in Argentina, but are inadvertently taken around the globe by people (seems we have lots of effects on flora and fauna we don't know about or understand). Their colony may be the largest ever known, according to a BBC Earth News story by Matt Walker on July 1, 2009.

> *These introduced Argentine ants are renowned for forming large colonies, and for becoming a significant pest, attacking native animals*

and crops. In Europe, one vast colony of Argentine ants is thought to stretch for 6,000km (3,700 miles) along the Mediterranean coast, while another in the US, known as the "Californian large," extends over 900km (560 miles) along the coast of California. A third huge colony exists on the west coast of Japan.

But it now appears that billions of Argentine ants around the world all actually belong to one single global mega-colony. Researchers in Japan and Spain led by Eiriki Sunamura of the University of Tokyo found that Argentine ants living in Europe, Japan and California shared a strikingly similar chemical profile of hydrocarbons on their cuticles. But further experiments revealed the true extent of the insects' global ambition.

The team selected wild ants from the main European super-colony, from another smaller one called the Catalonian super-colony which lives on the Iberian coast, the Californian super-colony, and from the super-colony in west Japan, as well as another in Kobe, Japan. They then matched up the ants in a series of one-on-one tests to see how aggressive individuals from different colonies would be to one another. ...

Whenever ants from the main European and Californian super-colonies and those from the largest colony in Japan came into contact, they acted as if they were old friends. These ants rubbed antennae with one another and never became aggressive or tried to avoid one another. In short, they acted as if they all belonged to the same colony, despite living on different continents separated by vast oceans. The most plausible explanation is that ants from these three super-colonies are indeed family, and are all genetically related, say the researchers. When they come into contact, they recognize each other by the chemical composition of their cuticles. "The enormous extent of this population is paralleled only by human society," the researchers write in the journal Insect Sociaux, in which they report their findings.

How about that? Here's another thing that'll blow your mind. Slime mold cooperate. You know, *slime mold*. They spend most of their life as single celled, uh, yucky things. But if times get tough and they can't find food, the slime mold get together – literally.

Elephants, Slime, Mold, and Worms - Oh, My!

They coagulate and form streams of cells, maybe a hundred thousand. Then, *they start to behave as one organism!* They start their slimy moving around in unison, looking for food. It's like they all clasp hands and start acting as one being. That's not all. This big (well, only a few millimeters, but relatively speaking) hulking slime mold thing can *stand up*!

> *It stands on end to create a CN Tower-like structure, a slender stalk supporting a mass of spores on top. Eventually the spore case ruptures, releasing individual spores to the wind. When they hit the ground, (hopefully in a place where food supplies are adequate) they germinate, creating, once again, a single-celled amoeba, and the story begins again.*
>
> *This remarkable process has intrigued biologists for fifty years, because it appears to be a simple, almost transparent example of the process underlying our own embryonic development. How does the single human egg produce the hundreds of specialized cells in our bodies? How do single unspecialized amoebas become stalk cells or spores?* http://www.softhawkway.com/fram_self.htm

Well, I don't know. I've seen mold in the shower, but didn't know all this was going on. Seems to be saying the slime mold get together and stand up, like maybe a whole bunch of human cells could get together and just stand right up on two legs and start walking!

And who woulda known that some slime mold *cheat*? Yea, when they stand up, the generous cells in the stalk harden and then die, essentially giving up their lives for the sake of the dudes on top that get to have their offspring continue. They're like the Marines of slime mold. The thing is, if they're different strains of slime mold (I know, who knew there were different kinds of slime mold?), some of them will just never end up being the stalk. They somehow manage to Tom Sawyer the other strains to make the stalk. According to the article I mentioned before in *National Geographic*, titled, "Minds of their Own; Animals Are Smarter Than You Think," "Deceptive acts require a complicated form of thinking, since you must be able to attribute intentions to the other [slime mold] and predict the [mold's] behavior."

You know, I'd assumed the little slime dudes just acted out of pure instinct, but maybe that's wrong. Could they *talk* about what they want to do? Nah; mold, bacteria, and stuff don't talk, right? Well, who knew, but *yes*. Not, like, human language; they don't have mouths and vocal chords. They communicate *chemically*. For example, some marine bacteria emit a chemical glow to effectuate their public discourse. How do we know? Scientists like Bonnie Bassler, who has won a MacArthur genius award and been elected to the National Academy of Science, tell us so. From an NPR radio story on September 12, 2006 by Richard Harris, http://www.npr.org/templates/story/story.php?storyId=6061852:

> [S]he figured out how these bacteria talk to one another. They use not just one but two chemical signals. We head into a pitch-black room with a rack of Petri dishes, so she can demonstrate this bacteria's hidden talent. She shakes the dishes and they emit a blue glow. It turns out that when one of these bacteria is all alone, it doesn't glow. After all, that would be a waste of effort because nothing could ever see such a tiny amount of light. But it does send out chemical signals that say, hey I'm here ... and it listens back for other bacteria sending the same signal. When enough bacteria are doing this, they know they have a quorum. All of a sudden, they light up and do all sorts of other things to act in concert, like a super-organism.
>
> "So they turn on and off 100 different genes, to let them turn off behaviors that are good when you're alone and turn on genes that are good when you are a community. And for reasons we don't understand, the gene that lets them make this beautiful blue light is one of the genes they turn on," Bassler says. ... It turned out that communication is universal among all bacteria, including the nasty ones.
>
> "Melissa Miller, a graduate student in [Basssler's] lab, went on to show that cholera has a circuit like [the marine bacteria] and what it does, it doesn't make light, it makes toxins, and so that's what it does as a group," Bassler explains. "So they can't make you sick as one, but if they wait and they launch their attack together, it's fantastic."
>
> And what's more, Bassler came to realize that bacteria don't just talk to others of their same kind. One chemical she discovered serves as a universal language that all species of bacteria can understand. So all bacteria can talk to all other bacteria. She also realized that because bacteria evolved first, they must have invented chemical com-

> *munication, which is now used to organize all giant collections of cells into organisms.*
>
> *"This is what happens in your body," she says. "It's all this chemical communication. We call them hormones, and your kidney cells don't get mixed up with your heart cells because there are these different languages. And that's exactly what this bacterium is doing."*

Talking bacteria! I'm amazed by what science can teach us. Bassler is saying various parts of our bodies communicate with each other through some sort of universal bacterial/hormonal chemical language, like we speak in English. The heart says, "Yo, legs, stop running so fast or I'm gonna have an attack!" Stomach says, "Mouth, man, too much food comin' in; shut 'er down." Brain says, "I'm in charge, dudes!" Or something like that. It sounds crazy, but did you ever think about how the inner parts of your body *would* communicate? It must be some kinda way; they can't not work together, all squished inside there in the dark, digesting, pumping blood, white cells attacking viruses, yada yada.

Most people would probably say that bacteria and little funky things like that aren't important. Think again, people! Adrian Higgins in the gardening article from the *Washington Post* that I referred to above, mentioned a book by biologist James B. Nardi of the University of Illinois called *Life in the Soil*. Apparently, "a square meter of healthy garden soil is home to 10 trillion bacteria, 10 billion protozoa, 5 million nematodes, 100,000 mites, 50,000 springtails, 10,000 creatures called rotifers and tardigrades, 5,000 insects and arachnids, 3,000 worms and 100 snails and slugs." OMG. Right there in a suitcase or two of soil. All that stuff going on; eating, belching, defecating, and procreating like crazy. Man, I'm glad I don't have to live in the soil; sounds nasty. But Nardi says the microorganisms in the soil exert an "extraordinary influence on the health of the earth." According to the Higgins article, "For example, if you were to measure the metabolic activity of bacteria in the top six inches of an acre of fertile soil, [Nardi] writes, you would find that it exceeds the metabolic energy of 50,000 people."

Another book about soil is called *Secrets of the Soil*, by Peter Tompkins and Christopher Bird. Recognize those authors? They also wrote the book, *The Secret Life of Plants*, which came out in 1973. I remember reading it after college, but more on that later. Tompkins and Bird agree with Nardi and basically say soil is the key to human and planetary health. In *Secrets of the Soil*, it's stated that "more microorganisms germinate in half a cup of fertile earth than there are humans on the planet .... The combined weight of all the microbial cells on earth is twenty-five times that of its animal life; every acre of well-cultivated land contains up to a half a ton of thriving microorganisms, not to mention up to a ton of earthworms, which can daily excrete a ton of humic castings." Uh, castings are basically, well, worm poop. It's very beneficial to the soil and plant growth.

Charles Darwin thought earthworms were really special. Yea, I know, he's the father of the theory of evolution, but according to an NPR story on February 12, 2009, "his book, *The Formation of Vegetable Mould Through the Action of Worms, With Observations on Their Habits,* published in 1881, sold even better than *On the Origin of Species* during Darwin's lifetime." http://www.npr.org/templates/story/story.php?storyId=100627614. Darwin conducted extensive experiments on earthworms, and proved they were beneficial in moving soil, at a time when most people thought they were just garden pests.

According to *Secrets of the Soil*, Charles Darwin wrote, "It may be doubted whether there are many other animals which have played so important a part in the history of the world as have these lowly organized creatures." I only remember worms from fishing as a kid – I guess those were bloodworms – and they would wriggle wildly when you cut them to put on the hook (which I can sympathize with). Now, come to find out, Darwin also, "went so far as to suggest that earthworms are cognitive beings because, based on his close observations, they have to make judgments about the kinds of leafy matter they use to block their tunnels," as stated in the *National Geographic* article mentioned above, "Minds of their Own". According to Tompkins and Bird:

> *Earthworms are prodigious diggers and earthmovers, capable of burrowing down as deep as fifteen feet. They can squeeze between and push apart the soil crumbs, and one worm alone can move a stone fifty times its own weight. ... Experiments have shown that soils with earthworms drain from four to ten times faster than those without. ... With their mixing, digging, burrowing, fertilizing, and humus-making activities, the worms have an immense impact on the soil, its texture, its fertility, and its ability to support everything that lives in or on it, especially plants that form the basis for our food supply.*
>
> *Night crawlers, so named because they creep about at night on*

> *the surface of the earth, feed on leaves, which they drag down into their burrows, and even with their pin-head brains they have the wit to pull them down by the narrow end. ... In suitable weather, night crawlers can spend a goodly portion of their nocturnal activities in the pursuit of sex, even an entire night coupled to a willing hermaphroditic mate, each possessing both male and female organs.*

Goodness, life just gets curiouser and curiouser. Worm sex! Just never thought about it that much (I suppose that's a good thing). But all this information about slime mold, microorganisms, bacteria, and earthworms does change one's perspective. Even as recently as the 1970's, according to the *National Geographic* article, most scientists – as well as people generally – just thought animals and all these creatures were "incapable of any thought. They were simply machines, robots programmed to react to stimuli but lacking the ability to think or feel." I'm glad scientists are now studying things like rat laughter and worm poop. I think it's best to know the facts about life, rather than just cling to old ideas.

Scientists also know that we humans are full of it, bacteria, that is. In fact, according to the Kid's Post section of the *Washington Post* on December 15, 2008, "As far as humans go, there are more bacteria living on (and in) our bodies than there are people on earth. In fact, it is estimated that each of us carts around about a trillion of them." My goodness, a *trillion* bacteria on each of us?

According to Tompkins and Bird, "The intestinal tract is lined with vital microbes that alone enable food to digest. Microbes proliferate throughout the length of the intestine. And every inch of skin teems with friendly but unseen creatures by the billion." What? Then I read in the *Washington Post* on February 12, 2007 in an article titled, "Close Look at Human Arm Finds Host of Microbes," by Rob Stein, which asks, "What do you see on your arm?" I'd say, uh, nothing, maybe a few hairs.

> *But that's not what Martin J. Blaser of New York University School of Medicine sees. With the help of the latest scientific tools, Blaser sees a complex microscopic world teeming with a vast array of microorganisms. "The skin is home to a virtual zoo," said Blaser, a microbiologist who last week published online the first molecular analysis of the bacteria living on one small patch of human skin. "We're just beginning to explore it."*

> *The analysis revealed that human skin is populated by a diverse assortment of bacteria, including many previously unknown species, offering the first detailed look at this potentially crucial ecosystem.* [Note: they're saying the skin on your arm is an entire flippin' ecosystem!]
>
> *The work is a part of a broader effort by a small coterie of scientists to better understand the microbial world that populates the human body. Virtually every orifice of the digestive tract are swarming with bacteria, fungi and other microbes. By some estimates, <u>only one out of every 10 cells in the body is human</u>.* [Yea, my emphasis added!]

Well, that's some perspective. Ninety percent of our bodies are not our bodies! We each have in and around our bodies a trillion bacteria; doing all that stuff they do in the soil, right in and on us. But I don't feel all those little rascals crawling all over and in me. If they're there – and what are you supposed to do, *deny* the scientists who have all those special little measuring tools? – it means I don't perceive any difference at all from those trillions of beings that are living as part of me. No perception of separation at all, yet we know we're separate from those bugs. Or are we?

Eleven | People Are Naked

Hey, remember that show in the mid '70's called, "The Wild, Wild World of Animals?" Oh, you weren't born then? Some of you, anyway. I was in my twenties, and I remember watching it. It had a cool intro: http://www.youtube.com/watch?v=jDlKa G24CPU&feature=related. Lots of film of animals doing all sorts of things. Grazing, fighting, running, eating their prey, etc. William Conrad was the narrator.

Anyway, when I was living at the Island about that time, I wrote this as part of *Phantasmagoria*:

The Wild, Wild World of Animals

Good evening, nature lovers. I'm glad you can be with us tonight. You'll remember last week we took a look at the little-known habits of the African Swampwampus. What a strange creature! Tonight, we'll try to get to know a little better the most complex of worldly animals: Human Beings.

Humans inhabit the earth in nearly every nook and cranny. There are, to be sure, many areas of hostile climate where they do not regularly venture, such as the coldest mountains and hottest jungles. But humans have shown a capacity and ability to evolve and develop constantly, especially recently, in the last few thousand years.

There are many different breeds of human beings, just like the various strains of dogs, fish, birds, etc. There are different races, basically divided according to color and physical differences. Geographical and cultural characteristics further distinguish groups of human beings.

These differences divide many human groups into antagonistic countries and nations.

Often nations band together to oppose another segment of the human population, usually on the premise of achieving a certain goal or goals. These confrontations are most often begun because of the human desire to dominate as a means of achieving security, or a sense of security. Isn't it unfortunate that humans appear to be basically mistrustful of one another? This skepticism is most likely a trait developed by the need to survive in a scarce environment.

One observes, however, that men and women of this species possess a general desire to live together peacefully and amiably. Humans are not innately warlike or evil, but have not yet developed their whole family to the point where inter-human conflict and mistrust can be eradicated. We cannot venture to predict that such a time will come, but we know that its possibility cannot be denied.

Humans are the only creatures that have learned to clothe themselves as a means of protection from harsh climates. This is the prime reason for human's ability to thrive in many different environments and inhabit such a large spectrum of climes. Other creatures have learned to protect themselves from Mother Nature, to be sure, but humans can go from one element to the next and develop adequate garb to enable them to live comfortably. Humans also have produced superior shelter, communication, and transportation techniques. Their ability to learn and communicate knowledge to others is the most remarkable and important development in history. Comfort and security are the products of this faculty. Humans can learn from others and build on fellow humans' thoughts, actions, and experiences. No other animal possesses the physical endowments, especially intellect, that permit humans to mold their environment to such a great extent in order to fill their needs. There is even considerable merit to the argument that man has changed the world too much!

Humans have created machines to do work and these technological marvels produce quite efficient results. Machines to count, to load, to carry, to lift, and even to live for man. Certainly, our world would be much different if it weren't for the tools humans have created. But it is easy to make machines compared to the difficulty in realizing the long-range effects of these components. This is where humans will demonstrate whether they are capable of fulfilling their needs and goals without over-fulfilling some and neglecting others.

In other areas, the human species is also very remarkable. Humans are the only kind of animals that define much of their work as art. This work is above and beyond the normal needs of survival and has no real physical use or solves any problems. But man has the intellectual prowess to create in a more than functional manner. Humans appre-

ciate, and know that they appreciate, beauty in all its forms - touch, taste, smell, sight, and sound. It is a unique human trait that drives an individual to pursue something that can only be successful in a completely subjective manner. And then, usually for but a transitory moment in space and time. This artistic drive is another characteristic of man that seems to pick his face out of the worldly crowd of beings. Humans often work for pure mental enjoyment or, what the individual considers, gain.

One aspect of the human family that is similar to other animals is the division of sexes, male and female. One well-known thinker believes that all actions performed by humans throughout their entire lives are influenced by sexual needs and differences. This is not entirely true, but especially for males, it does not miss the mark by much. Males strive for recognition and to make love. Women strive to love and be loved. Certainly these are overlapping, socially instilled characteristics, but in a pinch they will suffice for an explanation. Of course, all animals desire to be fulfilled; the different ways in which fulfillment is pursued is what makes our world go round.

Here, we have a scene of a man and a women and their family. Human offspring are usually born one at a time, although births of up to eight babies at once have been recorded. The couple usually has between one and a dozen children, although the average born to each family now is about 2.5. A man and woman usually, or ideally, join together in marriage for as long as they shall live, but many times they prove incompatible and even after rearing children, separate in search of another companion. No one is perfect and it is understandable, given the vagaries of human love and compatibility, that life long togetherness is not always possible.

Humans have a very difficult time in understanding and coping with their imperfections. But amidst all these problems, there is greatness, realized and potential, in the human species; greatness in the sense of living comfortably, harmoniously, and amiably with nature. That is the task of this unusual, bipedal animal.

Well, it looks like that's all the time we have. I hope that you have enjoyed our brief glimpse of the Human Being. There is more, much more to be said. And we will say it, next time we look at - The Wild, Wild World of Animals.

Are we really just like animals? Or stated differently, *are we just animals?* Well, it's a fact that *people are naked!* We'd all be running around nude in the forest if we hadn't figured out clothes. Covering our nakedness sets us apart from animals, but isn't that an immaterial distinction? We think we're so handsome and good-looking in our $100 jeans, cashmere sweaters, and sparkling jewelry. We're better than spiders and worms cuz we are *so attractive*. Well, one might argue that many of us are *really ugly* without clothing. Just imagine if all the obese people you see walking around were naked? That would be so disgusting I can't imagine. What other animals are scared to be naked in their own skin like humans?

We used to go in stores at the beach where they sold shells, knick-knacks, and hermit crabs. My kids had hermit crabs when they were young. We'd keep them in a little cage, but they usually died. Probably didn't like being in a cage much. I can understand that. There are, like, 1100 different types of hermit crabs. They get in their shells as protection against predators. I don't think we have as good a reason. If you think about it, we didn't get designed too well if we have to wear clothes to stay warm. Interesting that hermit crabs are one of the only other species that wear "clothes" like we do.

It's funny, but we're actually close genetically to a lot of animals, not just the obvious ones like monkeys. Take sea urchins, for example. They're cousins of starfish and can live for *more than a hundred years*, according to an article by Rick Weiss in the *Washington Post* on November 13, 2006, titled, "Sea Urchins' DNA Code Mapped." That's older than my Mom! Anyway, *they're closer to us genetically than fruit flies*. The article said that an international team of 250 scientists has "determined the exact order of all 814 million letters of DNA code that carry the instructions for making and maintaining a sea urchin." No wonder I couldn't do it (make and maintain a sea urchin, that is)!

> *Sea urchins, it turns out, have 23,300 genes, only slightly fewer than humans have. Many are similar to genes that cause human diseases, including Huntington's disease and muscular dystrophy. ... Surprisingly urchins have 979 genes that, by the looks of them, are involved in sensing light or odors – not bad, considering they have neither eyes nor noses. Lots of those genes are most active in urchin "feet," suggesting those appendages are as important for vision and other sense as for movement.*

Well, don't that beat all? We have elephants grieving, slime mold talking chemically, and now sea urchins smelling with their feet. What will they think of next?

Another article by James Gorman proclaimed, "Fruit Flies Are Our Cousins, Too." That was in the *New York Times* on August 22, 2006, and said, like us, fruit flies begin having trouble sleeping as they get older! That study was published in the Proceedings of the National Academy of Sciences. The article also said, "We have a huge amount of DNA in common with yeast, also our distant cousins." And, who'd a thunk, but some animals may have mid-life crises, just like us! "[R]esearchers report that captive chimps and orangutans do show the same ebb in emotional well-being at midlife that some studies find in people." "Midlife crises may not be unique to humans," Associated Press, *Washington Post*, November 27, 2012. My goodness, where will these similarities ever end?

An article in *National Geographic* from February 2009, by Matt Ridley, titled, "Modern Darwins," says that, "over the past decade, as scientists compared the human genome with that of other creatures, it has emerged that we inherit not just the same number of genes as a mouse – fewer than 21,000 – but in most cases the very same genes." In that article, neurobiologist Constance Sharff says there's a thread running through all creatures; "The genetic hardware a bird uses to learn to sing probably isn't far from what a mouse uses to learn to run a maze, and what you use to learn to speak."

In fact, "Over the past two centuries, people have had to disabuse themselves about various ideologies asserting that humans are fundamentally different from animals. Biologists have shown that our arms and legs and organs have long evolutionary histories. Beliefs about the uniqueness of human behavior might well be the last bastion of our superiority complex, but ... even this redoubt may be crumbling." *Washington Post*, March 19, 2007. Oh, no, you mean the doctrine of "human exceptionalism" is on the wane?

An article entitled, "Our Inner Animal. Beasts are More Human – and Humans More Bestial – Than We Think," reviewed two books, *The Well Dressed Ape*, by Hannah Holmes and *Animals Make Us Human*, by Temple Grandin. *Time*, January 16, 2009. The review says that, "Both writers are after the same thing. They want to demolish the hard line that separates people from animals."

> We don't think of ourselves as poisonous, but our mouths are as full of noxious, infectious bacteria as is a Komodo dragon's, and a human bite can be seriously toxic. ... The premise of *The Well Dressed Ape* is that everybody knows human beings are really animals, but nobody cops to it linguistically. Just talking about ourselves the way we talk about animals is a step toward self-knowledge. ...

> There aren't many worse insults for a human than to be called an animal, but these books – which do that at great length – are instead strangely ennobling. They make you realize how much effort we expend every day convincing ourselves that we're different and what a relief it is to admit that we're not. It's lonely here at the top of the tool-using hierarchy - why don't we let down our fur and join the club? If they'll have us, that is. If animals could describe us in return, the results might not be so flattering.

I guess lots of folks don't want to admit we're like apes, fruit flies, or sea urchins, or that we're not much different from animals. Like I said before, I'm kinda proud of that, myself. There is a gorilla in Britain's Port Lympe Wild Animal Park that now walks upright, like a person; I'm proud of that dude, too. Here, watch: http://www.huffingtonpost.com/2011/01/27/gorilla-walks-like-human_n_814994.html?view=print. I don't necessarily want to have a lotta apes, sea urchins, or fruit flies sleeping in my bed, but it's nice to think we're not the only ones on the planet that are special. Takes the pressure off a bit. It does seem right to me, though, that many animals would *not* want to be compared to us. We can be pretty mean and nasty, and totally disrespectful of the earth that sustains us.

In all this cogitating about the similarities between animals and us, a logical next question would be, what about plants? I mean, plants are alive. They don't have heads, brains, and legs like we have, but certainly they have an awareness of being alive, don't you think? I know when I've forgotten to water the plants in our house, they look pretty sad. But they brighten up and almost smile in the sun. In *The Secret Life of Plants*, Tomkins and Bird say plants can perceive and understand many things around them, including human thoughts and actions! They recount numerous studies where scientists have measured plant responses to various stimuli through machines such as galvanometers (which measure electrical conductivity). So, when attached to a plant, it will show changes in the plant's internal energy. According to *Wikipedia*, early studies were done by:

> Indian scientist Sir Jagdish Chandra Bose, who began to conduct experiments on plants in the year 1900. He found that every plant and every part of a plant appeared to have a sensitive nervous system and responded to shock by a spasm just as an animal muscle does.

> *One visitor to his laboratory, the vegetarian playwright George Bernard Shaw, was intensely disturbed upon witnessing a demonstration in which a cabbage had violent convulsions as it boiled to death. Bose found that the effect of manures, drugs, and poisons could be determined within minutes, providing plant control with a new precision. In addition, Bose found that plants grew more quickly amidst pleasant music and more slowly amidst loud noise or harsh sounds. He also claimed that plants can "feel pain, understand affection etc.," from the analysis of the nature of variation of the cell membrane potential of plants under different circumstances.*

Cleve Backster, who worked with lie detectors (which utilize galvanometers), also did similar experiments in 1966. Tompkins and Bird say Backster attached the galvanometer to a plant and dunked one of its leaves into hot coffee, but the plant didn't react. So, he thought about trying something worse; he'd burn the plant's leaf.

> *The instant he got the picture of a flame in his mind, and before he could move for a match, there was a dramatic change in the tracing pattern on the graph in the form of a prolonged upward sweep of the recording pen. Backster had not moved, either toward the plan or toward the recording machine. Could the plant have been reading his mind? ... Later, as he went through the motions of pretending he would burn the plant, there was no reaction whatsoever. The plant appeared to be able to differentiate between real and pretended intent.*

Other studies reported in the book demonstrated that plants react to affection, joy, and pleasure, as well as to threats. Studies also showed that plants become fatigued when exposed to continuous stress; even that plants have memories. This is also from *The Secret Life of Plants*:

> *We had a man molest, even torture, a geranium for several days in a row. He pinched it, tore it, pricked its leaves with a needle, dripped acid on its living tissues, burned with a lighted match, and cut its roots. Another man took tender care of the same geranium, watered it, worked its soil, sprayed it with fresh water, supported its heavy*

> branches, and treated its burns and wounds. When we electroded our instruments to the plant, what do you think? No sooner did the torturer come near the plant than the recorder of the instrument began to go wild. The plant didn't just get "nervous;" it was afraid, it was horrified. ... Hardly had this inquisitor left and the good man taken his place near the plant than the geranium was appeased, its impulses died down, the recorder traced out smooth – one might say tender – lines on the graph.

I wouldn't be surprised if many people don't buy this about plants. It seems silly; plants reacting emotionally to human actions and thoughts? But one thing can't be denied; plants have sex. Yes, indeedy, boys and girls, they have sex organs and that's how they reproduce.

> [P]lants have female organs in the form of vulva, vagina, uterus, and ovaries, serving precisely the same functions as they do in a woman, as well as distinct male organs in the form of penis, glans, and testes, designed to sprinkle the air with billions of spermatozoa ... . Like animals and women, flowers exude a powerful and seductive odor when ready for mating. ... Flowers that remain unfertilized emit a strong fragrance for as many as eight days or until the flower withers and falls; yet once impregnated, the flower ceases to exude its fragrance, usually in less than half an hour. ... Pollen, which performs the same function in almost precisely the same manner as does the semen of animals and men, enters the folds of the plant vulva and traverses the whole length of the vagina, until it enters the ovary and comes in contact with the ovule.

When you come to think of it, those sexual parts of plants (the pistil being the female organ and the stamen being the male) are very attractive; a lot nicer than our own, I'm thinking. You know who else thoughts plants were cool? Yea, buddy, Charles Darwin. After he published the *Origin of Species* in 1859, and spent all that time studying earthworms, he devoted most of the rest of his life to studies of plant behavior. He wrote a 575 page book entitled, *The Power of Movement in Plants*, published just before his death. Though Darwin didn't conclude plants have a nervous system, like another scientist of his day, Gustav Fechner, "Darwin could not get out of his mind that plants must have sentient ability," according the Tompkins and Bird. Darwin indicated in his book that the plant's *radicle* (the first part of a seedling or plant embryo to emerge from the seed during germination) "acts like the brain of one of the lower animals."

Speaking of Darwin, I've always believed in evolution. I don't feel any less about myself because I might have been descended from chimps. I'm not going to get into a big exegesis of the *Origin of Species* (never read it), but it doesn't seem to be a theory in serious question. "Time and again, biologists are finding that Darwin had it right: evolution is the best way to explain the patterns of nature," said an article in *Time* dated January 23, 2009, by Carl Zimmer, titled, "Evolving Darwin." In the *National Geographic* article referenced above, "Modern Darwins," it says scientists are observing evolutionary changes that occur in a matter of a few years as opposed to centuries. For example, they are seeing changes in finches year in and year out, and confirming them through DNA analysis. David Resnick, a biologist studying guppies, said, "Darwin thought evolution by natural selection was too slow to observe, ...[b]ut we're watching evolution occur over just a few years."

We're seeing evolution in our daily lives, too. We know humans during, say, the days of the colonists were mostly smaller than modern people. If you go to Williamsburg and look at the clothes, you can tell. People are much bigger nowadays. I think kids and teenagers are larger now than when I was growing up. We also know that bacteria are changing to avoid the effects of antibiotics. These are just a few examples.

One of my favorite scientist-writers who provides a great perspective on man and the world is Jared Diamond. You remember hearing about him, from his *Why is Sex Fun?* book.

In his book, *Guns, Germs and Steel*, Diamond outlines the history of man and how we came to the global situation we have now. Diamond is a professor of physiology and geography, who has expanded into evolutionary biology and biogeography. I think of him also as an anthropologist, because a lot of the information he covers in his books is based on study of stuff found in the dirt. Things like old pots and clothes, bones, etc. Scientists like him analyze all sorts of little details and can figure out lots of things. Here's some information about us from Diamond that I find very interesting.

> *Human history, as something separate from the history of animals,*
> *began [in Africa] about 7 million years ago (estimates range from 5*

> to 9 million years ago.) About that time, a population of African apes broke up into several populations, of which one proceeded to evolve into modern gorillas, a second into the two modern chimps, and the third into humans. The gorilla line apparently split off slightly before the split between the chimp and human lines. Fossils indicate that the evolutionary line leading to us had achieved substantially upright posture by around 4 million years ago, then began to increase in body size and in relative brain size around 2.5 million years ago. ... Although Homo erectus, the stage reached around 1.7 million years ago, was close to us modern humans in body size, its brain size was still barely half of ours. Stone tools became common around 2.5 million years ago but they were merely the crudest of flaked or battered stones. In zoological significance and distinctiveness, Homo erectus was more than an ape, but still much less than a modern human.
>
> All that human history, for the first 5 or 6 million years after our origins about 7 million years ago, remained confined to Africa. ... At present, the earliest unquestioned evidence for humans in Europe stems from around half a million years ago, but there are claims of an earlier presence. ... By about half a million years ago, human fossils had diverged from older Homo erectus skeletons in their enlarged, rounder, and less angular skulls. African and European skulls of half a million years ago were sufficiently similar to skulls of us moderns that they are classified in our species, Homo sapiens, instead of in Homo erectus.

Then, there were the Neanderthals from about 130,000 to 40,000 years ago. Their brains were larger than ours and they were the "first humans to leave behind strong evidence of burying their dead and caring for their sick." We also have the Cro-Magnons, who did some great cave art, and finally, Diamond says, about 50,000 years ago was what he calls the "Great Leap Forward," from which scientists start finding garbage from our kinda modern people. They find more diverse and better tools, weapons, and things like that. It was somewhere around 11,000 years ago that North and South America were finally settled.

I find this fascinating. I'm just 60 years old, and it seems like there have been lots of changes in my life. My Mom saw even more, having been born in 1915. Just think of all the people who lived in various degrees of humanhood over the last 100,000 years or so. Did they think just like us? What in the world did they understand about life? They certainly didn't have the benefit of modern science to give them an accurate picture of things. By today's standards, most people who ever lived would be way *undereducated*.

What kind of crops do you think the Cro-Magnons used to plant and eat? Answer – none, nothing, nada. Until about 11,000 years ago, humans only hunted and ate wild animals and foraged for plants. Plants were not domesticated until about that time. You know, I think of dogs and cats, horses and cows, as domesticated, but not *plants*. I assumed we always knew how to plant corn, wheat, and tomatoes. How strange to think of domesticating plants. For example, it was only about 8,000 years ago that figs, donkeys, and cats were domesticated in Egypt and 4,500 years ago that sunflowers were first farmed in the Eastern United States. Some of the earliest food production came out of the "Fertile Crescent," which was southwest Asia, now including parts of Iraq, Iran, Lebanon, Syria, Israel, and parts of Turkey. Once plants and livestock were domesticated, that meant huge changes. People started living in one place, with their animals. Diseases from the animals migrated to humans, too. Some of the major infectious diseases that have plagued humans came from animals, such as smallpox, flu (like swine flu), malaria, measles, tuberculosis, cholera, and even AIDS, according to Diamond.

When people started staying put, because they could center their lives around farming for food, that resulted in towns, cities, and increases in population. That in turn led to a need for an organizing force to regulate behavior among strangers. I used to think of it as the "stop sign" theory. If there were two paths or roads where people crossed (with their cows, horses, wagons, cars, whatever), at first, there'd be no need for a stop sign. You'd just let the other person go first, or vice versa. But when you get more traffic, and people moving faster, then you have to figure out a system so there aren't crashes all the time. A stop sign makes a lot of sense. But, just like intersections in towns and cities all over the world, next more complicated mechanisms are needed; traffic signals with red, yellow, and green lights so everyone understands what to do. Then, adding pedestrians to the mix, you get the *Walk/ Don't Walk* signs, and now they even have a clock that tells people how much time they have to cross (because it's frickin' dangerous for people, kids, and dogs on our roads these days). You see, things get more complex basically because there are more humans.

Diamond goes into great detail in *Guns, Germs and Steel* about the manner in which people have organized themselves throughout history. We know that ants, slime mold, bees, elephants, and many other creatures live in complex societies. Humans certainly do, but when did we start getting sophisticated in our societal governance? Interesting to note that, "as recently as A.D. 1500, less than 20 percent of the world's land area was marked off by boundaries into states run by bureaucrats and governed by laws. Today, all land except Antarctica is so divided. Descendents of those societies that achieved centralized government and organized religion earliest ended up dominating the modern world." So, things have really changed in even the last 500 years.

Diamond identifies four categories of these human groupings: bands, tribes, chiefdoms, and states. *Bands* are not rock bands, but small, usually related, societies with between maybe 5 to 80 people. They are usually nomadic hunter-gatherers and don't farm. They're egalitarian in the sense that land is shared and there's not a formal leadership structure. Their conflict resolution is informal and they lack many types of institutions we have, such as lawmakers, judges, and businesses. Diamond says all humans probably lived in bands until 40,000 years ago, and probably most until about 10,000 years ago. Gorillas, chimps, and bonobos live in bands, which is another indication of our similarities with them. There are a few human bands left, but only in extremely remote areas, such as the Amazon jungle.

After bands, there are *tribes*, which typically would have perhaps a few hundred people, most of whom are related by blood or marriage. They also don't need police, laws, and other modern dispute resolution mechanisms, because most people know each another. The problem of conflict resolution becomes much more acute, says Diamond, among strangers where people don't trust one another. He relates that tribes are still common in some areas, such as New Guinea, where, "if a New Guinean happened to encounter an unfamiliar New Guinean while both were away from their respective villages, they engaged in a long discussion of their relatives, in an attempt to establish some relationship and hence some reason why the two should not attempt to kill each other."

Tribes were also generally egalitarian, although they might have a "big man," but even he would not have an inherited position. Diamond says neither tribes nor bands allow individuals to become disproportionately wealthy, since everyone has debts and obligations to many others in the community.

Chiefdoms are much larger, with thousands of people. They began probably about 7,500 years ago and, since most people were not related or knew one another, they had to learn not to kill strangers. One way the chiefdoms did that was to give one person – the chief, who had a permanent, often inherited position – a monopoly to use force for resolving disputes. The chiefdoms generally had organized food production and more division of labor, as well as a stratified society, luxury goods for the elite, public architecture, and sometimes slavery, because there were menial jobs to be done since it was not an egalitarian society. Chiefdoms have different exchange systems than bands or tribes; the latter two were mainly based on reciprocal trading, but chiefdoms involved more payments in the form of tribute to the chief and his class. The chief may redistribute some of the wealth to the people, but often retained significant amounts.

States are the forms of social groupings we're most familiar with, since basically the entire world is currently divided up into states (except Antarctica). States started coming into existence with groups of 50,000 or more people, and of course now China and India have over a billion. That's a lot of strangers! So, dispute resolution is managed through laws, judges, and governments. The exchange system of states is redistributive, meaning people pay taxes and the government decides how to spend those funds for the benefit of the public. There is intensive food production (like our industrial food production system nowadays). The classes are usually not egalitarian, with a significant wealthy class. States certainly have had slaves (witness our own country about a hundred fifty years ago).

Both chiefdoms and states permitted wealth to be transferred from the common people to the ruling class; in other words, they were *kleptocracies*. Thus, chiefdoms and states that transfer wealth to the upper stratum of people are much different than bands and tribes. But how do these kleptocrats stay in power? Diamond says there have been four different tacts: 1) take away the weapons of the common people and give them to the elite; 2) redistribute a lot of wealth for popular projects; 3) use the monopoly of force to maintain public order; or 4) gain public support by constructing an ideology or religion justifying the kleptocracy. He says there were supernatural beliefs present in bands and tribes, but not used to justify central power, the transfer of power, or maintain the peace.

Well, that's surprising, or maybe not. Controlling the people through religion; reminds me of the Emperor Constantine starting the Christian religion. Diamond says:

> *Besides justifying the transfer of wealth to kleptocrats, institutionalized religion brings two other important benefits to centralized socie-*

> ties. First, shared ideology or religion helps solve the problem of how unrelated individuals are to live together without killing each other – by providing them with a bond not based on kinship. Second, it gives people a motive, other than genetic self-interest, for sacrificing their lives on behalf of others. At the cost of a few society members who die in battle as soldiers, the whole society becomes much more effective at conquering other societies or resisting attacks. ...
>
> [T]he official religions and patriotic fervor of many states make their troops willing to fight suicidally. ... The latter willingness is one so strongly programmed into us citizens of modern states, by our schools and churches and governments, that we forget what a radical break it marks with previous human history. ... Such sentiments are unthinkable in bands and tribes. ... Naturally, what makes patriotic and religious fanatics such dangerous opponents is not the deaths of the fanatics themselves, but their willingness to accept the deaths of a fraction of their number in order to annihilate or crush their infidel enemy.

So, we go from Neanderthal-like naked people in bands to having state societies using religious and patriotic fervor to control the population and fight other states. I don't think slime mold or sea urchins do that. I'm not sure our way is better than bonobos. Wouldn't having sex be a better way to resolve conflicts than suicidal fighting? If bonobos can figure it out, why can't we?

If we're so smart, the most gifted "chosen" species, made in the image and likeness of God, how'd we get to this murderous stage? Diamond refers several times above to humans needing some reason, such as kinship or religion, *not* to kill one another. Why would humans kill one another, if not afraid of other humans? I just don't think they would. Do you?

Twelve | **Stardust**

I know, I know. We're supposed to be talking about love and oneness. But we *were*, and we *are*, I swear. The point of the last three chapters is that, *we're all the same*. We're just stardust.

And in some ways, homo sapiens aren't all they're cracked up to be, relative to other beings. Meaning, we're not so great or different as we think. My son Jay says, "Humans are overrated."

Part of recognizing oneness is understanding we're *not* special. *A Course in Miracles* teaches that believing one is special is *insane*. The *Course* says, "Pursuit of specialness is always at the cost of peace." If one thinks he or she's special, and separate from others, then one cannot truly understand oneness or, therefore, love.

Specialness *is* totally reasonable in another way; in knowing that *everything* is special, *including* me. That's an *entirely* different concept. *That*, is oneness. It's an appreciation of the gargantuanness and interrelatedness of everything. There's an *elation* in oneness and love. There's a, "It don't get any better than this," understanding. The perspective of knowing our bodies are composed of trillions of other cells that are not our own, and of the *vastness* of space.

I remember one night decades ago my father and I were lying on the dock at the Island, listening to the waves lap at the shore under us. We'd look up at the star laden sky, and try to circumnavigate the Milky Way with our gaze. Always pointing out the North Star, and the Big Dipper, my favorite constellation. We felt so small, yet so large.

One time, I asked, "Dad, how much does the world weigh?"

Dad was an engineer, and very practical. So, as we lay there on our backs, looking up at the stars, he proceeded to figure out how much our planet earth weighed. It went something like this:

> *Son, I suppose the first thing is to start with a small part. A cubic yard of dirt is about one and a half tons. I know that seems pretty heavy, and it is. Dirt is very heavy. So, we have that basic number to work with. Then, how many cubic yards of dirt in the world?*

I was already over my head, but Dad kept going.

> *Well, the circumference of the earth around the equator is about 25,000 miles, and the radius is about 4,000 miles.*

I had no idea how he knew this.

> *So, the volume of a sphere is equal to 4/3 Pi times the radius to the third, and Pi is about 3.14. ... Let's see 4/3rds of Pi is about 4 multiplied by 4,000 (that's the radius) to the third is like 64 squillion, or something like that. That's the volume, and if a cubic yard of dirt weighs one and a half tons, that means the world weighs about 96 squillion tons.*

I was dumbfounded. How did he know that? Was he right? And how many is a *squillion*?

Then, my father said, "Do you think if you put the world on a fly, it would squash it?"

"Dad, of course, it would squash it!" The idea of the entire planet being put on a fly just cracked me up. Poor fly!

There's a point here. Our planet is really big, as well as heavy. Even the comparison of a ginormous heavenly body to a fly is hard to get your head around. Yet, *our Universe is nothing but unfathomable.* Garnering the appropriate perspective to understand infinity is difficult. We're sort of like the fly in that respect.

Did you ever see the short video called *Powers of Ten*? It's a fascinating look at how reality changes by moving away from a picture of a couple having a picnic in Chicago out to the galaxy and beyond, and then moving closer in to the cellular level, each time by a power of ten. It's a 1968 American documentary short film written and directed by Ray Eames and her husband, Charles Eames, rereleased in 1977. Check it out: http://www.youtube.com/watch?v=0fKBhvDjuy0. The sheer size differential of things is amazing.

Or, how about Scale of the Universe, an online size calculator? It shows how big everything is in comparison to everything else. For example, a human at 2 meters tall is 46,500,000,000,000,000,000,000,000,000 times smaller than the estimated size of the *entire* universe, but only 600,000,000,000,000,000,000 times smaller than our entire *galaxy*. And atoms, once considered the smallest particle when Sir Isaac Newton named them in the 17th Century, are 7,000,000,000,000,000, 000,000,000 times larger than the planck, which is currently the smallest possible size we can measure. http://primaxstudio.com/stuff/scale_of_universe/

It wish Stephen Hawking could have been there on the dock with Dad and me. Hawking is the brilliant physicist who has ALS, Lou Gehrig's disease, that's left him almost completely paralyzed. He's in a wheelchair and speaks through a computer that recognizes his meaning through twitches in his cheek. He's highly regarded as one of the most influential scientists of our day. I picked up his book *A Brief History of Time* in an airport bookstore years ago, but it was incomprehensible. So, I was happy when he and Leonard Mlodinow published a simpler version, *A Briefer History of Time*, for lugs like me. Even that book is pretty hard to understand, but it's fascinating. If he'd been on that dock with us, he probably would have said something like this:

> *Today we know that stars visible to the naked eye make up only a minute fraction of all the stars. We can see about five thousand stars, only about .0001 percent of all the stars in just our galaxy, the Milky Way.*

Dr. Hawking, that's amazing, because when Dad and I look up at the Milky Way, it seems like there are, well, squillions of stars, and you're saying that's only like 1/10,000 of the stars in our galaxy? He'd reply, Yes:

> *The Milky Way itself is but one of more than a hundred billion galaxies that can be seen using modern telescopes – and each galaxy contains on average some one hundred billion stars. If a star were a grain of salt, you could fit all the stars visible to the naked eye in a teaspoon, but all the stars in the universe would fill a ball more*

> than eight miles wide. ... In one second, a beam of light will travel 186,000 miles, so a light-year is a very long distance. The nearest star, other than our sun, is called Proxima Centauri (also known as Alpha Centauri C), which is about four light-years away. That is so far that even with the fastest spaceship on the drawing boards today, a trip to it would take about ten thousand years.

For the serious scientists, Kid's Post in the *Washington Post* on December 8, 2010, said there are now thought to be about 300 sextillion stars in the universe, which is 3 followed by 23 zeros. That's three times more than earlier estimates (in case you were wondering). How many is 300 sextillion? "There are about 50 trillion cells in the human body, and there are about 6 billion people. Which comes out to about 300 sextillion. So the number of stars in the universe is about the same as the number of human cells on Earth."

Another perspective I find fascinating is speed (not the drug; actually never did it, but the speed of things in motion). Hawking mentioned the speed of light above. Consider this, sound travels at 760 miles per hour. That seems slow comparatively, but of course it's way faster than most people have ever gone, even flying in planes. A bullet leaves a gun at about 1,000 miles per hour. That's fast!

Now, think about this: the earth is rotating on its axis at about 1,000 mph. The earth orbits the sun at about 66,000 mph. Our solar system orbits around the Milky Way Galaxy at about 483,000 mph, and still it takes the sun and its planets about 225 million years to make a single revolution around the axis of the Milky Way. *The Milky Way itself is moving at about 1.3 million mph,* according to the Astonomical Society of the Pacific. http://www.astrosociety.org/education/publications/tnl/71/howfast.html. And you're sitting here reading this book. It just blows me away.

The key thing for me here is *perspective*. How would we know any of this if it weren't for scientists? And if we don't listen to what they say, we're basically staying in the dark ages. In many respects, we *are* still in the dark ages. In one of his new books, *The Grand Design*, Hawking, again with Leonard Mlodinow, says that humans are in a "primitive state of development ... ." Our cultural and personal philosophies continue to be shaped by religious beliefs and outdated understandings of the world we live in. I can honestly say I've learned much more about spirituality from science than from the *Baltimore Catechism* of my youth. In *The Grand Design*, Hawking also says, "[P]hilosophy is dead. Philosophy has not kept up with modern developments in science, particularly physics." He might have added that religion is dead too, as a reliable guidepost for our future world. I don't think Jesus ever said anything about quantum mechanics!

Here's an example of our social discourse not keeping up with new scientific knowledge. When was the last time you were at a cocktail party and people were discussing dark energy? Probably never. That's partly because *no one knows what it is!* Well, then, you say, it must be pretty insignificant stuff, right? Not according to Steve Nadis, Editor of *Origin and Fate of the Universe*:

> *The universe according to [the satellite orbiting the earth since July 2001 named the Wilkinson Microware Anisotropy Probe] contains 4 percent ordinary matter and just 0.4 percent luminous matter – stars, galaxies, and the like. All the rest consists of dark matter (23 percent) and dark energy (73 percent). [These results] reveal a sobering fact: 96 percent of everything is almost a total mystery because we can't yet describe the true nature of dark matter and dark energy.*

It hasn't been that long since scientists knew anything about this dark whatever-it-is. A *Washington Post* article by Guy Gugliotta and titled, "Astronomers Find More Evidence of 'Dark Energy,'" on Wednesday, May 19, 2004, reported that:

> *Astronomers announced new evidence yesterday that the mysterious force known as "dark energy" is causing the universe to expand ever more rapidly, perhaps eventually leading to cosmological "loneliness," in which galaxies grow so far apart that the heavens will appear empty." ... "This is clear and direct evidence that the expansion of our universe is accelerating," Cambridge University astronomer Steven Allen, leader of the Chandra study, said at a news conference. "Our results are consistent with the cosmological constant, and they have important implications for the fate of the universe."*

Another article in the *Washington Post* on November 17, 2006, titled, "Force that Counters Gavity Existed Early in Universe," by Marc Kaufman, said that a NASA study also showed that dark energy is pushing the expansion of the universe.

> *It also lends support to one of the most debated theories put forward by Albert Einstein, what the renowned physicist called a "cosmological constant" in the universe that works against gravity, creating equilibrium.*

However, despite calling dark energy a "constant," it seems to be acting like a universal accelerator. According to another article in the *Washington Post* titled, "Mysterious 'Dark Energy' Not as Ominous as Thought," by Joel Achenbach, on December 17, 2008:

> *The universe as we know it is believed to have sparked into existence about 13.7 billion years ago. From that initial big bang onward, it expanded, but until 5 billion years ago gravity dominated the show, gradually slowing the expansion. Then dark energy became the more powerful actor, and the expansion began to accelerate as surely as if someone had put the pedal to the metal. ... Hovering over the new research is the minor matter of the fate of the universe. If the universe continues to expand at an accelerated rate, [Lawrence M. Krauss, a physicist at Arizona State University,] has estimated, in about 100 billion to 1 trillion years, almost all the galaxies we see will be so far away they will vanish from sight. It will be a much darker universe.*

That's what they mean by 'cosmological loneliness.' Apparently, scientists had previously been worried the newly discovered dark energy as it took control would rip the universe apart. They're now feeling it won't, so we only have to worry about the sky being very dark in about a trillion years, whew! Glad I had that chance to check out the Milky Way with my Dad!

The ancients had much clearer skies than we do for looking at stars. There weren't big city lights to obfuscate the darkness. Our modern city skies are so bright no one seems to realize when the moon's rising or full, much less be able to see the Big Dipper. It's like we're losing sight of our true home. The humans living thousands of years ago didn't have a lot of what we have, and certainly science was very subjective back then. According to Hawking in *A Briefer History of Time*:

> *Aristotle thought that the earth was stationary and that the sun, the moon, the planets, and the stars moved in circular orbits around the earth. ... In Ptolemy's model [in the second century A.D.], eight*

> rotating spheres surrounded the earth. Each sphere was successively larger than the one before it, something like a Russian nesting doll. The earth was at the center of the spheres. [H]is model was generally, although not universally, accepted. It was adopted by the Christian church as the picture of the universe that was in accordance with scripture, ... Another model was proposed in 1514 by a Polish priest, Nicolaus Copernicus[, who] had the revolutionary idea that not all heavenly bodies must orbit the earth. In fact, his idea was that the sun was stationary at the center of the solar system and that the earth and planets moved in circular orbits around the sun. ... [N]early a century passed before the idea was taken seriously.
>
> In 1609, Galileo started observing the night sky with a telescope, which had just been invented. When he looked at the planet Jupiter, Galileo found that it was accompanied by several small satellites or moons that orbited around it. This implied that everything did not have to orbit directly around the earth, as Aristotle and Ptolemy had thought. At the same time, Kepler improved Copernicus's theory, suggesting that the planets moved not in circles but in ellipses. The true explanation for why the planets orbit the sun was provided only much later in 1687, when Sir Isaac Newton ... said that a particular force was responsible for [causing the planets to move around the sun], and claimed that it was the same force that made objects fall to earth rather than remain at rest when you let go of them. ... It was the first time in history anybody had explained the motion of the planets in terms of laws that also determine motion on earth, and it was the beginning of both modern physics and modern astronomy.

Wow, telescopes weren't invented until about 1600, and physics didn't arrive as a science until about seventeen centuries after the supposed life of Jesus. It puts a perspective on the level of knowledge at that time compared to now, and the framework for peoples' beliefs two thousand years ago. I mean, the people alive at the supposed time of Christ believed the earth was flat and that didn't change for seventeen hundred years.

Physics even up to the time of Einstein was much different than now. The new physics is pretty counter-intuitive and arcane sometimes, but Hawking does a good job explaining the key concepts in *A Briefer History of Time* and *The Grand Design*. Einstein's special theory of relativity was groundbreaking for the science of physics, and stated:

> [T]he laws of science should be the same for all freely moving observers, no matter what their speed. ... In other words, the theory of relativity requires us to put an end to the idea of absolute time. Instead, each observer must have his own measure of time, as recorded by a clock carried with him, and identical clocks carried by different observers need not agree. ... We must accept that time is not completely separate from and independent of space but is combined with it to form an object called space-time.

This is hard to understand. What's "an object called space-time?" Dunno. Hawking explains it by saying, we understand three coordinates of length, width, and height, also like latitude, longitude, and height above sea level, to describe the location of a point. So, if we add time, then we get a fourth dimension, called space-time.

Einstein combined his special theory of relativity with Newton's theory of gravity and in 1915 published his general theory of relativity. According to Hawking, Einstein's theory is "based on the revolutionary suggestion that gravity is not a force like other forces but a consequence of the fact that space-time is not flat, as had been previously assumed. In general relativity, space-time is curved, or 'warped,' by the distribution of mass and energy in it."

Well, again, it's hard to follow all this unless you're a trained physicist (which I am most definitely not), but the conclusions from these theories lead us to a new and different perspective on our world. Specifically, as one example, physicists now know that *time moves more slowly the closer one is to the earth's surface!* If you were on the top of Pike's Peak, and I was at the bottom, time would be moving more slowly for me than for you! According to Hawking:

> [E]ach individual has his own personal measure of time that depends on where he is and how he is moving. ... Space and time are now dynamic quantities; when a body moves or a force acts, it affects the curvature of space and time – and in turn the structure of space-time affects the way in which bodies move and forces act. <u>Space and time</u>

not only affect but are also affected by everything that happens in the universe. (My emphasis added.)

This means everything in the universe is totally *interconnected*. I've mentioned that before, as a way of describing *oneness*. It seems to be demonstrated as true by physics. We also learn this interconnection does not occur in a predetermined way.

Hawking discusses quantum physics in *A Briefer History of Time*, also known as quantum mechanics. It deals with the actions of energy and matter at the atomic and subatomic matter. Many brilliant scientists, such as Max Plank and Werner Heisenberg, contributed to the theory. Heisenberg developed the *uncertainty* principle, which says, "nature does impose limits on our ability to predict the future using scientific law. ... One of the revolutionary properties of quantum mechanics is that it does not predict a single definite result for an observation. Instead, it predicts a number of different possible outcomes and tells us how likely each of these is. ... Quantum mechanics therefore introduces an unavoidable element of unpredictability or randomness into science." Hawking says in *The Grand Design*:

> *For example, according to the principles of quantum physics, which is an accurate description of nature, a particle has neither a definite position nor a definite velocity unless and until those quantities are measured by an observer. ... Quantum physics might seem to undermine the idea that nature is governed by laws, but that is not the case. Instead, it leads us to accept a new form of determinism: Given the state of a system at some time, the laws of nature determine the probabilities of various futures and pasts rather than determining the future and past with certainty. ...*
>
> *Quantum physics tells us that nothing is ever located at a definite point because if it were, the uncertainty in momentum would have to be infinite. In fact, according to quantum physics, each particle has some probability of being found anywhere in the universe. ... It is important to realize that probabilities in quantum physics are not like probabilities in Newtonian physics, or even in everyday life. ... Probabilities in quantum theories are different. They reflect a fundamental randomness in nature. The quantum model of nature encompasses principles that contradict not only our everyday experience but our*

> *intuitive concept of reality. ... But quantum physics agrees with observation. It has never failed a test, and it has been tested more than any other theory in science.*
>
> *According to quantum physics, you cannot "just" observe something. That is, quantum physics recognizes that to make an observation, you must interact with the object you are observing. ... The universe, according to quantum physics, has no single past, or history. The fact that the past takes no definite form means that observations you make on a system in the present affect its past. ... [T]he universe doesn't have just a single history, but every possible history, each with its own probability; and our observations of its current state affect its past and determine the different histories of the universe.*

So, raise your hand if you ever woulda figured out that observation of the present could affect the past. Not I. This quantum physics stuff defies belief, but here the smartest physicists in the world tell us that all experiments prove its theories. Hawking says in *The Grand Design*, "[C]ommon sense is based on everyday experience, not upon the universe as it is revealed through the marvels of technologies such as those that allow us to gaze deep into the atom or back to the early universe."

Common sense also wouldn't say that there could be more than one universe. I certainly don't see more than one universe. Do you? I quoted Richard Carrier earlier about the possibility of more than one universe, if you recall. That is one of the tenets of what is called M-theory, as explained in *The Grand Design*. Hawking says it's the only current model that offers the possibility of a single comprehensive theory of the universe. M-theory essentially states:

> *[T]hat ours is not the only universe. Instead, M-theory predicts that a great many universes were created out of nothing. Their creation does not require the intervention of some supernatural being or god. Rather these multiple universes arise naturally from physical law. ... Each universe has many possible histories and many possible states at later times, that is, at times like the present, long after their creation. Most of these states will be quite unlike the universe we observe and quite unsuitable for the existence of any form of life. Only a few would allow creatures like us to exist. Thus our presence selects out*

from this vast array only those universes that are compatible with our existence. Although we are puny and insignificant on the scale of the cosmos, this makes us in a sense the lords of creation.

As I read these words, one lesson I get is that we can't be sure of anything. There is no definitive past. But we can understand that we don't understand everything there is to know, i.e., that it's OK not to have all the answers.

In the movie, *What the Bleep Do We Know?*, which I mentioned previously, some of our core beliefs and assumptions about life are shaken and stirred. It's an unusual film; part movie, part documentary. The website says the star, "Amanda, played by Marlee Matlin, finds herself in a fantastic Alice in Wonderland experience when her daily, uninspired life literally begins to unravel, revealing the uncertain world of the quantum field hidden behind what we consider to be our normal, waking reality." Fourteen different scientists and mystics serve as commentators. I noted physicist John Hegelin's comment that, at the atomic and sub-atomic levels, we are *not separate at all*. With more of the background of modern physics, as I've been boring you with for a while, you can see how this just might be right.

Think of this; we don't know what dark energy or dark matter are, and they comprise most of the universe. It makes sense that the rest of us, our energy and matter, are connected, because we comprise the same type of matter and energy. If you go way out into the universe, like in Powers of Ten, we on this planet look like, well, you can't even see us. We're just a speck, like a piece of dust in the air. I'd say a piece of dust is just one thing, not made up of lots of parts, but we know that's not the case. And it's the same for *everything* around us. Composed of the same elements; for goodness sake, *we're mostly water – H_2O!* There really isn't any separation, from that perspective. And what matters in terms of trying to understand it, to know truth - our limited perspective, or how we're viewed in context with the universe?

That's the message of the movie; we don't know shit. We can keep trying to find out the secrets of the universe, and maybe one day we'll find them. But one of the characters in the movie says that *not knowing makes us joyful!* Imagine that we're in such a fabulous place that we can't understand it; *how lucky is that!*

Some other quotes I wrote down when I watched the movie:

> *All we see is the tip of the iceberg.*
> *Atoms – electrons and protons – pop in and out of existence.*

Heaven is Everywhere

Particles can be in two or more places at once.
Atoms are not things; they're tendencies.
There is completely amazing magic under your eyes.
Every single one of us affects the reality that we see.
Thought alone can completely change the body.
There are different levels of truth.
Most religions have assumed God is separate, but that is a blasphemy.
The height of arrogance is the control of those who create God in their own image.
Each cell is definitely alive, and each cell has a consciousness.
Most people surrender and live their lives in mediocrity.
Setting up right and wrong is not a good way to look at things; there is no good or bad.
We are all interconnected.
I am connected to it all.

Sounds like a bunch of crazy new age stuff, doesn't it? But it feels *good*. Without the background of quantum physics, it may seem like ethereal, feel-good bullshit. But knowing what Stephen Hawking and all of the brilliant scientists of the world have to say lends credibility to it. It's not mumbo jumbo; it's *jumbo!*

Having been taking this voyage on earth for over half a century, I've learned enough to begin to be able to put together my own theory of existence and life. I'm beginning to know what's true for me. Here are some things I now believe:

Each of us is just like a flower. A flower grows, blooms, fades, and dies. But that time of being and blooming is the flower's perfection – it will not *be* again – and this life is our perfection, too. If there was only nothingness except a flower, that flower would be the antithesis of nothing; it would be everything. If there was only that flower, it would be perfect. We're like that flower, and all flowers. We're just like the slugs, worms, and birds. *Our lives are perfect; our planet is perfect; heaven is literally everywhere!*

Did you watch the series *Planet Earth* on television? In high definition, it's jaw-dropping spectacular. Scenes of places and animals that're hard to believe. Check it out: http://dsc.discovery.com/tv/planet-earth/. I get goose bumps just watching the scenery. Sometimes it's hard to believe we actually live here on this crazy ball of matter. The series is just gorgeous; if you haven't seen it, you really should.

If this isn't heaven, then what do you think heaven *should* be like? What would perfection be like? They always tell us that heaven would be sitting up on clouds with the angels in God's love. OK, take away fighting and killing and fear-based stuff, and what do you have here on earth? *Heaven, I'm tellin' you.* Humans cause most of the awful things that happen, which many misperceive as evil. Sure, there are natural disasters and freak accidents. But humans aren't evil; quantum physics doesn't include a law of evility. We humans get all angry, upset, and worried about threats like – taxes! Taxes? You'd think the government was trying to force us to eat dirt and dig ditches. There's nothing in quantum physics that says taxes are important. In fact, there is nothing in quantum physics to fear, because fear isn't real.

Think of it another way; have you ever been on a walk in the woods or stood on top of a mountain? What did you see and feel there? Sky, the sun, moon and stars at night, trees, birds. Animals, clouds, butterflies, plants, flowers. When I look down out the window of a plane, I'm telling you I see perfection, oneness, and love. *You don't see fear out there.* There is just – everything. You know, *this is that, that's that, all this is that.* It's only what is, and I believe it's all love. That's how I'm choosing to think of life and the energy. Just like my cousin Henry Glassie says, wherever there's art, there's love. So, to take that one step farther, wherever there is *anything at all*, that's love. And we are love, too, just like Jerry Jampolsky says in *Love is Letting Go of Fear*. If we're love and everything is love and we're all connected, then love really is onenesss.

We can open ourselves up to any or all possibilities, since truly *we are the lords of creation*. It's true from the perspective of quantum physics, which says the observer is crucial for existence. That's true for me in the practical sense that, what humans believe and think creates the social world we live in. What people think determines how people act. And if their thinking is fear-based, then we have a fearful, murderous, hateful world. Just like the elephants that were never hunted, or the animals on the Galapagos, things would be entirely different if we didn't fear one another. I talked a lot about fear earlier in the book, and so I'm thinking that by now you can at least tell what fear-based actions are, versus love-based actions. Can you?

We shouldn't allow ourselves to be held back by unreliable memes, particularly many of the dogmatic moral precepts that find their base in the subjective musings of ancients who didn't even know the earth was round. Just as the new physics says that unpredictability is a constant and there's no absolute time, there also are no absolute judgments about good or evil, or conclusions about God. We should just see what is!

And a belief that heaven is all around is even more special when understanding the transitory nature of life. Scientists are also telling us of all sorts of amazing, yet sometimes troubling, things. Did you know the moon might be shrinking? Should we be worried about that?

According to an Associated Press story written by Randolph Schmid, titled, "Moon may be Shrinking, But Very, Very Slowly," on August 10, 2010, "New research indicates that cracks in the moon's crust have formed as the interior has cooled and shrunk over the past billion years or so. That means the surface has shrunk, too, though not so much you'd notice from gazing at it." The moon has only actually shrunk about 300 feet, and it doesn't seem like it's going away any time soon, so that's good news. I suppose if it got real small, though, it would either fly off into space (and presumably our tides would get all weird with flooding and all) or be pulled by the earth's gravity and crash into us (which I'm guessing wouldn't be such a good thing).

Or did you know the sun hasn't had many spots in the last few years and also may be shrinking? According to a *Washington Post* article, titled, "Why is the Sun Losing its Spots?" by Stewart Clark, on June 22, 2010, "Sunspots and other clues [from modern space telescopes] indicate that the sun's magnetic activity is diminishing and that the sun may even be shrinking. ... Sunspots are windows into the sun's magnetic soul. They form where giant loops of magnetism, generated deep inside the sun, well up and burst through the surface, leading to a localized drop in temperature that we see as a dark patch. Any changes in sunspot numbers reflect changes inside the sun."

The apparent issue is that, though the number of sunspots fluctuates, scientists had thought a recent slow period would result in more spots in 2008, but it didn't happen. Though sunspot activity has increased recently, scientists are still concerned. Research indicates that over the centuries temperatures on earth have dropped when there's low sunspot activity. That could ultimately bring on a new little ice age, thereby taking us off the hook somewhat for manmade climate change. Not sure we want something like that to happen, but things will ultimately be much worse. Life as we know it will end, one way or another.

You've heard about the Hubble Space Telescope, I'm sure. We've come a long way since Galileo's first telescope. The Hubble images are fabulous; I love them. Eerie, yet heavenly. According to the Hubble website:

> *The Hubble Space Telescope's launch in 1990 sped humanity to one of its greatest advances in that journey. Hubble is a telescope that orbits Earth. Its position above the atmosphere, which distorts and blocks*

the light that reaches our planet, gives it a view of the universe that typically far surpasses that of ground-based telescopes. Hubble is one of NASA's most successful and long-lasting science missions. It has beamed hundreds of thousands of images back to Earth, shedding light on many of the great mysteries of astronomy. ...

Hubble's discoveries have transformed the way scientists look at the universe. Its ability to show the universe in unprecedented detail has turned astronomical conjectures into concrete certainties. It has winnowed down the collection of theories about the universe even as it sparked new ones, clarifying the path for future astronomers. Among its many discoveries, Hubble has revealed the age of the universe to be about 13 to 14 billion years, much more accurate than the old range of anywhere from 10 to 20 billion years. Hubble played a key role in the discovery of dark energy, a mysterious force that causes the expansion of the universe to accelerate.

Hubble has shown scientists galaxies in all stages of evolution, including toddler galaxies that were around when the universe was still young, helping them understand how galaxies form. It found protoplanetary disks, clumps of gas and dust around young stars that likely function as birthing grounds for new planets. It discovered that gamma-ray bursts — strange, incredibly powerful explosions of energy — occur in far-distant galaxies when massive stars collapse. And these are only a handful of its many contributions to astronomy. The sheer amount of astronomy based on Hubble observations has also helped make it one of history's most important observatories.

We are so fortunate to know all this, and see into the cosmos. For instance, have you heard of the Butterfly Nebula photograph taken by Hubble? Check it out: http://hubblesite.org/newscenter/archive/releases/2009/25/image/f/format/zoom/.

This is how it's described in the *Washington Post Magazine* of December 6, 2009:

The butterfly Nebula is the product of a star in its death throes. It's a star much like the one we see rising in the east every morning. This will happen to us. This is our future. The star is about 3,800 light years away, in the constellation Scorpius. Those wings are actually hot

> *streams of particles being ejected by the star into interstellar space. As the star starts to run out of hydrogen and helium fuel, its core contracts, and, simultaneously, the intense radiation of the star blows the outer layers into space. It's not an explosion, but more of a spewing. The universe, to be chemically interesting, has to have stars. And stars have to die.*

Of course, there's nothing at all we could do about any of this; shrinking moon, dying stars, whatever. There are so many things way beyond our ability to influence. So, the point is, no sense sitting around worrying about fewer sunspots, taxes, and stuff like that.

Quantum physics does not predict anything in particular, and doesn't say we'll be here again, or that we'll reincarnate into a cow. It doesn't say we'll go to heaven or hell. It says anything is possible. It says that everything is now, with no future or past. So, why not believe that this here is perfect? Why not live by the creed that *Heaven is Everywhere*, as opposed to God will send me to hell if I sin? It's a much more rational and positive viewpoint, IMHO.

I realize we've discussed many fearful things on our voyage through this book. If we reasonably don't have to get too worried about the sun quitting on us, can we also learn to let go of other fears we have? Can the knowledge we have of life, from our own direct experience and from what science teaches, help form a basis from which to get beyond our fears? I've tried to describe a physical cosmos and planet that from all detached perspectives could only be called perfect, just like the entire universes, yet we moan and groan about our condition in life all the time. Taxes! Traffic! The mortgage! How do we remove these fears and really get to the oneness, to the love?

Thirteen | Let It Be

I was in a war one time, in high school. The girls and guys were "rolling" each other's houses, you know, with toilet paper. The object was to sneak up to a girl's house at night, throw the rolls over the house and surrounding trees so the paper would stream out and cover them, and then run away before you could get caught by the girl's dad or the police. We knew where all the girls lived, and they knew where we lived. It was just a matter of execution, and having enough TP.

With about six guys, you could easily launch 30 or more rolls over the house before being discovered. If no one was home, we were *in like flint*. But if you hit three or four houses in a night, you'd need more than a hundred rolls. That would be a good night.

So, where to scrounge up that much toilet paper? I think we'd generally swipe some from home or buy a few rolls, though some of my friends weren't above pinching them from a store. I didn't do that, of course, but I didn't tell them not to. We needed a lot of rolls.

One night we heard that a boarding student at our high school, Georgetown Prep, knew where the janitors kept the TP and he could grab us a box. A box? We wondered how much was in a box. But hey, I drove my Dad's 1962 Cadillac out there and arranged to pick it up.

The boarder's name was Ted, but everyone called him "Banana Mouth." I think it was because he could fit a whole banana in his mouth, horizontally. Meaning, if he smiled, you saw the entire banana where his teeth would have been.

I'll never forget him running out of the dorm while we stood around the car waiting. He was carrying a *huge* box! It was so big he was sort of wobbly. We were stunned, but figured we'd be good to go for a while. Turns out the box had something like 120 rolls in it. We threw it in the trunk and skedaddled.

I don't remember how many houses we hit, but we did good. Still, when we were done, I had like 40 rolls left in the trunk. I kinda forgot about them until a few days later, when my Dad asked me how come there was so much toilet paper in the trunk. I answered as coyly as I could without lying, but it was OK because the family didn't have to buy TP for a while.

Actually, it was sort of a badge of honor, having your house rolled. That meant you were pretty cool. The girls thought enough of you to roll your house. You weren't a dweeb. So, I was not entirely disappointed one Saturday night when I heard my Mom yell, "They're out there!"

I scurried to the window and peered out. There they were! All the girls out in the driveway lofting toilet paper into the trees and up on the house. We banged on the window and they took off!

I scrammed out the door and ran down the street after them, my old man huffing behind me. I caught up with them after about a block. Then, I didn't know what to do. We were all out of breath and I couldn't think of anything to say, except, "Hey, you all oughta come back and clean up." I didn't think that was very likely, and when my Dad finally got to us, the girls raced off.

Well, Dad and I walked back to the house and surveyed the damage. Not too bad. We went inside and I started to call my friends (on one of those old black rotary phones). All of a sudden, my Mom yelled again, "They're back!"

Sure 'nuff. I looked out the window and there they were, about 8 girls. They had driven their car into our driveway, and some of them were starting to pull the paper out of the trees!

I didn't know what to do. But my Dad did. He went outside and called to them, "Hey, thanks for coming back, but don't worry about cleaning up. Why don't you all come in and have some ice cream?"

You could have knocked me over with a feather.

My father wasn't mad at these girls. He thought it was funny, and he liked them. He immediately let go of any sort of anger, and forgave them.

The girls all came in, relaxed in our kitchen, ate ice cream, and talked with my parents. I just sort of sat there. I called my friends and told them I knew where the girls were.

How could my Dad do that, just forgive them. I mean, they hadn't done any serious harm and the rain would eventually clean off the trees. Still, he just let it go. Does forgiveness involve absolving someone of a wrong they've done to you, or is it broader than that? Nearly forty years later, I find myself studying about forgiveness. Remember we talked some about forgiveness already, but there's more to learn.

I search amazon.com to find some books on forgiveness. Here's one: *Forgiveness is a Choice*, by Robert D. Enright, PhD. The cover of the book says it's "A Step-by-Step Process of Resolving Anger and Restoring Hope." The cover also says Robert Enright is "the forgiveness trailblazer." This should be a good book.

The first chapter is called, "Forgiveness: Path to Freedom." That's what I'm talkin' about! This must be the key. He describes forgiveness as a *process* and that great benefits come to people who have successfully completed the process of forgiveness. Specifically, they "have reduced or eliminated negative feelings, ... thoughts, ... and behaviors toward the offender." Also, that they "develop positive feelings, ... thoughts, ... and behaviors toward the offender." Hmm, that sounds good, but is there always an *offender*? And there hadn't been much of a process for Dad; he just forgave them in a snap.

Enright says it's clear that, "the more people forgave those who deeply hurt them, the less angry they were. ... [R]esearch on the physical effects of anger has produced significant evidence of the danger of holding onto anger." And lack of forgiveness is, "strongly related to coronary heart disease" and other physical problems. He cites studies saying, "those who forgave [] former spouse[s] were more emotionally healthy than those who chose not to forgive." He continues, "Forgiveness requires focusing less on the anger itself and more on the *source* of the anger. When we forgive, we face the fact that another person hurt us."

These are all good things about forgiveness. I'm troubled, though, by the apparent need for an "offender," that someone *hurt* the "offended." I don't know why, but is forgiveness limited to the narrow situation where there's a perceived offense or hurt? When the priest on behalf of Jesus forgave me in the sacrament of penance, had I hurt Jesus? Could puny little me, as a child, really hurt Almighty God?

I'm now interested in Enright's definition of forgiveness. In spite of being the "trailblazer" of forgiveness, he uses a definition developed by philosopher Joanna North of Great Britain, from her article entitled "Wrongdoing and Forgiveness" in *Philosophy*:

> *When unjustly hurt by another, we forgive when we overcome the*

> *resentment toward the offender, not by denying our right to the resentment, but instead by trying to offer the wrongdoer compassion, benevolence, and love; as we give these, we as forgivers realize that the offender does not necessarily have a right to such gifts.*

Enright continues:

> *Dr. North's definition makes it clear that forgiving begins with pain and that we have a right to our feelings. First, we are acknowledging that the offense was unfair and will always continue to be unfair. Second, we have a moral right to anger; it is fair to cling to our view that people do not have a right to hurt us. We have a right to respect. Third, forgiveness requires giving up something to which we have a right – namely our anger and resentment.*
>
> *Forgiving is an act of mercy toward an offender, someone who does not necessarily deserve our mercy. It is a gift to our offender for the purpose of changing the relationship between ourselves and those who have hurt us. Even if the offender is a stranger, we change our relationship because we are no longer controlled by angry feelings toward this person. In spite of everything that the offender has done, we are willing to treat him or her as a member of the human community. That person is worthy of the respect due to every being who shares our common humanity. ... I feel that the ultimate goal of forgiveness is that the forgiver experience positive feelings and thoughts toward the offender.*

Hold on. I don't like what he's saying. Yea, sure, we should treat all humans with respect (animals and the environment, too). Granting mercy to an offender, agreed. But, why *must* there be an offender, and why *must* we have been hurt by them to forgive?

Having his house teepeed by the girls didn't hurt my Dad. He wasn't offended. Neither was I. I didn't think they were wrong, because I'd done the same thing to them. Yet, I think it's fair to say we forgave them. We just decided to let go of any angry emotions or negative feelings we might have had. *Isn't that also forgiveness?* And in case you think my example is silly, realize that not everyone forgives so easily, even minor actions like that. A man shoveling snow in Manassas, Virginia attacked a 23-

year old guy with a box cutter and slashed his throat, because the dude threw a snowball at him. *Washington Post*, "Man Held in Attack on Snowball-Thrower," January 27, 2011. I think that man could've handled his anger better and just let it go, don't you?

Having been exposed to Jerry Jampolsky's ideas of forgiveness discussed in Chapter 4, where forgiveness is just *letting go*, the Enright/North view of forgiveness also seems *judgmental* to me. I mean, what if the "offender" doesn't think s/he did anything "wrong"? In another book I got on the subject, *Total Forgiveness,* by R.T. Kendall, he says, "It is my experience that most people we must forgive do not believe they have done anything wrong at all, or if they know that they did something wrong, they believe it was justified. I would even go so far as to say that at least 90 percent of all the people I've ever had to forgive would be indignant at the thought that they had done something wrong."

I think that's right; most people don't think they've done wrong. If they felt they'd done wrong, wouldn't that just be guilt? The problem with Enright's view of forgiveness is that it forces us to make judgments. We have to feel "unjustly" hurt. What does that mean? Can you feel "justly" hurt? I suppose when my father whipped me with the ruler for stealing money, I felt justly hurt. And I knew I'd done something society thought was wrong, but it didn't feel to me I was acting in a wrong or evil way from the standpoint of the universe. It's just what I did. I believe many people feel the same way about things they do, from Idi Amin to Adolf Hitler to the Pope.

Another problem for me with Enright and North's definition of forgiveness is that it claims we have a *right* to anger. What kind of "right" would that be? Who gave us this right? If people feel they have a right to anger, that means it's OK to be angry? And what is anger? As we have learned, it's basically fear. So, then we have a right to fear and be afraid? And everyone can have the right to be fearful and angry and that will be conducive to bringing peace to the world? *Not.*

Let's think more about that concept of having a *right* to be angry. What rights do I have in the first place? Well, here in the United States, I have the right to free speech and religion, and to vote, things like that. Where did I get these rights? From the government, isn't that so? From all of us citizens buying into the concept that we all get theses rights.

Thomas Jefferson in the Declaration of Independence famously wrote: "We hold these Truths to be self-evident, that all Men are created equal, that they are endowed by their Creator with certain unalienable Rights, that among these are Life, Liberty, and the Pursuit of Happiness." Pretty good stuff. But do we really have those rights?

Jefferson said these were *self-evident truths*. Where did they come from? If you believe God gave us those rights, that's fine, but what if there isn't a God, or a God who gives out rights?

Isn't our situation more like other earth denizens? For example, does a pig or a worm have rights? I don't think a duck has any rights. Doesn't seem to me quantum physics says we have any rights like that. Do we have the right to live? No, I don't think we have that right, except as humans collectively through government give us the right. I don't think we have a universal right to happiness. The universe doesn't give us rights like that. The universe just is, and somehow enables our life, but there's no such right inscribed in the dark energy. Likewise, there's no right to anger. So, I think it's clear we *don't* have any *inherent* rights.

Another aspect of this definition that confounds me is "hurt" – what is *hurt*? Maybe, emotional pain? What is emotional pain? That would just be a thought, wouldn't it?

If my Dad had run out of the house and yelled at the girls and called them evil wenches, they may have felt hurt. What would that mean to them? I suppose they'd feel less than good about themselves. They might feel unlovable and not worthy. They might have felt guilty, like they were wrong and bad to have rolled our house. That would be *their* reaction, the thought they had.

Or, they could feel my Dad had unjustly hurt *them*. They might feel they did what was asked – coming back and cleaning up the toilet paper – and he had no right to yell and make judgments about them, and that he hurt them. In which case, they'd be angry, which Enright says they'd have a right to be. Would they really have a right to anger? They'd rolled our house! Enright says they could retain that anger forever if they wanted, when or until deciding to give my father the "gift" of forgiveness. That's what Enright and North call it, a "gift" to the offender. North would say Dad didn't have a right to that gift of being forgiven if he'd yelled at them, unless the girls decided to give it to him. Of course, if the girls later forgave him, how would Dad even know? Would they send a letter saying they'd forgiven him? How would he even know he'd received their gift otherwise?

The whole thing makes your head swim, trying to figure out who was wrong, hurt, or angry. It's not simple to do; kind of a morass, actually. Enright basically acknowledges this problem, by saying people who thought they were wronged often find out that, *oops*, they weren't. "Sometimes people discover as they reach the last phase of the forgiveness process that they were not victims of a terrible injustice, but instead were victims of their own lack of understanding, something they couldn't see at the beginning."

Exactly! They *incorrectly* perceived wrong or hurt. From the perspective of the universe (any one of them), a perceived hurt or wrong *simply can't be that*. As I study these things, I've come to see that *there's no right or wrong, no evil or hurt, in the universe*. I just don't believe there is. Sure, people do all sorts of unloving and awful things; killing, murdering, and raping. These acts must be addressed under the laws of the government. As a society, we have to protect against uncivil behavior to the extent possible, to preserve the peace. But, there just can't *ultimately* be right or wrong.

Here's another example of why I think there's no right or wrong. I have a half-brother, Don. He was my father's son from his first marriage. Born in 1934, he was about 20 years older than me, so I never lived with him and didn't know him that well.

Except that I did, because he was a larger than life figure. He just wanted to have fun with life. He skied, sailed, travelled around the world, built and owned hotels and boats, and loved women. He had two wonderful kids with his first wife, who is a great lady. Then, since he had that "wild male Glassie gene," he partied and eventually married a woman who I think was about 28, and he was much older. They had three wonderful girls. Then, when he was 75, he was diagnosed with stage 4-pancreatic cancer, which can take you in a few days or a month. He did chemo, and hung in for 21 months. He visited India and China after he'd had the cancer for over a year. He was skiing in Aspen a few weeks before he passed. Whenever I was with him, *I had a blast*, and he was a great older brother at those times. http://www.jamestownpress.com/news/2011-02-10/Obituaries/Donelson_Don_C_Glassie_Jr.html

Thing was, he didn't tell hardly anyone when he got diagnosed with cancer, not even his children. We all knew something was up, but he only told his sister, best friend, and girlfriend at the time. She wanted to have a baby, so they did *in vitro* fertilization and had a son when he was 75 years old! When he finally succumbed and was in the hospital during his last days, he didn't want anyone around him. Didn't want to see anyone.

Some would say that was wrong; he should've told his family and let them share their love for him. He didn't. Some would say it's *wrong* to have a child at age 75 when you know you're going to die - very soon. He provided for them.

So, I thought to myself, you know, *there's no right or wrong way to die*. We all go, some more painfully or uglier than others, but there's no right or wrong way to die, or to live, for that matter.

Using Enright's definition of forgiveness is a *very* slippery slope. Once you define forgiveness as requiring a wrong or hurt, you're immediately making a judgment. Dad didn't make a judgment as to whether the girls were good or bad, right or wrong; he let it go. That's what Jampolsky describes as forgiveness. "Simply stated, to forgive is to let go," he said in *Love is Letting Go of Fear*. That definition seems cleaner, fresher, and not contingent on making judgments.

Another of my favorite writers and theorists is Eckart Tolle. I don't know if you've seen him. Sort of a weenie little guy; reminds me of an elf. But he's a profound thinker and successful author. When it comes to forgiveness, Tolle says in his book, *The Power of Now*, "Forgiveness is to offer no resistance to life – to allow life to live through you." Hey, that's really letting go! It's not based on right or wrong or judgments, it's just letting things be; you know, like the Beatles song, *Let it Be*. Tolle continues:

> *Nonforgiveness is often toward another person or yourself, but it may just as well be toward any situation or condition – past, present, or future – that your mind refuses to accept. Yes, there can be nonforgiveness of the future. This is the mind's refusal to accept uncertainty, to accept that the future is ultimately beyond its control. Forgiveness is to relinquish your grievance and so to let go of grief. It happens once you realize that your grievance serves no purpose except to strengthen a false sense of self. ... The alternatives are pain and suffering, a greatly restricted flow of life energy, and in many cases, physical disease. ... [Acceptance] is an essential aspect of forgiveness.*
>
> *Forgiveness of the present is even more important than forgiveness of the past. If you forgive every moment – allow it to be as it is – then there will be no accumulation of resentment that needs to be forgiven at some later date. ... [A]nger or resentment strengthen the ego enormously by increasing the sense of separateness, emphasizing the otherness of others and creating a seemingly unassailable fortresslike mental position of "rightness."*

I think Tolle gets to the same point – forgiveness is good and nonforgiveness can make you sick, but it's more, I don't know, cerebral or maybe less dogmatic. A more sophisticated sense of letting go. And he confirms what I've been saying about anger creating a sense of separateness from others, which makes one feel "right."

Sure, I think it's great one can have positive feelings toward someone who has offended or hurt them; I'm all for that. But why not have positive feelings about other people *all the time*? Isn't that one of the most important teachings of the alleged Jesus, *love your enemy*? Or is it just easier to say, the enemy is evil so I don't have to love them?

I've had lots of these conversations with people. A friend said to me, "Of course, there's evil in the world." We discussed that a lot in Chapter 5. I think evil is just a way that people try to rationalize the horrible things humans do to one another out of fear. There is no *thing* that is evil. It's only a perception, an illusion, of separation. It also is dependent on a frame of reference that is by its own nature limited.

Besides, who decides what's evil? Eckart Tolle addresses evil in his book, *A New Earth*.

> *On a collective level, the mind-set "We are right and they are wrong" is particularly deeply entrenched in those parts of the world where conflict among two nations, races, tribes, religions, or ideologies is long-standing, extreme, and endemic. ... Both sides believe themselves to be in possession of the truth. Both regard themselves as victims and the "other" as evil, and because they have conceptualized and thereby dehumanized the other as the enemy, they can kill and inflict all kinds of violence on the other, even on children, without feeling their humanity and suffering.*
>
> *Here it becomes obvious that the human ego in its collective aspect as "us" versus "them" is even more insane than the "me," the individual ego, although the mechanism is the same. By far the greater part of violence that humans have inflicted on each other is not the work of criminals or the mentally deranged, but of normal, respectable citizens in the service of the collective ego. One can go so far as to say that on this planet "normal" equals insane.*

So, Tolle is saying that simple distinctions of right and wrong are not only false, but lead to devastation on the planet. Surely, though, there's right and wrong, I'm asked. Well, yes and no. Here's a way to look at it. There's a right answer on a test, and a right play to make in baseball – throwing to first on a ground ball, runner at third, with two outs. Of course, there are decisions that are wrong in terms of our civil laws – speeding, stealing, stalking, and killing. These are right or wrong from the standpoint of baseball or the legal system. What I'm talking about is *right or wrong from the perspective of the universe.*

Tolle says, "The deeper interconnectedness of all things and events implies that the mental labels of "good" and "bad" are ultimately illusory. ... There's nothing that strengthens the ego more than being right. Being right is identification with a mental position – a perspective, an opinion, a judgment, a story. For you to be right, of course, you need someone else to be wrong ... ." So, thinking one is right is actually a way of feeling superior or different; a perception of separation. It's based on fear, and making those distinctions is not so simple as it seems.

Let's look at an example from the last chapter. Say that the sunspots slow down, finally cease, the sun explodes, and every single bit of matter on the earth is fried in seconds. All of the lovely mountains, streams, birds, and children – flash – gone instantaneously. Would that be wrong? It wouldn't be so fabulous for us, but it would not be wrong from the perspective of the universe.

Say a tree falls down in a windstorm and smashes my car; wrong or evil? No. What if I was in the car? Dead as a doornail. Is that wrong? What if the Washington Nationals never win a World Series? What if I get the epizootic and die tomorrow? What if your child is killed in a car crash? Or maybe a huge meteor hits and we all die in a cold and dust covered planet?

How about if I step on an ant on the sidewalk? What if I deliberately step on the ant? What if some loon kills a loved one? What's the difference in those acts from the standpoint of a rapidly expanding universe? They are not wrong or right, bad or good; they're just what *happens*, like black holes and dying stars.

Of course, there can be right or wrong from our limited human perspectives. But is our perspective the only one? Of course, *not*. No matter what you might think of all the quantum physics stuff we talked about, there are infinitely different perspectives, some of which might evoke, "That was right," and others, "That was wrong." And neither might be right, or both may be right, just as particles can be in two places at the same time. But ultimately, they're in no particular place at all, and nothing about them is right or wrong.

I've realized there's no sin, either. I know a lot of religious teaching is focused on sins, how we can do bad and sin against God. But I think *humans* created sin, and essentially use it to control behavior, like Jared Diamond was saying. Think about it in the same way we looked at good or bad, right or wrong. If an anteater eats an ant, that's not a sin. If I step on an ant, that's not a sin. If a bird craps on my head; not a sin. If I fall down the steps; nope, not a sin. (I did and it was a shame and a pain, but not a sin.) If a raging river takes out a beaver damn, not a sin. If someone calls me an idiot, not a sin.

If someone steals from me, that's not a sin. It may be against the law or support a legal claim against the thief, but not a sin. The person was not acting in a loving way, but still no sin. Remember Lesson 6 of the *Baltimore Catechism* said there were two kinds of sin, mortal and venial sins? A mortal sin is a "grievous offense against the law of God [that] takes away the life of the soul." Nothing in quantum physics or any other serious discipline I know of declaring that some act or omission takes away the life of the soul. And what are God's laws anyway? The Ten Commandments? Nothing in quantum physics about the Ten Commandments. Quantum physics basically says there aren't any laws, so I'd say modern science also doesn't support the theoretical concept of sin.

Again, people can do some awful things, and from a religious frame of reference they could be called sins. Killing someone, stealing, raping, and dumb stuff like that would be sins. Most people would agree on that. But the slope gets slippery here, too. I don't think masturbating is a sin. If it is, a lot of us have got some black splotches on our souls. I don't think sex for purposes other than procreation is a sin. (What other purposes; um, well, fun?) But the Catholic Church says it is: sex for purposes other than procreation is sinful. Any kind of sex with contraception is a sin. That includes a vasectomy. I'm a sinner. Drat. According to the website http://www.catholic.com/library/Birth_Control.asp:

> *Contraception is "any action which, either in anticipation of the conjugal act [sexual intercourse], or in its accomplishment, or in the development of its natural consequences, proposes, whether as an end or as a means, to render procreation impossible" (Humanae Vitae 14). This includes sterilization, condoms and other barrier methods, spermicides, coitus interruptus (withdrawal method), the Pill, and all other such methods. …. . Contraception is wrong because it's a deliberate violation of the design God built into the human race, often referred to as "natural law." The natural law purpose of sex is procreation.*
>
> *The pleasure that sexual intercourse provides is an additional blessing from God, intended to offer the possibility of new life while*

> *strengthening the bond of intimacy, respect, and love between husband and wife. The loving environment this bond creates is the perfect setting for nurturing children. But sexual pleasure within marriage becomes unnatural, and even harmful to the spouses, when it is used in a way that deliberately excludes the basic purpose of sex, which is procreation. God's gift of the sex act, along with its pleasure and intimacy, must not be abused by deliberately frustrating its natural end—procreation.*

Really, this is just insane. I doubt bonobos think they're sinning when they have sex. This position represents a serious *judgment* based on church dogma, and further shows how sin is so subjective and an unreliable determinant for assessing behavior. It falls into the natural law trap, too. Just as quantum physics shows there are no such "laws," attempting to make judgments such as alleging sin based on natural law is a fatally flawed analysis.

Many Christians and Muslims believe homosexuality is a sin. Listen to what cheery minister Joel Olsteen said on the *Piers Morgan Tonight* show on CNN, as reported in the *Washington Post*, "Under God," by Elizabeth Tenety, on January 29, 2011:

> *Yes, I've always believed, Piers, the scriptures show that homosexuality is a sin. But I'm not one of those who is out there to bash homosexuality and tell them that they're terrible people and all of that. I mean, there are other sins in the Bible, too. And I think sometimes the church – and I don't mean this critically – but we focus on one issue or two issues, and there's plenty of other ones. So I don't believe that homosexuality is God's best for a person's life – sin means to miss the mark.*

He's saying the Bible declares homosexuality a sin, so it's a sin; he doesn't bash gays and say they're terrible, but clearly he would be forced by the scriptures to make that judgment even if he didn't condemn them publicly. He is morally judging someone else based on what kind of sexual activity they like. Alleging sinfulness is just an arbitrary judgment. And remember what they say about judging: judge not, lest ye be judged. (The Bible, of course, has *some* good advice, in spite of its underlying faulty foundations.)

It may be hard to get your head around, because we've been brought up believing in right and wrong, good and evil, and that we're sinful creatures. It's just *bunk* and a childish perspective, frankly. We're no more sinful, evil, good or bad, than a June bug. As I said, the problem with these concepts is that they mandate judgments. Making judgments is perceiving separation, so is fear-based. Jerry Jampolsky in *Love is Letting Go of Fear* says, "Not judging others is another way of letting go of fear and experiencing Love."

Of course, we have to observe and make decisions about our actions all the time. We must evaluate and decide what we think is best for us. Making moral judgments is not the same thing. Tolle says in *A New Earth*, "[W]hen I criticize or condemn another, it makes me feel bigger, superior." Criticizing and condemning are judgments, versus observing or perceiving.

I will admit, though, that forgiveness of a tragic perceived wrong or hurt can be very difficult. We somehow have developed a belief that, when someone deeply does cause us loss, people deserve retribution. When the 9/11 attackers brought down the Twin Towers, many Americans seethed with anger and a desire for revenge. The feeling of an eye-for-an-eye was very strong in this Christian country, even though the mystical Jesus countenanced against it. But there are people who do forgive in extremely difficult situations. Enright acknowledges that, "Forgiveness is possible even under the most brutal and unfair circumstances." Here are some examples:

Audrey Kishline was the founder of Moderation Management, a nonprofit organization that preached responsible drinking in moderation. Audrey had problems with drinking in the past, and reformed. However, she went back to secret binge drinking and drove drunk on March 25, 2000 the wrong way down I-90 in Washington. She killed the daughter and husband of Sheryl Davis, and was sentenced to prison for vehicular homicide.

Audrey had told Sheryl she was sorry in the courtroom, but one day Sheryl came to visit Audrey in jail. Here was the transcript of their meeting as reported by Dennis Murphy on MSNBC's Dateline, http://www.msnbc.msn.com/id/14627442/ns/dateline_nbc/:

> **Audrey Kishline:** *I see them come in. You know, they've already been through all the checks and stuff. We're brought into a glassed-in room where the guards can completely watch us, but can't listen.*

> **Davis:** And I looked at her lawyer. And I asked, I said, "Is it okay if I hug her?"
>
> **Kishline:** And all of a sudden her arms were around me, and my arms were around her.
>
> **Murphy:** And you expected perhaps a slap across the face.
>
> **Kishline:** That's what I was expecting when she—but when she touched me, and I was gonna try to say, "I'm sorry," I got the - "I am," she goes, "Audrey, I forgive you." Now those were the very last words I ever, ever expected to hear from her, out of her lips.
>
> **Murphy:** She was the person that drove drunk and killed your daughter. Didn't you have every reason to say, "How dare you be a drunk driver?"
>
> **Davis:** And what good does that do? If you don't forgive somebody you're going to live with that turmoil for the rest of your life.

I remember hearing about an 18-wheeler truck that plunged off the Chesapeake Bay Bridge in August 2008. Tied up traffic for hours, and what a horrible image. That huge truck smashing through the guard rails and flailing a couple hundred feet down to the water. Turns out the driver didn't cause the wreck. John Short, Sr. was a safe driver, called in at the last minute to carry a load of chickens. He swerved his truck to avoid hitting the Chevy Camaro driven by Candy Baldwin that swerved into his lane at 4:00am. She said she'd fallen asleep.

You'd expect the family to be pretty angry at Candy, huh? No, they forgave her. Here's what John Short, Jr. said. "[My father] said, 'I've made mistakes in my life and the only thing I can do is ask for people to forgive me. … He just demonstrated that with himself, so that's how we learned about forgiveness and not holding grudges." As reported by Neal Augustine of WTOPnews.com on August 14, 2008.

Here's another story of forgiveness from Rawanda, where genocidal violence between the Hutus and Tutsis took the lives of nearly a million people. This is the story of a woman named Mukantabana, in a CNN story "Woman Opens Heart to Man Who Slaughtered Her Family," by Christiane Amanpour http://articles.cnn.com/2008-05-15/world/amanpour.rwanda_1_hutu-gitarama-tutsis?_s=PM:WORLD:

> In 1994, Mukantabana's husband and five of her children were hacked and clubbed to death by marauding Hutu militias. Among her

> family's killers was Jean-Bosco Bizimana, [a friend's] husband. "In my heart, the dead are dead, and they cannot come back again," Mukantabana said of those she lost. "So I have to get on with the others and forget what has happened. ... Women and girls were raped, and I saw it all ... . The men and boys were beaten and then slaughtered. They told others to dig a hole, get in, then they piled earth on top of them, while they were still alive."
>
> Yet today, Mukantabana shares her future and her family meals with Bizimana, the killer she knew, and his wife, her friend Mukanyndwi. ... Mukantabana admits that it was difficult to forgive. She said she did not speak to Bizimana or his wife for four years after the killings. What put her on the road to healing, she said, was the gacaca process. "It has not just helped me, it has helped all Rwandans because someone comes and accepts what he did and he asks for forgiveness from the whole community, from all Rwandans," she said.
>
> Bizimana said he did just that. "You go in front of the people like we are standing here and ask for forgiveness," he said. But despite his confession and apology, Mukantabana said, reconciliation would not have happened unless she had decided to open her heart and accept his pleas. In Gitarama, Bizimana said, "It hurts my heart to see that I did something wrong to friends of my family, to people who we even shared meals with," he said. "I am still asking for forgiveness from the people I hurt." Amazingly, many seem to have forgiven.

This type of reconciliation process was used to help South Africa get through its post-apartheid era. Nelson Mandela has said, "Forgiveness liberates the soul. That's why it's such a powerful weapon." "Three Men and a Dream," by Janice Kaplan, in *Parade*, December 6, 2009.

These are just a few examples of forgiveness in difficult circumstances. There are many more. The sister of one of the Lockerbie airplane bombing victims meeting with Libyan leader Moammar Gadhafi as an "ambassador of reconciliation" and forgiving him for his role in the murder. "Sister of American Lockerbie Victim Visits Gadafi," by Joe Sterling, CNN, September 25, 2009: http://ac360.blogs.cnn.com/2009/09/25/sister-of-american-lockerbie-victim-visits-gadhafi/. Then, there were the parents of two girls who were murdered by Matthew Murray in a shooting spree at a church, who forgave the killer and embraced his parents when they met. "Killer's Parents Hug, Cry with Parents of 2 Slain Teens," January 09, 2008, by Eric Marrapodi and Wayne Drash, CNN: http://articles.cnn.com/2008-01-09/justice/colorado.shooting_1_murrays-senior-pastor-parents?_s=PM:CRIME.

A lot of people don't buy forgiveness and would prefer vengeance. They don't believe humans can change and heal from tragic events, like 9/11. In comments addressed to the *Washington Post's* On Faith section on September 11, 2010, "9/11: Have We Healed," some show they don't see a way to forgive. Here are some of the comments.

> *Wounds of 9/11 will never heal: The wounds of 9/11 will never heal because we lack genuine commitment to find common ground.*
>
> *We cannot heal without justice: The nation has not healed from the attacks on 9/11, because an act of war is not a wound or disease. Terrorists attacked and killed Americans on Sept. 11 motivated by a radical form of Islam. Until we win the war against terrorism and make radical forms of Islam as unattractive as we have made Nazi ideology, we will not have finished the job that 9/11 started.*
>
> *The human condition makes healing impossible: It is ugly and it is dangerous, given the capability of enemies to not only hate each other but annihilate each other, but it is part of the human condition.*

All I can say is – *baloney*. Forgiving can be hard for some, but it does not mandate a process. It's just *letting go!* If someone put a hot coal in your hand, you'd drop it right away. Forgiveness can be that easy. Like Nike says, you *just do it*. Here's what I said in my book *Peace and Forgiveness*:

> *The greatest lesson in forgiveness is so misunderstood. They said he said, whatever is loosed in heaven is loosed on earth. This means simply that, if you let it go, it is gone. If you do not let it go, it is bound to you.*
>
> *It does not mean that some man in a black robe decides whether or not god forgives you. God and the Universe have already forgiven you. The key is you forgiving. If you forgive, the burden of fear is removed and love replaces it. If you do not forgive, you bear the weight and burden of fear yourself and you keep it in your heart and your soul.*

Let It Be

Louise Hay in her book, *You Can Heal Your Life*, says it well:

> *Many people come to me and say they cannot enjoy today because of something that happened in the past. Because they did not do something or do it in a certain way in the past, they cannot live a full life today. Because they no longer have something they had in the past, they cannot enjoy today. Because they were hurt in the past, they will not accept love now. Because something unpleasant happened when they did something once, they are sure it will happen again today. Because they once did something that they are sorry for, they are sure they are bad people forever. Because once someone did something to them, it is now all the other person's fault that their life is not where they want it to be. Because they became angry over a situation in the past, they will hold on to that self-righteousness. Because of some very old experience where they were treaded badly, they will never forgive and forget.*
>
> *Because I did not get invited to the high school prom, I cannot enjoy life today. Because I did poorly at my first audition, I will be terrified of auditions forever. Because I am no longer married, I cannot live a full life today. Because I was hurt by a remark once, I will never trust anyone again. Because I stole something once, I must punish myself forever. Because I was poor as a child, I will never get anywhere.*
>
> *What we often refuse to realize is that holding onto the past, no matter what it was or how awful it was, is ONLY HURTING US. … We are only hurting ourselves by refusing to live in this moment to the fullest. The past is over and done and cannot be changed. This is the only moment we can experience. Even when we lament of it in this moment, we are experiencing our memory of it in this moment, and losing the real experience of this moment in the process.*

Wouldn't it have been great if Eric Harris and Dylan Klebold had realized this before killing their classmates and teachers at Columbine? We would not have lost 32 more students and teachers at Virginia Tech if Cho Seung Hui had understood this perspective. Too bad they didn't know that, when you let go of the past and the future, you're in the *now*.

Fourteen | **Now, My Personal Allegory**

I don't remember the guy's name. He was working on seawall construction at the Island when I lived there. The contractors stayed on the Island during the week while they worked, then took off for home on Friday. For some reason, this guy stayed Friday night. We didn't have anything to do on the Island, which I was used to, but he wanted some action. He said, "Let's go over to a titty bar in Lexington Park." I hadn't been to the Park much, and didn't know the local scene enough to know any of those sorts of establishments. But I was 23, so said, "Sure."

It was a winter night, but I'd driven the Seahawk, our 42-foot wooden boat, back and forth from the Island a jillion times. So, it wasn't a problem to go across in the dark. We crossed the Sound, tied up the boat on the mainland, and drove to Lexington Park.

I don't remember much about the bar. The chicks weren't so hot. We drank a bunch of beers and left. It had started raining, and when we got back to the dock, it was coming down *cats and dogs*. The wind was out of the east, and you know what that means. Cold, wet, yucky. The eastern flow pushed water up the River and lifted logs and things off the shore, where they lurked just under the water, hard to see. The rain and clouds made it pitch black. No moon. But we wanted to get back to the Island to sleep.

The wind was blowing hard and the rain was sideways. We untied the ropes and pushed off from the dock. My senses were heightened and I didn't feel the beer at all.

We motored out the creek and into the Sound, and the lights from shore disappeared. It was dark as a dungeon, wipers barely clearing the windshield before water covering it again. We were rockin' and rollin' big-time with the waves. I grabbed the wheel and held tight. Had to flex my knees to balance and stay upright. The other guy held on for dear life.

I couldn't see where we were. The compass didn't work. All I had to go on was keeping the same angle to the waves, as we crashed through them and they washed over the boat. We had to go about a mile into the storm.

I knew we didn't want to miss the Island to the south and head into the River. We also didn't want to hit a sandbar or run aground in the shallows.

But I was totally calm. Not scared in the least. No panic, no fear. Concern, yes; I knew it would take skill and intuition to land us safely at the Island pier. I was just in a zone, in the moment.

I've had that experience other times driving a boat, or a car through snow. Sometimes writing, or playing music. Freakin' calm. Still. Peaceful.

The first thing I saw after about fifteen minutes of intense focus was … the *Island dock!* I'd bee lined straight to the pier! We tied her up and raced to the house. In a few minutes, we were safe and warm by the fire.

What *is* that state of mind? That sense of total harmony with your surroundings? Being in the zone?

> *One of the daredevils in "Steep," a documentary about extreme skiing, insists he is not hooked on the adrenaline rush of this death-defying sport. His denial is hard to believe. In one scene after another of this movie, written and directed by Mark Obenhaus, skiers hurtle down slopes at angles of 55 degrees or greater and leap off precipices into the unknown. The tiniest miscalculation, we are told, could result in death.*
>
> *The sport's practitioners - addicts might be a better word – would rather talk about how extreme skiing puts them totally in the moment and gives them an appreciation of life so acute it makes this sport, which has a high fatality rate, worth pursuing at all costs.*
> "Talk About Slippery Slopes," by Stephen Holden, *New York Times*, December 21, 2007.

Ah, "in the moment." "An appreciation of life so acute," it's worth risking one's life. Eckhart Tolle explains it this way in *The Power of Now*:

Now, My Personal Allegory

The reason why some people love to engage in dangerous activities, such as mountain climbing, car racing, and so on, although they may not be aware of it, is that it forces them into the Now – that intensive alive state that is free of time, free of problems, free of thinking, free of the burden of personality. Slipping away from the present moment even for a second may mean death.

"These are moments in which your mind becomes so entirely absorbed in the activity that you 'forget yourself' and begin to act effortlessly, with a heightened sense of awareness of the here and now (athletes often describe this as 'being in the zone')." This is from a blog on the *Huffington Post* on January 22, 2011, written by Lance P, Hickey, Ph.D, senior editor and contributing writer of www.pursuit-of-happiness.org. Hickey calls it *flow*.

He says it's been found "that people who experience a lot of flow in their lives also develop other positive traits such as high concentration, high self-esteem, and even greater health." He recounts how Zhuangzi, Chinese philosopher of the 4th Century BCE, taught "the 'ultimate happiness' is gained when we have learned to 'let go,' engaging in activities for their own sakes without any ulterior motives. In such a state all human actions become spontaneous and fresh, childlike in their intensity. On the highest level we transcend our egos and merge with the Dao, or *the way*, the underlying unity that embraces all things in the Universe. … In the state of flow with the Dao there is no 'me' and no 'it.'"

Hickey also says, "This idea of flow as a kind of 'caring for life' is a major theme in Robert Pirsig's popular work *Zen and the Art of Motorcycle Maintenance*, for example, when he writes, 'When you're not dominated by feelings of separateness from what you're working on, then you can be said to 'care' about what you're doing.'"

I experienced flow that night driving to the Island. And you know you've felt it before, or perhaps seen it in sports when Michael Jordan scored at will. The comments above also echo a now familiar theme; being at one versus perceiving separateness. I didn't feel any fear, just heightened awareness of the moment. I felt almost at one with the storm.

When I was driving the boat, I wasn't afraid, but very focused on the boat, water, rain, wind, and everything around me so I could achieve my (our) goal of reaching the Island safely. I wasn't worried about not landing at the dock on the Island. It wasn't possible for me to miss the dock, even though I didn't have any idea how

I would actually find my way without being able to see. I wasn't worried about the future, only each wave right then and there.

Is that potential fear of a dangerous situation the same as fear in other circumstances? Tolle again:

> *The reason why you don't put your hand in the fire is not because of fear, it's because you know that you'll get burned. You don't need fear to avoid unnecessary danger – just a modicum of intelligence and common sense. ... The psychological condition of fear is divorced from any concrete and true immediate danger. It comes in many forms: unease, worry, anxiety, nervousness, tension, dread, phobia, and so on. This kind of psychological fear is always of something that might happen, not of something that is happening now. You are in the here and now, while your mind is in the future. ... A great deal of what people say, think, or do is actually motivated by fear, which of course is always linked with having your focus on the future and being out of touch with the Now. As there are no problems in the Now, there is no fear either.*

The *Now*. What the heck does that really mean? Well, it's hard to disagree with Eckhart on this:

> *Nothing ever happened in the past; it happened in the Now.*
> *Nothing will ever happen in the future; it will happen in the Now.*

That's right when you think about it. Nothing but now. Tolle says;

> *The oak tree or the eagle would be bemused by [the] question, "What time is it?"*
>
> *"What time?" they would ask. "Well, of course, it's now. The time is now. What else is there?" ...*
>
> *The more you are focused on time – past and future – the more you miss the Now, the most precious thing there is. Why is it the most precious thing? [B]ecause it is the only thing. It's all there is. The eternal present is the space within which your whole life unfolds, the*

> *one factor that remains constant. Life is now. There was never a time when your life was not now, nor will there ever be. ...*

Why is it helpful to see everything from the perspective of the present? Eckhart replies:

> *All negativity is caused by an accumulation of psychological time and denial of the present. Unease, anxiety, tension, stress, worry – all forms of fear – are caused by too much future, and not enough presence. Guilt, regret, resentment, grievances, sadness, bitterness, and all forms of nonforgiveness are caused by too much past, and not enough presence.*
>
> *Most people find it difficult to believe that a state of consciousness totally free of negativity is possible. And yet this is the liberated state to which all spiritual teachings point. It is the promise of salvation, but right here and now. ...*
>
> *A great deal of what people say, think, or do is actually motivated by fear, which of course is always linked with having your focus on the future and being out of touch with the Now. As there are no problems in the Now, there is no fear either.*

Ah, ha!! That's it. Since there's no fear in the now, all one has to do is let go of perceptions of the past and future, and fear disappears. Much as light drives out darkness, presence in the moment eliminates fear. All the fears we discussed, all the random paranoias, just evaporate in the now. When I let go of my fears, and disregard the past and future, *nothing's left but now*. If I do that, for example, while looking out my window at a tree this very moment, I'm in the now. And in the now, there's nothin' but peace. There's nothing but a realization of the oneness and interconnectedness of everything. I look out the window and *I see only love*.

Eckhart says, "When you act out of present-moment awareness, whatever you do becomes imbued with quality, care, and love – even the most simple action."

And, *finally*, I realize the importance of quantum physics in all this. Remember that Stephen Hawking and other modern quantum physicists say there's no past or future; that everything is happening at the same time. They actually say there's not even a distinct present, that now is always. There is only the now.

> *Physicists will argue that we should not give any preferential status to what we call "the present." To a physicist, "now" is a subjective concept that just doesn't show up in the equations of nature. This defies common sense, of course, but no amount of protestation, arm-waving and spluttering will conjure from the physical laws any evidence that any one point in time exists differently than any other.* "The Wow Factor," by Joel Achenbach, *Washington Post* Magazine, December 6, 2009.

Friends ask what quantum physics can mean to us in our daily lives. (Well, not that many friends, but you know, a couple people did recently.) I think it's this: if according to science, there isn't really a past or future, then letting fears from the past and anxieties of the future determine our state of mind is silly. If the past and future aren't really in the past or future, then they're illusions. There's no separation even of time and space. Seeing anything other than the *now as everything* seems to deviate from an accurate perspective of the truth.

Tolle's explanation of the now shows that being in the now is subjective, that everything just depends on how you look at things. In our role as observers of the universe, we help create reality, as quantum physics teaches, so it's critical that we don't look at things in an unhelpful way. Viewing life as awful and humans as inherently vile does *not* promote peace. Observing life as perfect does.

Do you remember how Tom Harpur said in Chapter 2 that the only way to describe the divine nature of our world is through "myth, allegory, imagery, parable, and metaphor?" So, adopting a personal allegory or belief system is important, I believe, to help lead me through life and give it purpose. No one perspective is right or wrong, but some views are more nurturing, positive, and helpful for a happier and more peaceful society. So, as an example, this short writing by Thich Nhat Hahn from *Prayers for Healing,* by Maggie Oman, is as cogent an allegory as one could have:

> *Our true home is in the present moment.*
> *To live in the present moment is a miracle.*
> *The miracle is not to walk on water.*
> *The miracle is to walk on the green earth in the present moment.*
> *To appreciate the peace and beauty that is available now*
> *Peace is all around us – In the world and in nature –*
> *And within us –*

> *In our bodies and our spirits.*
> *Once we learn to touch this peace, we will be healed and transformed.*
> *It is not a matter of faith;*
> *It is a matter of practice.*

It *is* a miracle to walk on the earth, to *be* at all. We needn't try to imagine a creator of the universe to explain things for us. Just look at *what is*. That's it. Thich Nhat Hahn is saying we can *learn* how to see the miracle of life. One has to be conscious to see it, since our perspective determines how our lives play out. From quantum mechanics, we know that our observation of life creates it. We create by thought, as my teacher Avery Kanfer loves to say.

So, what's my personal allegory; what's my goal? Here, this is what I've decided for myself. *My goal is peace.*

As Avery also says, if you have peace of mind, you can live under a cardboard box in the woods and be happy. Peace of mind is internal peace. Not being worried about lots of dumb inconsequential stuff. Sure, I'm concerned about the mortgage and relationships with my wife and kids, and things at work. But I've gotten pretty good at letting go of the negative stuff. I'm no guru, but I appreciate this life.

Eckhart Tolle says in *A New Earth*, "Many poets and sages throughout the ages have observed that true happiness – I call it the Joy of Being – is found in simple, seemingly unremarkable things." I agree. I often wander around our house on a Saturday night, and look at the things I've called into my life, or that called me into theirs. I could go on about this for a while, but just as an example: I sit in the living room in the dark with a candle lit, and gaze at each of the paintings, pieces of furniture, and knick-knacks. I know each one. They're part of me, and part of my life. They're like my siblings in existence. I take pictures of things around the house, of stuff in the rooms. If I have to leave here some day, meaning before I die, I'll be able to look back at my time with this house and my roommates, so to speak. I really love my home, and I have a feeling that this house loves me, too.

OK, so now you think I've gone off the deep end. How could my house love me. Well, why not? According to Tolle in *A New Earth*:

> *All things are vibrating energy fields in ceaseless motion. The chair you sit on, the book you are holding in your hands appear solid and motionless only because that is how your senses perceive their vibrational frequency, that is to say, the incessant movement of the*

> *molecules, atoms, electrons, and subatomic particles that together create what you perceive as a chair, book, a tree, or a body. What we perceive as physical matter is energy vibrating (moving) at a particular range of frequencies.*

That makes sense. We know from scientists, as we've discussed, that we're made up of all sorts of particles, even cells that aren't human. So, in this respect, everything is alive. We talked a lot before about how plants and animals are so much like us; why not everything else, too? We're pretty sure rocks and gravel aren't dark energy or dark matter; they're clearly made of the same things we're made of, you know, chemicals and molecules and stuff like that. They're part of the same energy we are. Tolle explains it this way in *The Power of Now*:

> *Everything that exists has Being, has God-essence, has some degree of consciousness. Even a stone has rudimentary consciousness; otherwise it would not be, and its atoms and molecules would disperse. Everything is alive. The sun, the earth, plants, animals, humans – all are expressions of consciousness in varying degrees, consciousness manifesting as form.*

Sometimes, as I walk around the house or back yard, I have the distinct feeling that *everything is one*, in all places and time; that I'm looking at the *entire universe* through my puny eyes. And what I see in my mind is love; *love everywhere*. In my personal allegory, there ain't nothin' *but* love. Here's a little verse I wrote in *Poems of Peace and Forgiveness*:

> **From Above**
> *When you look*
> *down*
> *from above,*
> *don't you see love?*
>
> *From a plane*
> *in the sky,*
> *as far as the eye*

can see,
there's love.

Mountains tall
valleys wide
rivers run
through
farms, side
by side.

The sun beams
down
A heavenly light
the earth is
peaceful, and
all is right.

That's the way I think about things. Everything is love. Love is oneness. I believe there's no better way to look at life. I can't believe how lucky I am to be, here, now, in love.

Eckhart Tolle has an interesting story about this perspective. Turns out he was a student at Cambridge when Stephen Hawking was teaching there. He knew of Hawking only as an emaciated and paralyzed person in a wheel chair. One day, Eckhart held the door open for him and wrote in *A New Earth*:

> [O]ur eyes met. With surprise I saw that his eyes were clear. There was no trace in them of unhappiness. I knew immediately that he had relinquished resistance; he was living in surrender. ... A number of years later when buying a newspaper at a kiosk, I was amazed to see him on the front page of a popular international news magazine. ... There was a beautiful line in the article that confirmed what I had sensed when I had looked into his eyes many years earlier. Commenting on his life, he said (now with the help of the voice synthesizer), **"Who could have wished for more?"**

Heaven is Everywhere

Yes. No one could've ever imagined more to life than this. It is much more wonderful, complex, and exhilarating than anything I could conceive.

If I don't feel this way sometimes, I just remind myself it's what I believe. I witness what I'm feeling or thinking about, evaluate it, and then let go of any fear-based emotional content. I remember what I've called this book: *Heaven is Everywhere*.

I believe that. I choose to take the approach there will *not* be another life for me (at least not one here on earth). Maybe there will be, but I don't have actual evidence of that; perhaps I will someday. I've heard of the people who say there's a white light. A friend of mind had a heart attack and said he saw it – the white light. But I've decided to look at life the way Richard Carrier does; *this is a one shot deal*. I'm not going to hope for anything else. This is my perfection, just like the rose. Heaven is all around; *you just have to think of it that way to make it true*.

Maybe life just reboots when I die, like Dr. Robert Lanza said in a blog on *Huffington Post* on June 11, 2010, in an article titled, "What Happens When You Die? Evidence Suggests Time Simply Reboots." He references Einstein's comment that, "The distinction between past, present, and future is only a stubbornly persistent illusion." Lanza then says, "Death is a reboot that leads to all potentialities. That's the reality that the experiments mandate." That'd be cool; I could just be recycled or reincarnated. It would be fabulous to see the white light, and hang out with my father and his friends in their angel garb. But, I don't know for sure.

A friend who lost her husband said she felt his warmth near her shoulder for some time after he passed. Grant said Eileen hung around for a while after she died; he felt her beside him when they spread her ashes in Seattle. So, maybe we do go on. I hope so. But I think it's more conducive to treating this life as special to think of it as a single affair; kind of like a *one-night stand*.

Then, *life becomes so very, very precious*. It's like everything you've ever wanted, but more. Have you ever thought about it that way?

It pains me to think of anyone killing another person, or taking his own life. Of course, there's no right or wrong in any of it. But there's so much to love; stealing another person's chance at existence because of some random insecurity is *insane*.

I've also been asked, how can everything be perfect? What about a pile of dead babies? How can *that* be perfect? I reply that it's awful -- *from our perspective*. Truly terrible. We catalogued some dreadful conduct by humans earlier in this book. I'm not saying any of that is perfect from a social standpoint. I *deplore* all that crazed violence. But a pile of dead human babies is just the same *from the perspective of the universe* as a pile of dead ants, birds, or skunks, or maybe even a pile of plumbs or oranges. The universe doesn't judge such things; everything just is, and it's all perfect.

Here's another way to look at it. Tolle says in *A New Earth* that the ego in people can't distinguish between a situation and *their reaction* to a situation.

> *You might say, "What a dreadful day," without realizing that the cold, the wind, and the rain or whatever conditions you react to are not dreadful. They are as they are. What is dreadful is your reaction, your inner resistance to it, and the emotion that is created by that resistance. In Shakespeare's words, "There is nothing either good or bad, but thinking makes it so."*

I like to say the *moon is full all the time*; it's just that sometimes it doesn't seem that way *to us*. And *everyday is a beautiful day*, though often it's raining, snowing, cold, or torridly hot and humid. Tolle says detachment helps to not see things as good or bad, right or wrong. How one thinks affects outlook entirely. Detachment is like forgiving, letting go of an emotional attachment to a thing or a situation. Doesn't that mean one becomes immune to life, jaded about life; wouldn't that be boring? Tolle says, No:

> *Being detached does not mean that you cannot enjoy the good the world has to offer. In fact, you enjoy it more. Once you see and accept the transience of all things and the inevitability of change, you can enjoy the pleasures of the world while they last without fear of loss or anxiety about the future. ...*
>
> *You are able to live with uncertainty, even enjoy it. When you become comfortable with uncertainty, infinite possibilities open up in your life. It means fear is no longer a dominant factor in what you do and no longer prevents you from taking action to initiate change. ... If uncertainty is unacceptable to you, it turns into fear. If it is perfectly acceptable, it turns into increased aliveness, alertness, and creativity.*

This is analogous to quantum physics and the uncertainty principle we discussed. If nothing is certain, then resisting uncertainty will be counter-productive and uncomfortable. But accepting what *is* eliminates fear. "The underlying emotion that governs all the activity of the ego is fear," says Tolle. "[A]nger and resentment strengthen the ego enormously by increasing the sense of separateness, emphasizing the otherness of others and creating a seemingly unassailable fortresslike mental position of 'rightness.'" See, we keep coming back to identification of fear as a sense or perception of separateness. What would Tolle say love is then?

> *In the stillness of presence, you can sense the formless essence in yourself and the other as one. Knowing the oneness of yourself and the other is true love, true care, true compassion. ... So, love is the recognition of oneness ... .*

So, my personal allegory claims its main goal as peace, for the world, and peace of mind, for me. How do I get there? I adopt a construct like Jerry Jampolsky that views only two sources of our actions; love and fear. You recognize that as a recurring theme of this book. And you've seen that I've cogitated on this big time, so that I define love as oneness and fear as the perception of separation. From there, I try to observe and evaluate everything from that rubric. Not from right or wrong, good or evil, or compliant with the Bible or not. I don't want to judge. I just try and ask; is that action, principle, or conduct based on love or fear? Does it represent and engender a feeling and sensitivity to oneness? Or does it reflect a sense of separateness?

From there, I'm drawn to the love, and repulsed by the fear. I try to think and act in a manner that's conducive of love, and try not to become hostage to fear. I want to foment oneness and minimize fear.

So, when I read, see, or hear about murder, rape, killing, terrorism, hatefulness, discrimination, anger, genocide, tyranny, repression, despotism, kleptocracy, rage, fury, self-righteousness, slaughter, intolerance, bigotry, hate, greed, lying, defensiveness, meanness, stealing, guilt, hopelessness, depression, jealousy, blame, attack, humiliation, dishonor, entitlement, stress, control, and similar things, I equate them with fear. They're not good or bad, but they are *not* based in love. There are a gazillion shades of fear, but they're not supportive of a loving world. The humans who find themselves being fearful see themselves as separate, and not the same as the rest of the world. Still, they're blameless, and *never* not worthy of love.

On the other hand, when I perceive happiness, gratitude, thankfulness, sharing, trust, kindness, charity, generosity, openness, reconciliation, honesty, sincerity, spontaneity, laughter, joy, bliss, tenderness, compassion, politeness, reverence, truth, trust, creativity, art, music, acceptance, unity, forgiveness, and similar manifestations of interdependence, I know they're from love and oneness. And the more we open ourselves up to love, unity, and interdependence, the more we can *eradicate* all of the things that go with fear.

How are we gonna figure out how to do that? If we could just learn how to capture the love and let go of the fear, darn it. I think there are many ways, but here's one; the process described in *The Breakout Principle*, by Herbert Benson, M.D., and William Proctor. Benson is from Harvard Medical School, so you know he's an intellectual stud. Here's what it's about:

> *The Breakout Principle refers to a powerful mind-body impulse that severs prior mental patterns and – even in times of great stress or emotional trauma – opens an inner door to a host of personal benefits, including: greater mental acuity, enhanced creativity, increased job productivity, maximal athletic performance, spiritual development.*
>
> *The most significant phrase in the above definition is "severs prior mental patterns." Many if not most of the problems we face in terms of blocked creativity and productivity, subpar athletic performance, flawed health, or even stunted spirituality can be traced back to unresolved destructive or negative thought patterns – such as nagging anxieties, stress-related emotional baggage, or circular, obsessive "mental tapes."*

Daggone. Sounds like these "prior mental problems" have to do with fear. Benson says, " You can actually learn to turn on a natural inner switch to sever those past mental patterns and activate Breakouts that will transform your daily life. ... Learning to activate this trigger can also provide you with what superior athletes call the *physiology of zoning*, or *getting in the zone*." So, this must mean there are ways to learn how to let go of fear. And why wouldn't sports and spirituality have similar principles? Thich Nhat Hanh says we *can learn* how to perceive the world as a miracle; why can't we learn to get into the flow in a similar way?

One of the first things Dr. Benson says is, "to break mental patterns, it will often be necessary to *quit concentrating on yourself*. You must disregard the way *you* are performing, or the way you are appearing to others – and dwell instead on what's happening *outside* your narrow orbit of self-interest. ... [I]t may simply be necessary to shift your focus from old, negative mental habits to promising new possibilities – even if you still don't quite know the exact nature of those possibilities." In other words, let go of the past and embrace the uncertainty of what's now. Wouldn't it be cool if we could use some sort of rubric like this to help us figure out how to let go of the past and be in the love in the now?

Benson says there are four distinct stages to advancing, which reminds me of my Dad saying we learn in steps. Our improvements are not just gradual inclines, but more like going along horizontally and then jumping up vertically to the next level.

The first stage, according to Benson, is a process that "begins with a hard mental or physical struggle. … The person on a spiritual quest may plunge into concentrated study, a 'dark night of the soul,' or intense, prayer, meditation, or soul-searching." I feel like I've been soul-searching all my life; haven't you? Certainly, I've been very focused on trying to learn what's up for over a dozen years. In some respects, I'd say living can be a struggle, if only because we don't know what's really going on with life. It also seems humans as a species have been searching and struggling for *a long time*. First, finding enough food and shelter; recently, having good jobs and stable societies. So, yes, I'd say I've been engaged in the first stage; most of us, really.

"The second stage involves *pulling the Breakout trigger*." Huh, what's that mean? Benson describes this as "letting go," backing off," or "releasing" from the struggle stage. But the most important part of this second stage is you "*must completely sever prior thought and emotional patterns*." Well, severing thought patterns could be pretty hard for a lotta people. Many folks don't want to change, or change their mind. Dr. Benson seems to be saying that you *must* sever prior ways of thinking to reach the higher level you're seeking. In other words, *one must change his or her mind* to advance.

Now, the cool part is that there isn't any particular trigger you have to employ. Dr. Benson lists some sample triggers, which include:

- *Prayer, as defined by your religious tradition*
- *Meditation, as understood by your tradition*
- *Contemplation, as understood by your tradition*
- *"Eastern" triggers, such as tai chi, chi gong, or yoga*
- *Repeating for several minutes any positive or meaningful word or phrase*

- *Sitting quietly by yourself or with a group in a chapel or house of worship*
- *Listening to your favorite music*
- *Playing or singing music with which you are familiar*
- *Viewing a work of art, such as a painting or sculpture*
- *Soaking in a bathtub or hot tub*
- *Grooming with a repetitive routine*
- *Walking, jogging, bicycling, or performing any other repetitive exercise for at least fifteen minutes*
- *Becoming absorbed visually in a sport*
- *Regular, conscious breathing*
- *Sitting quietly in a garden*
- *Gazing over a seascape or a mountain range*
- *Strolling silently through the woods*
- *Relinquishing control over a person or job problem*
- *Imagining and accepting a "worst-case scenario"*
- *Eating at a quiet restaurant*
- *Sitting quietly with your pet*

Though these seem very different, they all have the power to scrap prior mental patterns. This begins to resonate with me. I've experienced "breakouts," or I might call them mind-changing experiences in these types of situations. First, though, I have to be *open to change*. I have to be willing to accept something new, to embrace creativity or let go of my mental hang-ups.

The third stage, according to Benson, is breaking out with a "peak experience." "Invariably, the peak will involve something unexpected – a surprise that produces unanticipated new ideas or higher levels of performance." So, the peak experience is when you reach that next step in whatever you're doing or searching for. If you put in the work trying to accomplish something, after some sort of a trigger, you can catapult up to the next level. You can learn to be in that zone where you want to be. There's no timetable. It can happen on any schedule, and may be fast or slow. That doesn't matter (for everything is happening at the same time anyway, remember?).

I've never read any books by Abraham Maslow (I mean, one can't read *everything*), though I'm familiar with his "hierarchy of needs." Benson references Maslow in *The Breakout Principle* and his work on *peak experiences*. Through the wonders of the Internet, I go to a website devoted to Maslow, which says, "Peak experiences are especially joyous and exciting moments in the life of every individual. Maslow notes that peak experiences are often inspired by intense feelings of love, exposure to great art or music, or the overwhelming beauty of nature." http://www.abraham-maslow.com/m_motivation/Peak_Experiences.asp. I honestly think I've had some of those.

The fourth stage of the Breakout principle, says Dr. Benson, is achieving a "new-normal state – including ongoing improved performance and mind-body patterns." Here's where the old ways of doing things become no longer acceptable. Here's where you've reached the next step and can continue on your journey to your goal. Dr. Benson says you reenter the world of struggle for the next step. But I'd say that each step is really achieving a goal, and makes the next easier to achieve. After I learned to pilot the boat and experienced several challenging situations on the water, I was better able to handle the next one. And after one learns to deal with and understand life better through this sort of learning process – which might be triggered by a divorce, birth, death, injury, or anything – one can know how to accept life better.

Dr. Benson discusses Maslow further and says:

> *Maslow argued that the greatest personal creativity, well-being, and philosophical understanding became possible when a person moved up to the highest, self-actualization level. Furthermore, at the level of self-actualization, the individual was more likely to have a "peak experience," characterized by great insight, freedom from fear and anxiety, and a sense of being unified with an infinite or eternal dimension of reality.*

So, the point to me is that, through a process or mechanism of searching, letting go, and achieving new experiences, we can evolve to a more advanced state. That doesn't sound like rocket science, or quantum physics for that matter. The more I try and develop my own world-view and personal allegory, the more satisfied and self-actualized I can become.

I've talked about Dr. Benson's Breakout process just because it's one I've read about and fits in with the direction I'm taking with developing my personal allegory. There are an infinite number of paths, and none are right or wrong, but I can only reference the road I've been taking. I find this Breakout principle useful, partly because it reflects my experience, and partly because it gives me hope that humans can breakout on the path to peace and love.

I've told you about the basics of my personal allegory. Now, I'm going to tell you about the thought process I've discovered for myself that's conducive to finding and learning about peace. I call it *NWAL*.

Yep, you got that right. NWAL. Or N'wal. Might be pronounced like the first syllable of the Big Easy – NWAL-ins. Could also be thought of as *Know-All*.

Seriously. It's just an acronym. I like to use them. I mentioned I'm a lawyer. I developed this little mnemonic to help me remember the key aspects of a legal contract if I was put on the spot and had to give advice to a client really quickly. WIPIT. Warranty, Intellectual property, Payment, Indemnification, Termination. Pretty clever, huh?

Anyway, here's what NWAL stands for:

- **Now**
- **Witness**
- **Assess**
- **Love**

You're thinking; "I read all these pages for this?" OK, I'll explain.

NOW. If I want to consider something, any action or emotion; I first must get myself in the Now. It's hard to objectively evaluate something important if I'm all freaked out and anxious. I need to be thinking with a straight head, so to speak. Not worrying about the past or future; not prejudiced to any particular result or outcome. Not all emotiony. Just being in the present.

WITNESS. Once suitably situated in the now, I simply have to observe. I look at the action or emotion objectively. I don't try and judge or evaluate, but just see *what is*. Look at the behavior. Look at the facts. Avery Kanfer once told me to witness my emotions as a way of understanding what I was feeling. Take the emotional content out. That's what I'm talking about.

Witness implies a bit more of a serious scrutiny than just to use the word "observe." Not a religious meaning, just a significant examination of what I'm considering. Witness also requires a solemn look; for example, paying enough attention that one could be a witness in court. Also, it means to take off the blinders and look at the big picture objectively as possible.

ASSESS. Dictionary.com defines *assess* as, "to estimate or judge the value or character of." Evaluate has a similar meaning, and we tend to use that word more commonly in speech. I'd say assess is less frequently employed, and I wanted a word that conveyed an important role in making a determination about the subject matter of the inquiry.

This is the most critical step in the process. It involves evaluating whether something is based in love or fear. It does not involve judgment. I just ask, does the thought, action, behavior, conduct, or principal primarily evoke oneness and unity, or difference and separateness?

I don't ask if it's good or bad, right or wrong; just if it supports love or fear. Like this: war is not based in love, murder is not based in love; giving to the poor *is* based in love; supporting a healthy and diverse environment is based in oneness.

It's also very important in making an assessment to use the best and most reliable information and evidence I can. As you know, I try and rely on scientists, because they are subjected to rigorous analysis and peer review. In other words, in exercising my (what should be) fiduciary-type duties in life, I want to act on an *informed basis*. Of course, no one can know or read everything, but I put a value on trying to expand my knowledge and experience by pursuing an eclectic array of sources so I can think outside of the box.

LOVE. Once I've determined whether it's love or fear, then I have to make a decision how to act or what to think. Do I decide to cast my lot with love or fear?

Hey, you guessed it. I always try to choose love. I think that makes sense, don't you? Just think if everyone always chose love instead of fear; that would be awesome! No more wars, hate, anger, rape! Yay!

Whataya mean that'll never happen? We've got NWAL now, and so we can start looking for some ways to use it in our search for peace.

Fifteen | **Is it supposed to be?**

One Sunday after my Mom and Dad had gotten together (before I was born), they went to church. I'm sure Mom wore her veil, black dress, high heels, and looked pretty sharp. Dad, likely in a brown suit, maybe two-tone shoes, and a hat (which he would have taken off in the church).

My father wasn't much of a Christian. I don't remember him going to church at all, except maybe once in a blue moon with us. You remember me saying Mom went all the time, even when her soul was all black 'cuz of the mortal sin of marrying my father and she'd been excommunicated. My Mom used to tell this story:

The Mass proceeded. The congregation started out standing for the Introductory Rights. Then they sat for the Liturgy of the Word, stood for the Gospel, sat for the Homily, stood up for the Profession of Faith, sat for the Liturgy of the Eucharist, stood for the prayer over the gifts, knelt for the "Holy, holy, holy" part, stood for the Communion Rite, knelt after the Lamb of God response and during communion, sat for the Song of Praise, and stood for the rest. Here's the order of the mass in detail. http://catholic-resources.org/ChurchDocs/Mass.htm. Remember the mass was in Latin back then, with the people responding in dead-language unison.

Well, at one point, Dad must have gotten tired. He pulled out his handkerchief as he sat in the pew, wiped his brow, and put the handkerchief across his lap. Mom turned and said, "What's the matter, is your fly down?"

Dad looked at her and asked, "Is it supposed to be?"

Cracks me up. *Is it supposed to be?* Yea, Dad, right after you sit after the Gospel, all the men pull their zippers down. Seriously, many people around the world believe that God *wants* them to do stuff like that. Well, maybe it would be more accurate to say God wants them to pull their zippers up, because he's pretty particular about sex. Actually, from what I read, God is *extremely* particular about some things. Here are some examples in clippings from 2006-08:

- God doesn't want people to dance. Yep, if you're in Mogadishu, Somalia, you better not dance, because Islamist insurgents might come and whip you, like they did to 32 people taking part in traditional dances. Islamist spokesman Sheikh Abdirahim Isse Adow said, "We arrested 25 women and seven men who were dancing near Balad (a town). We released them after whipping them. … The dancing of men and women together is illegal and totally against Islam. We neither killed them nor injured them, but only whipped them in accordance with Islamic law." That was after they had stoned to death a 13-year-old woman the month before who had been "raped by three men while visiting her grandmother." I guess God doesn't want either dancing or little girls to have the audacity to get raped. "Islamist Rebels Whip 32 Dancers," by Abdi Sheikh, Rueters, November 15, 2008.

- God doesn't want people to use ringtones from the Koran. That's according to an AP article titled, "Muslim Clerics Decry Ringtones Using Koran," in the *Washington Post* on December 2, 2006. "One should hear the complete verse of the Koran with a pious mind and in silence. If it is used as a ringtone, a person is bound to switch on the mobile, thus truncating the verse halfway," said Mufti Badru-Hasan. "This is an un-Islamic act." I used to have an Egyptian song as my ringtone, but I don't think it was playing any Koranic verses.

- God wants women to work at home. He doesn't want them out running for public office. This was a big issue for Southern Baptists when Sarah Palin ran for Vice-President, but I guess they may have eased up on Michelle Bachman. The Apostle Paul in the Bible says in 1 Timothy 2:12: "I permit no woman to teach or have authority over a man." Ephesians 5:22 demands, "Wives, submit to your husbands as to the Lord." The Southern Baptist position was that a wife can work outside of the home only if her husband agrees. Whew, thank God Todd Palin "has long backed his wife's career in public service." "Southern Baptists Adjust to Idea of Palin," by Mike Baker, *Washington Post,* October 4, 2008.

- God wants Muslims to have nuclear weapons. None other than Osama bin Laden in 1998 in response to a question asking if he had nuclear or chemical weapons said that, "Acquiring weapons for the defense of Muslims is a religious duty. If I have indeed acquired these weapons, then I thank God for enabling me to do so." David Ignatius writing in an article titled, "Portents of A Nuclear Al-Qaeda" in the *Washington Post,* October 18, 2007.

- God does not want university students flirting. In Pakistan at Punjab University, if the Islami Jamiat-e-Talaba (a student morals organization) finds this sort of behavior, they will physically assault the guilty students. "We are compelled by our religion to use force if we witness immoral behavior," says Rana Naveed, 22. Outside of classes, there is strict segregation of students; no co-ed dining. *Time* magazine, "No Dates, No Dancing," by Aryn Baker, October 16, 2006.

- God wants fathers to kill their daughters if they have sex out of wedlock or otherwise dishonor the family. Faleh Hassan Almalecki ran over his daughter with a jeep and killed her because he thought she'd become too westernized. Terry Tang, reporting on February 22, 2011, "Faleh Hassan Almalecki Guilty: Jury Convicts Iraqi Immigrant for 'Honor Killing,'" in the *Huffington Post.* This was like one of about 20 estimated "honor" killings in Jordan each year, such as when a man fatally shot his 17 year old daughter because he thought she'd had sex, even though doctors had conducted a medical exam proving that she was still a virgin – which he knew before killing her - and an autopsy also confirmed those findings. "Father Kills Daughter; Doubted Virginity," by Shaffika Mattar, Associated Press, January 25, 2007.

- God does not want Richard Gere kissing a Bollywood actress. That was clearly the message around India after Richard Gere kissed popular actress Shilpa Shetty during an AIDS awareness event to highlight the epidemic among India's truck drivers. "In front of a cheering crowd, Gere kissed the giggling Shetty on the hand, then kissed her on both cheeks before bending her in a full embrace to kiss her cheek again." That resulted in angry crowds in several Indian cities burning them in effigy and shouting, "Down with Shilpa Shetty!" Public displays of affection are taboo in the Hindu country, where a spokesman for the Hindu nationalist party also condemned the kiss. "Protestors in India Burn Gere Effigies," by Sam Dolnick, Associated Press, April 16, 2007.

And here are some random examples from 2012:

- God doesn't like music, apparently. When Islamists took over part of northern Mali in the spring of 2012, they started shutting down music as against Islam. "Renowned singer Khaira Arby fled her home in Timbuktu after Islamists destroyed her instruments." *Washington Post*, September 1, 2012, "Islamists silence northern Mail's music tradition, by Sudarsan Raghavan. "They told my neighbors that if they ever caught me, they would cut my tongue out," said Arby. "They have installed an ultra-conservative brand of Islamic law in this moderate Muslim country, reminiscent of Afghanistan's Taliban and Somalia's al-Shabab movements. Now, women must wear head-to-toe garments. Smoking, alcohol, videos and any suggestions of Western culture are banned. The new decrees are enforced by public amputations, whippings and executions, prompting more than 400,000 people to flee." No singing and dancing for you!

- God doesn't like a 14-year-old Christian girl, if someone says she's been carrying "burnt pages of the Koran in a trash bag in her village outside Islamabad." *Washington Post*, September 8, 2012, "Judge grants bail for Christian girl in blasphemy case," by Michelle Leiby and Shaiq Hussain. She could be sentenced to life in prison for such crimes against Allah, who I'm sure was really offended by what she did, if she did it. Fortunately, the article says the judge in the case granted her bail and it was expected she'd be released from the *high security prison* where she'd been held.

- God doesn't want Jews wearing Crocs on Yom Kippur, the Day of Atonement, because they would be too comfy for the "somber atmosphere of the day of fasting and repentance." This was one of the rulings made by Rabbi Yosef Shalom Elyashiv, "revered by Jews worldwide as the top rabbinic authority of this generation for his scholarship and rulings on complex elements of Jewish law, ..." He passed on July 18, 2012 in Jerusalem at the age of 102, according to his obituary in the *Washington Post*, by Ian Deitch on July 19, 2012. "In another ruling, he banned ultra-Orthodox women, who covered their hair for modesty, from wearing Indian-made wigs in case the wig hair may have previously been used in Indian worship, which would disqualify them for Jewish ritual. The edict prompted the burning of thousands of wigs in Israel as well as in Jewish communities in Brooklyn and elsewhere." Of course, it did.

- God definitely does not want women to masturbate. We know this because when Sister Margaret A. Farley, of the Sisters of Mercy and a professor at Yale University, wrote her book in 2006 titled, "Just Love; A Framework for Christian Sexual Ethics," saying masturbation by women "usually does not raise any moral questions at all" and that it "actually serves relationships rather than hindering them," the Vatican said the Catholic Church teaches that masturbation is "an intrinsically and gravely disordered action." I'm not making this up. It was in the *Washington Post* article titled (sort of appropriately), "Vatican attacks U.S. nun over book on sexuality," by Philip Pullella on June 5, 2012. I'm thinking there are a lot of gravely disordered people out there!

What's the world coming to? Women want the right to drive in Saudi Arabia. Christians want to have sex. Homosexuals want the right to marry. God says No! I say God's a buzz-kill! Wait a second, though. How does everyone know what God wants?

Being the pseudo-scholar that I am, I decided to consult a book on the subject, aptly named, *What God Wants*, by Neale Donald Walsch. He's the guy who wrote all those *Conversations with God* books (that I never read; my bad). Walsch goes through a litany of things that humans believe God wants.

> *Many humans have been told that What God Wants is for humans to understand that God is the Supreme Being, ... Omnipotent, ... a jealous God, God is a vengeful God, God is an angry God, ... One result of this teaching: Many human beings are afraid of God. They also love God. So, many humans confuse fear and love, seeing them as connected in some way. Where God is concerned, we love to be afraid (we have made it a virtue to be "God-fearing") ... .*
>
> *Many humans have been told that What God Wants is for people to live good lives, and for good people to go to Heaven or Paradise after their deaths, while bad people go to Hell, Gehenna, or Hades. ... Many humans have also been told that it's when life ends that the real*

joy begins. ... One result of this teaching: Humans believe that life is not easy, nor is it supposed to be. It's a constant struggle. ...

Many humans have been told that What God Wants is for humanity to understand that God is male. ... One result of this teaching: Males are considered superior to females in nearly all of the world's cultures. In some of those cultures this manifests as cultural norms that do not allow females to go to school, to hold jobs of authority or responsibility, to leave the home without being in the company of a blood male relative, or to permit any part of their body to be seen in public, requiring them to be covered from head to toe. (Many of these cultural restrictions are justified as honoring and protecting women, or as protecting men from "temptation.") ...

Many humans have been told that What God Wants is for sexual unions to be experienced only with one's spouse for the purpose of procreation and the expression of love. One result of this teaching: Millions of people believe that sex may absolutely never be experienced in any way that deliberately prevents conception, and that while sex is wonderful, to experience sex simply for pleasure with no possibility of procreation is against the will of God and, therefore, "unnatural," immoral, shameful, and a giving in to baser instincts. ... As with the combining of love and fear in the earlier understanding of God, the combining of pleasure and shame in this construction has produced chronic emotional confusion: wonder, excitement, and passion, yet embarrassment fear, and guilt about sexual desires and experiences. ...

Many humans have been told that What God Wants is for marriage to be an everlasting union between a man and a woman. ... One result of this teaching: In most religious cultures ending a marriage for whatever reason, including mental or physical cruelty is deeply discouraged, and one major religion tells its followers that they may never divorce, may never remarry in the church nor receive the church's sacraments if they do divorce, and may never marry another person who has been divorced. ... [Like my Mom! Further,] human beings may not have sex with one outside of marriage, and therefore,

> *should they never remarry, they may have sex at no time during their entire lives.*

I'd never thought of it that way. An old maid or bachelor would be expected by many religions *never* to have sex. Pretty harsh. But, really, how many people have these sorts of ideas? I haven't found a survey on these particular topics, but here are some statistics on what people do believe:

According to a Pew survey in 2009, 55 percent of folks say a guardian angel has protected them from harm. "Mixing Their Religion," by Cathy Grossman, *USA Today*, December 10, 2009. A Gallup poll determined that 48 percent of Americans believe in creationism, while only 13 percent believe evolution most accurately explains human development. "Eager to be Reagan's Heir, but Not Bonzo's Cousin," by Chris Cillizza and Shailagh Murray, *Washington Post*, May 6, 2007.

Over a third read the Bible or other holy scripture once a week, but almost half say they seldom or never read other books about their religion, and 70 percent seldom or never read books about other religions. "Religious Literacy: Americans Don't know Much About Religion," by Rachel Zoll, *Huffington Post*, September 28, 2010. That study also found that over 40 percent of Catholics don't know their church teaches that the bread and wine actually are the body and blood of Christ. I knew that! A *Newsweek* poll reported in an article, "Special Report: Spirituality," by Jerry Adler, on September 5, 2005 found that 80 percent of people believe God created the universe and 67 percent believe their souls will go to heaven or hell when they die. I wonder if that's belief or certainty?

It's interesting the way religious people think about topics that appear contradictory to what their religions teach. Another Pew forum survey result showed that about 60 percent of white evangelical Protestants think the use of torture is often or sometimes justified, versus 50 percent of white Catholics, 45 percent of Protestants, and only 40 percent of the religiously unaffiliated. "Fiery Response to Pew's Torture Analysis," *Washington Post*, May 9, 2009. Since one of the most important teachings of Jesus is to love your enemies (rather than torture them), you'd think Christians would be totally against torture. And against the death penalty, however, "more than 7 in 10 Protestants (71%) support the death penalty, while 66% of Catholics support it. Fifty-seven percent of those with no religious preference favor the death penalty for murder." http://www.deathpenaltyinfo.org/religion-and-death-penalty#intro

That seems so peculiar to me. You'd think our religious brethren, Christians in particular (since Jesus was the Prince of Peace and told everyone to love one another), would be fanatically against war, torture, the death penalty, etc., and would know the most about religion. In fact, according to another Pew Forum survey dated September 28, 2010, "Atheists/agnostics, Jews and Mormons have the highest level of religious knowledge about world religions other than Christianity, though they also score above the national average on questions about the Bible and Christianity." http://pewforum.org/Other-Beliefs-and-Practices/U-S-Religious-Knowledge-Survey.aspx. How could this be?

On this point, I'll reference an article written by Michael Gerson called, "A Searcher With Faith in Mind," from the *Washington Post* on April 15, 2009. He was writing about a book by Andrew Newberg and Mark Robert Waldman titled, *How God Changes Your Brain*. Here's what the book says, according to Gerson:

> *Contemplating a loving God strengthens portions of our brain – particularly the frontal lobes and the anterior cingulate – where empathy and reason reside. Contemplating a wrathful God empowers the limbic system, which is "filled with aggression and fear." It is a sobering concept: The God we chose to love changes us into his image, whether he exists or not.*

This is a telling conclusion. Newberg and Waldman are basically saying that *we create God in our own image and worldview*. If we believe in a loving God, we strengthen the empathetic and reasonable parts of our minds. But thinking of a mean and nasty God leads to aggression and fear. I'd say the God of the Bible is *not* a nice God. In my *Baltimore Catechism*, I was told God was loving, but everything else about him was angry, judgmental, and vindictive. I mean, he would send people who never heard of Jesus to hell; what fault was it of theirs? The Bible is full of stories in which God condemns humans for one thing or another. How about Sodom and Gomorrah? Lot's poor wife looked back at the wicked cities being destroyed by God with fire and brimstone, and she was turned to a pillar of salt! Seriously, it's pretty sick to have even thought of that.

And, pardon me, but our Taliban brethren see a God who is angry and vengeful, who doesn't like young children dancing or a woman's shoulder exposed. According to Gerson, their concept of God changes them! They see a mean and nasty god, and so *they become mean and nasty*. How mean and nasty? According to Fareed Zakaria in an article in *Time* magazine titled, "A Moment for Moderates," on October 1,

2012, "The story seldom varies: a Westerner, or a handful of them, does something that attacks Islam (mishandles a Koran, attacks the Prophet). The episode is virtually unknown until radical Islamists publicize it to whip up frenzy, hatred and intolerance. Crowds gather outside U.S. embassies, and violence ensues. The regime disperses the crowds with tear gas and bullets. Order is restored, often by brute force, but the rage endures. ... As people watch the crowds and the violence, surely they must be thinking, Why is there so much anger in the Arab world?" According to Newberg and Waldman, those folks create an angry god and so they get to be angry as a result.

The contradictions in religion also are acute. You remember Mother Teresa. She was one of the holiest and most admired women on the planet. She was a champion for the poor and impoverished, *a true saint*. Turns out, after she died, a lot of her secret letters were found in which she comes across as desolate and aching from the absence of inner love.

She'd thought Jesus spoke to her on September 10, 1946, when she was working in Calcutta as a nun and teacher. According to a story in *Time* magazine on September 3, 2007, "Her Agony," by David Van Biema, Christ "called her to abandon teaching and work instead in 'the slums' of the city, dealing directly with 'the poorest of the poor' – the sick, the dying, beggars and street children. 'Come, Come, carry Me into the holes of the poor,' he told her. 'Come be My light.'"

Then, he never talked to her again. It was like not calling back after a first date. She had visions of herself talking with Christ on the cross. She said, "I want to love Jesus as he has never been loved before." Is that really a good goal? Isn't that a desire for specialness that *A Course in Miracles* condemns?

> *Although perpetually cheery in public, the Teresa of the letters lived in a state of deep and abiding spiritual pain. In more than 40 communications, many of which have never before been published, she bemoans the "dryness," "darkness," "loneliness" and torture" she is undergoing. She compares the experience to hell and at one point says it has driven her to doubt the existence of heaven and even of God.*
>
> *"Jesus has a very special love for you," she assured [Rev. Michael] Van der Peet. "[But] as for me, the silence and the emptiness is so great, that I look and do not see, Listen and do not hear- the tongue moves [in prayer] but does not speak ... I want you to pray for me- that I let Him have [a] free hand."*

> *[She wrote the local Archbishop Ferdinand Perier], "Please pray for me that I may not spoil His work and that Our Lord may show Himself – for there is such terrible darkness within me, as if everything was dead. It has been like this more less from the time I started 'the work.' ... "The more I want him – the less I am wanted, ... Such deep longing for God – and ... repulsed – empty - no faith - no love – no zeal."*

Poor chick. What a sad story. We thought she understood love better than anyone. But it turns out she didn't know *anything* about love in herself. Yes, she loved all the poor people she worked with; I don't doubt that. But she devoted her entire life to a myth. If there's no Jesus (and I'm sure of it), then who talked to her? Who was she praying to?

Basically, *herself*. And the big problem, for her and many others, is that what she got from religion was just separation and not love, not oneness.

If someone could have said, "Hey, Teresa, Babe. You can do whatever you want in life. Don't worry about God and Jesus, just do something you love and feel the fullness of life and the human community. Don't believe in a God who refuses to talk with you, because you'll end up living the reality of an absent God and feeling the emptiness. Just feel the oneness." Here are a few more examples about religion's contradictions.

I don't really have a problem with the United States Senate having a morning prayer. Whatever. And I think it's good if they shift around among different clergy. So, in 2007, Senator Harry Reid, the Majority Leader, invited Rajan Zed, a Hindu clergyman, to say the prayer. He began saying, "We meditate on the transcendental glory of the Deity Supreme, who is inside the heart of the Earth, inside the life of the sky and inside the soul of the heaven. May He stimulate and illuminate our minds."

During his prayer, he was interrupted by two women and a man in the gallery shouting, "This is an abomination," and other stuff like that. The protestors were led away, and the male cried out, "We are Christians and patriots!" They did this at the urging of the American Family Association, which had called on its members to object because the Hindu would be, "seeking the invocation of a non-monotheistic god." "Senate Prayer Led by Hindu Elists Protest," *Washington Post*, July 13, 2007.

Seriously, are they kidding? They're so sure there's only one God, the one they believe in (who wants humans to do or not do all that stuff Neale Donald Walsch talked about), they call another's beliefs an "abomination." Talk about the pot calling the kettle black. And the *American Family Association* thinks it's copasetic to scream degrading comments at Hindu priests?

Although, if there's a religious group that might actually deserve to be tagged an abomination, it would be the Westboro Baptist Church. I'm sure you've heard of them. The church is in Topeka, Kansas and its head is Fred Phelps; it only has a few members, and most are part of Phelps family. Here's what they think:

They believe homosexuality is a sin and should be a capital crime. They believe God hates homosexuals and imposes his justice on those who support homosexuality. They believe America supports homosexuality, so God punishes America. One way God does this is by killing soldiers. They believe God is punishing us for supporting gays by killing soldiers with improvised explosive devices ("IED's"). Church members picket funerals of soldiers and others with signs that say *God hates fags, God hates America, Pray for dead soldiers*, and similar jolly statements. http://en.wikipedia.org/wiki/Westboro_Baptist_Church. I guess they must know what God wants, huh? Interesting to have a group of religious people who believe in a God that hates. If you believe God hates, I suppose that means it's OK for you, too.

Research has shown that the self-righteous are more prone to extremes. Scott Reynolds and Tara Ceranic of the University of Washington concluded based on their study that, "people with exceptionally strong convictions about their moral goodness are likely to follow extreme courses of action because they can convince themselves that whatever they do is good." "Self-Righteous Prone to Extremes," by Shankar Vedantam, *Washington Post*, November 5, 2007. Sounds like the Christians who led the Inquisition, and the Taliban of today. Even as they're murdering those who teach girls, they bask in self-righteousness. Tell you what; a lot of folks need to be brought into the Twenty First Century.

So many people do want that certainty, even if it defies logic. They want someone to tell them what to think about life and that everything is going to be OK, that we're not really gonna die and never exist again. That we can live again together with our loved ones and God, and have as many virgins as we can handle. They need to know what God wants them to do.

So, what does Neale Donald Walsch say God wants?

Nothing.

Absolutely nothing at all.

What would a God really need humans to do? If there's a Supreme Being who figured out how to put the stars in the sky, why would he care about little us? I learned early on that God wasn't going to tell me what to do; he didn't stick the letters down on my dresser. It wasn't too bad realizing that. I didn't get all worried. You just have to get used to the idea.

Walsch does believe in God. "Nothing stands outside of God. Nothing exists without God, and God does not exist if nothing exists. You are the expression of God itself. So is everything around you. Even so-called inanimate objects are found, when examined under a microscope, to consist of particles constantly in motion. These particles and their movements are all part of God. Indeed, everything in the observable universe is God, in some form."

Walsch creates a good God. Osama bin Laden (rest his soul) created an angry god. Others, like me, do not create God for ourselves.

In my personal allegory, the concept of God is Oneness. But Walsch also says God is Oneness, so maybe we're just talking semantics. All of us humans talking past each other, searching for the same thing. Not realizing that *we are one*. Walsch has insightful thoughts about this:

> *Why have humans denied their Oneness? Because humans have confused oneness with sameness. We have not understood that no two fingers are alike, even though all are on one hand.*
>
> *Afraid of losing individuality, desperately fearful of disappearing their own identity, human beings have tenaciously clung to their <u>illusion of separation</u> from each other, from all things in life, and from God. Especially from God. For if humanity is not separate from God, not only do people fear losing their individual identity, oneness with Deity suggests a whole new way of acting, a whole new way of being, for which religions have left humans woefully unprepared.*
>
> *Yet it is not necessary to prepare yourself to love. Love is what you are, and so, loving comes naturally to you. Stopping yourself from loving is what is hard. Love of everyone and everything in life comes easily when fear of anyone or anything disappears. And fear of anyone or anything disappears when you realize that you don't need anything from anyone or anything, because everything you thought you needed to get from something or someone outside of yourself is available within you. ... In the end, <u>love and fear are the only feelings there are</u>. Life brings you a constant stream of opportunities to choose between*

> the two. ... Yet you never need fear the loss of self or love, <u>because you and love are one</u>, and even if everything else and everyone else was gone, you would still be able to feel love. Perhaps more so.
>
> <u>You and love cannot be separated in any way</u>, for love is what God is, and you are not separate from God, nor were you ever, nor will you ever be. This is so very difficult for so many people to believe, because religion today teaches exactly the opposite. It tells us we have been separated from God, and it fills us with the fear that we could be separated forever. Yet one day, religion will teach this no more. One day, very soon, all religions will speak of the unity of God and humanity, and indeed, of the unity of all of life, ...
>
> [The problem is that] the largest number of humans have refused to give up their belief in separation. Ironically, it is the idea of separation that holds the greatest attraction - and the greatest fear. And so we fear what we believe in. Yet the message in this book comes to tell you that <u>separation does not, and cannot, exist</u>. It is in no way an aspect of life. It is always an illusion. Always.
>
> Life is a unified expression of the only thing there is: life itself – which may also be called God. Our constant yearning for union is the outward worldly expression of the innermost knowing of the soul; we are one with all things and with life itself ... . (My emphases.)

Right on, Neale! Eckhart Tolle in *A New Earth* understands what Walsch is talking about in terms of the separation. He says it this way:

> And so religions, to a large extent, became divisive rather than unifying forces. Instead of bringing about an ending of violence and hatred through a realization of the <u>fundamental oneness of all life</u>, they brought more violence and hatred, more divisions between people as well as between different religions and even within the same religion. ... Through them, they could make themselves "right" and others "wrong" and thus define their identity through their enemies, the "others," the "nonbelievers" or "wrong believers" who not infrequently they saw themselves justified in killing. Man made "God" in his own image. The eternal, the infinite, the unnamable was reduced to a

> *mental idol that you had to believe in and worship as "my god" or "our god." ...*
>
> *In fact, the more you make your thoughts (beliefs) into your identity, the more cut off you are from the spiritual dimension within yourself. Many "religious" people are stuck at that level. ... Used in such a way, religion becomes ideology and creates an illusory sense of superiority as well as division and conflict between people. (My emphasis.)*

Neale Walsch may be on to something in another way. Maybe religions *can* change. Instead of proselytize the illusion of separation, maybe we could promote unity and oneness and unity of all people. Or maybe – just maybe -*we don't need religion*. Maybe there's another way?

It might be a good time to pull out NWAL. We can try and NWAL (using it as a verb) religion. OK, let's do it.

We'll sit quietly for a moment and relax. Let all the doings of life melt away, until we're just sitting and thinking of – nothing. It's like meditating, no matter what your mantra is. If you don't have a mantra, you can try one of these:

Love.

Oneness.

Shring.

Be bop a dip dop doop dope doo.

Doob.

Get in the *Now*. Feel the peace and oneness.

OK, good. Now (get it), we need to *witness* and observe religion, as it exists. I'm talking about the main religions of the world, generally, and particularly Christianity and Islam. I jotted down some things about religion as I observe it.

- Religion is based on old stories from a time when people didn't understand the world as we do. Religion is not based on science, such as quantum physics, and is not consistent with biology, astronomy, etc.

- Religion focuses on mythical characters, whose stories were meant to be allegories, but that many humans took literally.

- Religion makes people think they're small, insignificant, unworthy, sinful, and separate from God or the energy, rather than being part of it.

- Religion cajoles people to fear God.

- Religion makes people feel they need to live by arbitrary rules, or be punished and go to hell. It preaches right and wrong.

- Religion has generated wars and inquisitions, killing, and torturing in the name of God.

- Religion does not focus on peace or peace of mind as a main goal for society and individual humans, respectively.

- Religion wants people to think this life is just a prelude to a better life after death in heaven, so doesn't emphasize the beauty of the now.

- Religion teaches that miracles are only done by the superhuman, not that living itself is a miracle.

- Religion wants us to believe we're separate and better than all other species.

As an illustration of this last point, a former teacher and friend of mine, Father George Williams, wrote an article in my high school (Georgetown Prep) alumni magazine about the spiritual exercises of St. Ignatius. The highlighted part of the article said:

> *Human Beings are created to praise, reverence, and serve our Lord and by this means to save their souls. The other things on the face of the earth are created for the human beings, to help them toward the end for which they were created.*

This might be called *human* exceptionalism. I don't buy it. It assumes our souls will be lost unless we praise, revere, and serve the Lord in the way God wants, and tells us we're different and better than *all other animals or parts of creation* - which are just here to *serve us*.

Sorry, Father. I don't think so.

But, hey; here are some things that *are* extremely positive about religion.

- Religions preaches love and forgiveness. That is probably the best thing about religions. They teach that love is the goal.
- Religions do lots of good things for people; relief efforts, teaching children, caring for the sick, all that.
- Religion also feels good. Why? Because, I think, it can help us feel safe. It tells us hope is not lost; we can go to heaven. That can give folks optimism.
- Religion provides community. I always felt part of a community in religion. We were the Blessed Sacrament Bombers (sort of ironic, huh?); that was my team. We felt good that everyone believed the same things, and so we rooted for one another.
- Religion, lots of people say, is good for your health, can extend your life, and helps make you happier.
- Religion provides an avenue for spiritual expression. It's hard for many people to even begin to know what to think about life. Religion can provide some certainty in that regard.
- Religion provides meaningful rituals, such as around births, deaths, marriages, etc. Many religious holidays bring families together in celebration.

When I was young, I went to church every week. Being in a solemn church with music, singing, praying, and community made me feel peaceful and safe. I liked the feeling. During my mid-life crisis, I went to mass almost daily. It made me feel so good, like I was in a refuge. I prayed Hail Mary's on my little thimble rosary and thought deep thoughts. My contemplation through religion was a significant part of my decision-making and healing process.

I also checked in with Peter Aislie about this; you remember him, he taught me about Jesus being a mythical rather than historical figure. He says one of the important things about religion, which secular humanism doesn't address, is making the transcendent personal. What he means by this is, religion is meant to grapple with the spiritual aspect of things through myth and alegory. He says some of the early mystery religions understood the meaning of the teachings.

Unfortunately, many religions have misperceived the mythical, and thus make the transcendent separate. Rather than bringing God or the energy within, the literalists and fundamentalists put God outside us. That turns religion on its head, according to Peter.

He says some religions seem to understand this, such as the Quakers and some monks, but most of the religions are more of the literal variety. I might be painting with too broad a brush here, but I think this will suffice for our general purposes.

Now, we need to *assess* religion with NWAL. This is where we evaluate whether religion is love-based or fear-based. Most of the items on my first list above promote separation, rather than oneness. Promoting fear of God, and extolling God-fearing behavior, is not consistent with love or oneness. So, that means the underlying premise of religion as it applies to those aspects is fear. Fear of death, of God, etc.

What occurs to me also is that all of the "good" things about religion in the second list can be had or accomplished *without religion*. But most of the fearful things about religion don't seem like they can be extracted from religion. In other words, if you took those God-fearing aspects out of the major religions, what would you have left? You would not have anything resembling the religion, though you would have a focus on love without the fear.

I believe we can achieve community, good health, generosity, and ritual without being tied to any particular dogma. We can look at what science tells us about our world, rather than what old (really old) books tell us. We can also develop philosophies that make the transcendent personal, as Peter would say, without religion. Doesn't understanding oneness get you there?

We obviously also can have love and forgiveness without religion. That's really important, but as I've tried to explain, the concepts of love and forgiveness from the perspective of mainstream religion are different than mine and more limited, if not entirely incorrect. As Freke and Gandy say, once you understand oneness, that's love. Once you see what love is, it's hard to go back to the fear-based perspectives.

My assessment is that religion is predominately fear-based, rather than love-based. And so, it doesn't make much sense to support religion. The more I think about it, I believe our human population can't make the progress toward love that it must with religions as they are.

Walsch is on to something when he predicts that religion will change, and can begin to teach love and oneness. He says, "*Getting to heaven* will no longer be the ultimate purpose in life. *Creating heaven* wherever you are will be seen as the prime objective."

That's so logical. Why believe life revolves around something that isn't even part of this life and *may not be at all*? I've changed my mind to see heaven here now. Even if I may not have evolved enough to be a yogi and experience full immersion in the *bliss* of the now, in the all-encompassing energy, I can and do intellectually understand this perspective of heaven and miracles being here all the time. Can you?

Why have religion at all? What would be wrong with a new paradigm, a new thought process to help guide humans? Maybe metaphysical naturalism, as espoused by Richard Carrier? Maybe just a version of NWAL as a guiding faith principle? Or, heck, just make up your own.

There's a doctor in Alabama named Lee Baumann who's written a book called *God at the Speed of Light*, in which he asks whether *light* and *God* are the same? What would be wrong with that? Baumann apparently read books on quantum physics while studying to be a physician. According to an article in the *Washington Post* on January 10, 2009, "Interest in Faith and Science leads Doctor to Divine Light," by Gregg Garrison, Baumann said, "Einstein in his theory of relativity says that time stops at light speed. It's theoretically possible that any light wave can travel the entire universe and no time has elapsed. In a sense, light can be everywhere in the universe at once. ... like God." In a way, that sort of belief is allegorical too, just like my beliefs.

In truth, I hope that in a few decades hardly anyone will still believe literally in any mainstream religious doctrines. John Lennon sang it well in *Imagine*:

> *Imagine there's no Heaven*
> *It's easy if you try*
> *No hell below us*
> *Above us only sky*

Imagine all the people
Living for today.

I think we'd all be much better off with *love* as a guide rather than religion. The way my mind works, it seems the whole population could just slap its collective forehead one day and say, "What were we thinking with all that inane stuff that divided us? We're all in this together. Thank goodness we finally understand love!"

I know, you think I'm crazy, again. But, listen, peoples minds *are changing*! The percentage of Americans identifying themselves as Christian has dropped from 86 percent to 76 percent since 1990, according to a study conducted at Trinity College in Hartford Connecticut and reported in a *Washington Post* article on March 9, 2009, titled, "15 Percent of Americans Have No Religion," by Michelle Boorstein. The number saying they were Protestant went down from 17 million in 1990 to 5 million today. The only group that grew since that time is the one saying they had no religion. By 2012, a Pew Research Center study said that the 'nones' were up to 20 percent. In fact the 'nones' are now at 24 percent of the Democratic Party, making it the largest 'faith' constituency in the Party, with strongly liberal social views. As we shall see, there is a correlation between political views and religion.

This is good, but those folks who are turning away from traditional religion now do need something to *believe in*. What I'm saying is that *we can believe in love*, and *in ourselves*, and that's a perfectly fine belief system. We also could find faith in our wonderful world, *in nature*.

Many of those who still say they're in a religion are now mixing and matching from other sources. "Elements of Eastern faiths and New Age thinking have been widely adopted by 65% of U.S. adults, according to a survey by the Pew Forum on Religion & Public Life," as reported in the article, "More U. S. Christians Mix in 'Eastern,' New Age Beliefs," by Cathy Grossman, from *USAToday* on December 10, 2009. The survey said the findings point to a "spiritual and religious openness – not necessarily a lack of seriousness," said Pew researcher Greg Smith. The study showed that 25 percent of Americans find "spiritual energy" in physical things (welcome to my world). And 23 percent say yoga is a "spiritual practice."

The article mentioned a woman named Julia Jarvis, who was in graduate school at Southern Baptist Seminary in the 1980's. She since became an ordained minister in the United Church of Christ, leads an interfaith family project, and studies with Buddhist teacher Thich Nhat Hahn and finds a spiritual dimension in yoga. Her mother was a staunch Baptist all her life in Birmingham, Alabama but, like 49 percent of adults in the Pew survey, said she had a moment of "religious or spiritual awakening." Julia said, "My mother feared for years that I was no longer saved, but just two days before she died, she had an epiphany. ... She said she was 'told' in a spiritual experience to put aside all religious and political differences and just love each other."

That *is* what I'm talking about. It's just not that hard. All one has to do is seriously look at the belief system one has, and put it to the NWAL test. If the meter turns to love, then, *do it*. If the dial shows fear, *fuggedaboutit!*

It's OK to take bits and pieces of religions, science, even athletics and make them into a personal allegory. The *USAToday* story said so many Catholics have stopped paying attention to the Pope's edicts on "birth control, divorce, premarital sex that they don't think they are unCatholic when they believe and do what they please." So, they pick and choose, and if they really faced the contradictions and downright unbelievabilities of their religion, they wouldn't still call themselves Catholics. People just like to hold onto their historical identities. Eckhart Tolle would say those things aren't really who we are. They're just handles. Illusory descriptions of ourselves.

It's difficult for humans to let go of the things they've identified with for so long. I think the process of *letting go* in this regard is really crucial. Don't stop there! Look deeply at what you believe, test yourself and your faith. I ask myself, "Do I still really believe that?"

Many people are finally saying, "No." A *Washington Post* story, "In Major Poll, U.S. Religious Identity Appears Very Slippery," by Michelle Boorstein and Jacqueline L. Salmon, on February 26, 2008 reported that more people are exploring different religious and spiritual concepts, based on results of another Pew Forum poll, with more than 40 percent saying they have changed their religious affiliations since childhood. In the article, a woman named Anh Khochareun, who was raised Buddist in Vietnam, converted to Catholicism as a teen, and now doesn't identify with any religion, was quoted as saying about she and her husband, "We make our own faith within what we can do for ourselves in our own lives right now."

Another *Washington Post* article, "In Europe and U.S. Nonbelievers Are Increasingly Vocal," by Mary Jordan, on September 15, 2007, said nonbelievers are coming out of the closet, so to speak. One man, Graham Wright in England, just stopped praying and believing one day.

> *Wright, 59, said he was overwhelmed by a feeling that religion had*

> become a negative influence in his life and the world. Although he once considered becoming an Anglican vicar, he suddenly found that religion represented nothing he believed in, from Muslim extremists blowing themselves up in God's name to Christians condemning gays, contraception and stem cell research.
>
> "I stopped praying because I lost my faith," said Wright, a thoughtful man with graying hair and clear blue eyes. Now I truly loathe any sight or sound of religion. I blush at what I used to believe." ...
>
> New groups of nonbelievers are sprouting up on college campuses, anti-religious blogs are expanding across the Internet, and in general, more people are publicly saying they have no religious faith.

According to a *Washington Post* article on April 28, 2009, "Study Examines Choice of Religion," by Jacqueline Salmon, "more than three-quarters of Catholics and half of Protestants currently unassociated with a faith said that over time, they stopped believing in their religion's teachings." Young people are really good at this. A greater percentage of 18 to 24 year olds today than in previous generations say they have no religious affiliation, according to a February 17, 2010 article in *USAToday*, "Young Adults Today Are a Less Religious Bunch," by Cathy Grossman. Many are still very spiritual and pray to God, but they don't affiliate with a particular church.

This isn't just happening in America, Europe, and the developed world. It's happening, and much faster because of the influence of electronic communication, throughout the developing world, even in the Muslim countries. An article in the *Washington* Post, on July 22, 2007 was titled, "Losing My Jihadism." Written by Mansour Al-Nogaidan in Buraidah, Saudi Arabia, he says:

> Islam needs a Reformation. It needs someone with the courage of Martin Luther. This is the belief I've arrived at after a long a painful spiritual journey. It's not a popular conviction – it has attracted angry criticism, including death threats, from many sides. But it was reinforced by Sept. 11, 2001, and in the years since, I've only become more convinced that it is critical to Islam's future.
>
> Muslims are too rigid in our adherence to old, literal interpretations of the Koran. ... It's time to accept that God loves the faithful of all religions. ... It's time for Muslims to question our leaders and

> their strict teachings, to reach our own understanding of the prophet's words and to call for a bold renewal of goodwill, of peace and of light.
>
> I didn't always think this way. Once, I was one of the extremists who clung to literal interpretations of Islam and tried to force them on others. I was a jihadist. ... I joined a hard-line Salafi group. I abandoned modern life and lived in a mud hut, apart from my family. Viewing modern education as corrupt and immoral, I joined a circle of scholars who taught the Islamic sciences in the classical way, just as they had been taught 1,200 years ago. My involvement with this group led me to violence, and landed me in prison. In 1991, I took part in firebombing video stores in Riyahd and a women's center in my home town of Buraidah, seeing them as symbols of sin in a society that was marching rapidly toward modernization.
>
> Yet all the while my doubts remained. Was the Koran really the word of God? Had it really been revealed to Muhammad, or did he create it himself? But I never shared these doubts with anyone, because doubting Islam or the prophet is not tolerated in my country.

Mansour began to see the contradictions in his faith, in part by the way Islamic teachers molested their students and many devout Muslims lied to others. He began to wonder why the people could not question the religious leaders, and wrote about that, which eventually got him banned from writing. He's now looking for a Martin Luther type person with courage to challenge the status quo and "scholars who can convince Islamic communities of the need for a bold new interpretation of Islamic texts, to reconcile us with the wider world."

This is great. I wouldn't say a new view of Islamic texts is all that's needed, but it's a step in the right direction. As are the recent "Arab Spring" uprisings in Middle Eastern countries. And even when Islamists in the Arab Spring countries try to move things in their direction, such as the Muslim Brotherhood in Egypt, the people say, we don't want that stifling ideology to govern us. "Islamist setback could resonate far and wide," by Liz Sly in the *Washington Post*, on July 4, 2013. But we all can come together to move beyond our past history. We *have to* let go of our prior separateness. As Stephen Prothero says in his book, *Religious Literacy, What Every American Needs to Know – and Doesn't*: "The Fall into religious ignorance is reversible."

We need to develop a new faith for current Earth denizens that's more consistent with who we've become, rather than who we were. The human species is gradually coming out of its mid-life crisis. We can emerge into a new age of appreciating one another. Maybe it's the dawning of the Age of Aquarius. Or, we could just call it, the *Age of Love*.

I wonder if that'd be politically correct?

Sixteen | **Politics, as Usual**

My Dad was a Democrat; Mom, too. I remember Dad talking about going to see Franklin Roosevelt campaign for President. He said people didn't know Roosevelt wore steel braces on his legs because of polio. The press respected politicians and presidents back then, even though they might have written some tough stuff about them.

Dad thought Roosevelt was great. "He gave us back our whiskey," Dad said. Meaning that Prohibition was repealed right after Roosevelt took office. I remember Dad saying Prohibition had made ordinary citizens into criminals.

How about that? Being a criminal in your own country for drinking a beer or glass of wine. And that's after the economy went down the crapper and folks were *really* poor. The unemployment rate soared from 4% in 1929 to over 25% in 1932 during the Great Depression.

What did Roosevelt do? In his first inaugural address to the nation in 1933, he said:

> *So, first of all, let me assert my firm belief that the only thing we have to fear is fear itself - nameless, unreasoning, unjustified terror which paralyzes needed efforts to convert retreat into advance. In every dark hour of our national life, a leadership of frankness and vigor has met with that understanding and support of the people themselves which is essential to victory. ...*
>
> *Happiness lies not in the mere possession of money; it lies in the joy of achievement, in the thrill of creative effort. The joy and moral stimulation of work no longer must be forgotten in the mad chase of evanescent profits. ... If I read the temper of our people correctly, we*

> *now realize as we have never realized before our interdependence on each other; that we can not merely take but we must give as well … .*

This was a repudiation of fear and a call for oneness, for everyone to come together for the common good. He told the people that profits were of fleeting value, and that working together was the true joy and reward of life. He didn't rail against government; he knew government had an important role.

The importance of the "fear" line was psychological. It would take World War II to finally get the country out of the Depression; but if people could lift up their heads and move forward with hope, that was half the battle. A secret service agent said that, as a result of Roosevelt's inaugural speech, he felt "an injection of adrenalin in the veins of public morale. So far as the spirit of the thing was concerned, the Depression ended right there." When he went back to the White House that day, "it had been transformed during my absence into a gay place, full of people who oozed confidence and seemed unaware that anything was wrong with the United States." These quotes were from the book, *The Defining Moment, FDR's Hundred Days and the Triumph of Hope*, by Jonathan Alter.

In his first nationwide radio address, which became known as "fireside chats," FDR spoke to the people as if they were friends and emphasized the theme of the inaugural address:

> *Confidence and courage are the essentials in carrying out our plan. Let us unite in banishing fear. … We have provided the machinery to restore our financial system; it is up to you to make it work. It is your problem no less than it is mine. Together, we cannot fail.*

The many new programs of the New Deal improved the outlook of the nation, and made some important structural changes. For example, the New Deal brought in: increased regulation over the financial markets in efforts to protect the public against swindlers, a prohibition on child labor, limits on hours of the work week and a minimum wage law, and Social Security. The latter is a crucial tool for social justice. Old people didn't have to live in poorhouses anymore; they could have a safety net to get them through their golden years. A nation that takes care of its old folks and each other demonstrates compassion. My mother relies on her monthly Social Security check now. It's probably the same for millions of Americans.

According to Alter, one of the best examples of how Roosevelt changed the nation through his leadership was launching the Civilian Conservation Corps. The CCC hired young out-of-work men and put them in the national parks planting trees, building trails, and developing park land. Alter said, "Roosevelt's point was plain: Government counts, and in the right hands, it can be made to work. Strong federal action, not just private voluntary efforts and the invisible hand of the marketplace was required to help those stricken in an emergency."

Roosevelt was not afraid of government, and pushed on with a "spirit of bold experimentation" in trying to get the country out of the Depression. He said, "The test of our progress is not whether we add to the abundance of those who have much. It is whether we provide enough for those who have too little."

My parents believed that; we're all in it together and the government needs to help. I believe that, too. So, why is any of this important or relevant?

The reason is that politics can also be looked at through the NWAL lens. What types of political thought is based on love and which on fear? To me, the underlying spirit of Roosevelt's politics was the interdependence of the people in the nation and an obligation to help each other out. Many people hated Roosevelt and his policies and considered him a socialist. But isn't that really the type of government we have today, and what's wrong with that? We support the poor and sick, and why shouldn't we?

In a real sense, since giving is receiving, when we as a people acting through our government give to others in our society, we actually are *rewarded* with a more civil society. If there were only a few have's and many have not's, like in many Middle Eastern countries, we could have a revolt of the poor. Occupy Wall Street is a peaceful version of what could happen. Our streets could become war zones; we'd become a police state just to keep the order. It wouldn't be because of government, it would be government's failure to take care of its people. It's in our collective best interests to have a strong government that protects us, maintains our infrastructure, keeps the water clean, reduces pollution, makes sure our food and consumer products are safe, and helps ensure there are jobs so people can live decently - even if sometimes the government has to create the jobs, like the CCC.

These, of course, are liberal philosophies. It's clear that liberals tend to be Democrats in the United States and conservatives to be Republicans. We can look at the policies of the parties to help us determine which philosophy is more love-based and which more fear-based. In other words, which is based on oneness, and which on the perception of separation? Here are some musings on the topic from www.difference-between.com:

> A Democrat and a Republican differ in many ways, especially in their philosophy, ideas, worldview and politics. ...
>
> While a Democrat generally believes in a larger federal government, the Republican do not believe in such a concept. The Democrat believes that the government should look towards the greater good for the welfare of the people, irrespective of individual interests. They always tend to see the people as equal.
>
> The Republican holds economic equity on top more than any other thing. They believe that all answers lay with the people rather than with the government. The Republican wants less governmental interference and strongly believes in property rights rather than believing in welfare rights.
>
> A Democrat looks towards equality under a federal government and a Republican looks towards people looking after themselves. A Democrat stands for government-supported programs like healthcare and giving government subsidies to business, schools and hospitals, which means levying more taxes. A Republican wants decisions to be made at the state level with less government involvement.
>
> When the Democrat thinks federal government to be supreme and believes that every idea should originate from it, the Republican believes in capitalism and free market.
>
> A Democrat wants the right to legislate equality and a Republican want the right to earn equality. ...
>
> While the Democrat stands more for community, community responsibility and social justice, the Republican is for individual responsibility, individual rights and individual justice.
>
> Another difference between a Democrat and a Republican is that the former as a party tends to be pro-choice and the other one tends to be pro-life. When a Democrat favors social policies at the federal level, a Republican favors the policies at the state level. ...
>
> One can also come across many ideological differences between the two. Some examples are that when the Democrat believes in

decreasing military finance, the Republican wants to increase military budget.

I'm not sure Democrats think that *every* idea needs to come from government, but otherwise these points seem to be generally accurate, don't you think? Clearly, there are different worldviews happening here, so it's fair to ask what are the underlying differences. A few articles address this issue.

Here is what George Lakoff said in a blog titled, "Conseratism's Death Gusher," on the *Huffington Post* on July 16, 2010, writing about the Gulf Oil spill of 2009:

> *The conservative worldview says man has dominion over nature; nature is there for human monetary profit. Profit is sanctioned over the possibility of massive death and destruction in nature. Conservatives support even more dangerous drilling off the coast of Alaska and are working to repeal the President's moratorium on deep water drilling. Nature be damned; the oil companies have a right to make money, death or no death.*

That concept of nature as being only to serve man is the same my friend Father George Williams mentioned in his article I noted in the last chapter. It's consistent with the perception that man is separate and different - *entitled*. Interesting how religious and political views can be rooted in the same perspective. So tied together, in fact, that some politicians think they're inseparable. Mitt Romney, the Mormon presidential candidate said, according to E.J. Dionne writing in an article titled, "Boldness, Watered Down, in the *Washington Post* on December 7, 2007, "Freedom requires religion." Just think of that; no freedom without religion. I think it's really the opposite.

And here's a random definition of liberals in a review in the *Washington Post* titled, "Will the Real Liberals Please Stand UP," on March 1, 2009, by Thane Rosenbaum, about a book titled, *The Future of Liberalism*, by Alan Wolfe:

> *To be liberal, according to Wolfe, is to be optimistic about human purpose and potential, open to new ideas and strongly in favor of individual choice. Liberals, he writes, are inclined "to include rather than exclude, to accept rather than to censor, to respect rather than*

stigmatize, to welcome rather than reject." Liberals also see enormous social costs in inequality and value means over ends, procedure over passion, constitutional legitimacy over executive privilege.

A true liberal, Wolfe contends, is pragmatic, sober, skeptical and emotionally detached. Both the political right and far left, in contrast, are romantic at heart; they impulsively rush into military adventures and domestic crusades, yet often display defeatist tendencies premised on the belief that human beings, cursed by nature, cannot change their fate.

Fareed Zakaria wrote in an article titled, "The World Isn't So Dark," in the *Washington Post* on September 15, 2008, what answers he thought then Presidential candidates John McCain and Barack Obama might have to the question, "What kind of world do we live in?"

> We live in a very dangerous world, John McCain would respond. In his eyes, Islamic extremism is the transcendent challenge of the age. Jihadist warriors—funded and supported by states that adhere to their views—pose the central threat to the United States. In the rise of China, Russia and India, McCain sees turbulence. Russia and China, being autocracies, represent a special danger. Moscow's attack on Georgia was, for McCain, the "first serious crisis since the end of the cold war." The role for America, in such an environment, is to aggressively use its power—hard power—to fight evil, spread freedom and defeat the enemy. Otherwise we will lose the struggle for the 21st century.
>
> Obama's sense of the world is more optimistic. The dangers are real but not so all-encompassing. Obama speaks less of Islamic extremism in general and more of Al Qaeda and its affiliated groups specifically. He points out that compared with the cold war—when thousands of Soviet nuclear missiles were pointed at American cities—the threats we face today are reduced. He argues that most people in the Islamic world want development and a better life, not jihad. America's promise remains alive even in these countries.
>
> America's role, for Obama, is to restore its military strength, fight

Al Qaeda and its ilk, and deter rogue regimes like Iran. But it is also to stay calm, because in overreacting to dangers, we often cause new problems and crises. To lump together all Islamist groups is to exaggerate and misunderstand the threat. The Iraq War, for Obama, is a prime example of an alarmist overreaction, one that had the United States launch an unprovoked invasion of a country and rack up huge costs. If America can keep its cool and provide the help that countries really seek—in development, modernization and democracy-building—then we will gain in both security and legitimacy.

There is some truth to both visions of the world, but in my view the reality is much closer to Obama's—more so than most American politicians seem willing to admit. We live in remarkably peaceful times.

Eugene Robinson wrote a similar article in the May 26, 2009 *Washington Post* titled, "Worlds Collide." He asked the question, "Which reality do you inhabit; Obama World or Cheney World?" Here's how he answered:

In Obama World, human beings are flawed but essentially decent and rational. Most will behave in a way consistent with enlightened self-interest. In Cheney World, humanity's defects are indelible and irredeemable. Absent evidence to the contrary, evil should be assumed to lurk in every heart. Better to do unto others before they have a chance to do unto you.

In Obama World, Americans have a sense of community and shared purpose. Those upon whom fortune has smiled -- through accident of birth, educational opportunity, career-enhancing connections or any other kind of "right place, right time" serendipity -- recognize that extending a hand to those who do not enjoy such advantages is not just morally right, but ultimately beneficial to all. They believe that Henry Ford was right to pay his workers the shockingly high sum of $5 a day -- so they could afford to buy the cars they were making.

In Cheney World, sharing is for saps. Obtaining great wealth and power has nothing to do with being "fortunate," whatever that means.

> *It's all about preparation, focus and hard work. The idea that luck or connections could possibly have anything to do with, say, becoming the lavishly compensated chairman and chief executive of a megacorporation such as Halliburton? Preposterous and un-American.*
>
> *In Cheney World, ideals are nice and all that, but might makes right. We do what we want. Because we can. You got a problem with that?*
>
> *Obama World is an exciting place to live right now -- not perfect, to be sure, but full of energy and hope. If Dick Cheney wants to stay in his bunker, that's his business. Others might want to come up for some fresh air.*

I think these assessments are accurate. The underlying outlooks of conservatives exacerbated the reaction of Americans to the September 11, 2001 attacks. It was difficult in those years to even whisper anything questioning the steps taken by the Bush administration, even the attack on Iraq that was so damaging. Conservatives branded any such concerns as unpatriotic and un-American. There was a pervasive sense of fear. Fareed Zakaria said in an article titled, "We're Safer Than We Think," in the *Washington Post* on September 13, 2010, that this was a serious problem of the right, "where it has become an article of faith that we are gravely threatened by vast swarms of Islamic terrorists, many within the country. This campaign to spread a sense of imminent danger has fueled a climate of fear and anger."

That's an understatement. It's really no surprise that conservatives react so strongly to threats; they *startle* more easily, too. Seriously. Shankar Vedantam wrote an article in the *Washington Post* on September 19, 2008 entitled, "Startle Response Linked to Politics."

> *People who startle easily in response to threatening images or loud sounds seem to have a biological predisposition to adopt conservative political positions on many hot-button issues, according to unusual new research published yesterday.*
>
> *The finding suggests that people who are particularly sensitive to signals of visual or auditory threats also tend to adopt a more defensive stance on political issues, such as immigration, gun control, defense spending and patriotism. People who are less sensitive to potential threats, by contrast, seem predisposed to hold more liberal*

> *positions on those issues. ... Startle responses [] cannot be used to predict the political views of any one individual – there are many liberals who startle easily and many conservatives who do not. ... [But] the research suggests that people who adopt political views you disagree with are not stupid or irrational. Rather, they may arrive at their positions in part because they are predisposed to be more or less worried about risk.*

Being worried about risks is fear-based. Being afraid of change - which is the definition of conservatism in my mind - likewise. Richard Cohen writing in the *Washington Post* in a March 23, 2010 article titled, "It's Progress, Not Socialism," about the health care bill said:

> *Anger comes from fear. What was once a white Protestant nation is changing hue and religion. It is no accident that racial epithets were yelled at black law makers on Saturday in Washington and a kind of venom even gets exclaimed from the floor of the Congress: "You lie!" "Babykiller!" The protestors were protesting healthcare legislation. But they feared they were losing their country.*
>
> *Ever since the New Deal, the GOP has been the Party of the Past. It said no to the New Deal. It said no to Social Security. Important leaders – Barry Goldwater, for instance – said no to civil rights, as they now say no to gay rights. The party plays the role of the scold, the finger-wagger who warns of this or that dire outcome – not all of it wrong – and then gets bypassed by progress.*

Many people fear they're losing their storybook country. Life isn't the same as when I was growing up going to good old Blessed Sacrament School. It's not the same as when my Dad was a young man voting for Franklin Roosevelt. There are risks in these societal changes, to be sure, but we also have changed *so much for the better*. My friend Neal says that change is the *only thing* you can count on in life, but it's very fearful for many. Very often, it's an irrational fear.

I read an article in *Mother Jones* magazine, January/February 2010, written by David Corn, titled "Too Sane for Congress." It was about Rep. Bob Inglis, (R-S.C.) and an encounter he had with some constituents during the 2010 campaign:

> *In the middle of his primary campaign, Inglis had convened a small meeting with donors who had contributed thousands of dollars to his previous campaigns. This year, they hadn't ponied up, and during the meeting it became clear why:*
>
> *"They say, 'Bob, what don't you get? Barack Obama is a socialist, communist Marxist who wants to destroy the American economy so he can take over as dictator. Health care is part of that. And he wants to open up the Mexican border and turn [the U.S.] into a Muslim nation." Inglis didn't know how to respond.*

I wouldn't either. These citizens are terrified of socialism, communism, Marxism, economic ruin, dictatorship, health care, Hispanics, and Muslims, just to name a few fears they have. That's a heavy weight, I must say. That's a lot to be afraid of all the time. But, jeez, they can't be that afraid, can they? And don't they have their religion to fall back on?

It's true, Republicans are more religious than Democrats. According to another *Washington Post* article by Richard Cohen on January 25, 2011, titled, "The GOP's Lobotomy," 50 percent of evangelicals are Republican and only 34 percent are Democrats, "but the more telling figure is this one from [another] survey: Republicans are twice as likely as Democrats to believe Satan is a real spiritual entity."

So, if a person's deepest spiritual beliefs are fears of the devil, that infiltrates into what he or she thinks about other things. In other words, we can see a strong connection between conservatives and religious folks – their worldviews are rooted in fear, a perception of separation from God and the world. Actually, it *defines* many of them.

Take Mike Huckabee, for instance. An article from the *Washington Post* on December 20, 2007, "Is This Heaven? No, It's Iowa," was about campaigning in that state. Huckabee had increased his lead in the polls at the time, partly because of his 30-percentage point advantage among evangelicals. He commented, "There's only one explanation for [this lead in the polls], and it is not a human one. ... It's the same power that helped a little boy with two fish and five loaves feed a crowd of 5,000 people." One of his ads said, "Faith doesn't just influence me, it really defines me." If belief in a mythical God-man defines someone, what does that really mean?

Here's one of those positions that defines someone. When Keith Ellison, the Minnesota Democrat who was elected to Congress as the first Muslim in Congress in 2006 (as reported in a *Washington Post* article, titled "Coservatives Attacke Use of Koran for Oath," dated December 9 of that year), announced that he would take his oath of office on the Koran, it evoked a response from Dennis Prager, a conservative talk show host in Los Angeles. Prager wrote, "America is interested in only one book, the Bible. If you are incapable of taking an oath on that book, don't serve in Congress." Didn't know that was a condition for serving in the House or Senate. Seems like a pretty harsh and irrational judgment, actually.

Beliefs about the history of humans on the planet show similar trends between the religious and conservatives. A Gallup poll released December 17, 2010 and reported in an article titled "40 Percent of Americans Still Believe in Creationism," *Huffington Post* on December 20, found that:

> *40 percent of American still believe that humans were created by God within the last 10,000 years. ... A mere 16 percent of respondents subscribed to the belief of 'secular evolution:' that humans have evolved with no divine guidance. ... The poll also revealed that beliefs in creationism and evolution are strongly related to level of education attained. When results are narrowed to those with college degrees, only 37 percent of respondents maintain beliefs in creationism. Meanwhile, the belief in evolution without the aid of God rises to 21 percent. With regards to political affiliation, a majority of Republicans (52 percent) subscribe to creationist beliefs. This is compared to only 34 percent of Democrats and Independents.*

So, conservatives are generally more religious. I suppose that means liberals are more irreligious? My Dad was not religious and my Mom was, so obviously one can't make hard and fast conclusions as to individuals, just as with the startle reflex. But I don't remember perceiving such clear underlying bases for Democratic and Republican principles before. Am I just now understanding this, or is it a new phenomenon?

A good example of how political trends in society also move along the oneness and separateness axis comes from the developments leading up to the 2008 elections. David Broder wrote in the *Washington Post* on March 25, 2007 in an article titled, "An Opening for Democrats," that polls showed voters in the United States had become disenchanted with the Republicans and President George Bush because of Iraq and the perceived quagmire in the nation's capital. At the same time, there was significant growth in the support for liberal measures. For example, Broder said a Pew Forum poll showed that:

> [T]wo-thirds of those surveyed said they favored government-guaranteed health insurance, even if it means higher taxes. Seven out of 10 agreed with the general proposition that government has a responsibility to take care of people who can't care for themselves, and a similar percentage says government should guarantee every citizen a place to sleep and enough to eat. Those percentages have grown significantly since 2002, along with the belief that income inequality has increased in that same period.

Broder concluded that people generally were against Republicans and in favor of Democratic ideals in the spring of 2007. About six months later, polls showed similar results. In a *Washington Post* article titled, "Permanent Republican Majority? Think Again," on August 19, 2007, Andrew Kohut and Carroll Doherty of the Pew Research Center wrote that half the public favored Democrats and only 35% favored Republicans, reversing the trend that had put George Bush into office. They wrote:

> Here's something Democrats can really take heart from: Public support for more government aid to the poor and needy is back. The percentage of those who say that "it is the responsibility of the government to take care of the people who can't take care of themselves" has gone up 12 points since 1994, the pivotal year when Republicans took control of Congress … . Support for more government involvement in dealing with social problems is on the upswing overall. More Americans now subscribe to the sentiment that "the rich get richer while the poor get poorer." Another bad sign for Republicans, the party of staunch religious values; Most Americans remain religious, but the number expressing strong beliefs has dropped since the mid-1990s.

Interesting they note Republicans are the "party of staunch religious values." No wonder there seems to be a correlation as to the underlying nature of those belief systems.

Steve Pearlstein wrote in the *Washington Post* on December 14, 2007, in an article titled, "Business Over a Barrell," that major business organizations overreaching in

Washington had contributed to the change of attitude toward more liberal ideals. He said organizations like the Chamber of Commerce and the National Association of Manufacturers had "decided to bring the same inflexibility, partisanship and religious fervor to economic issues that Christian conservatives have brought to social issues. Their relentless crusade against taxes and regulations has damaged financial markets, weakened the economy, poisoned the political atmosphere and eliminated any possibility of effectively representing their members' interests with a Democratic Congress or White House."

These writings show a strong trend toward more liberal principles in 2007, the year before the Presidential election. Barack Obama took advantage of these beliefs that government should do more to win the Presidency in 2008. But that was just about the time that the deficits - and the rampant real estate and financial shenanigans - came home to roost. In the final months of the Bush presidency, the government had to bail out the economy. Barack's hope campaign reached deep into the nation's wounded psyche and I remember feeling as his inauguration approached: thank goodness, we can try and do good in the world again.

I went down to the Mall on that cold January inaugural day. It was really cold. We planned to get up – my son Jay and I - about Oh-dark-thirty. But he'd been awakened about three am by police activity down on Canal Road. Jay's a light sleeper, so heard it all and began watching the TV reports. He burst into our room and said, "I think we need to leave now! There's going to be a mass of people down there, so we have to go! The TV is showing people already waiting in the cold outside the metro stations to go down to the Mall!"

I got up and looked out the window. We saw the police lights, but didn't know what was going on at the time. "Jay, we have to get a little more sleep; get me up at five am."

Sure 'nuff. Five AM in the morning, Jay's ready to go. 'Course I had to grab a cup of coffee and perform my ablutions. So, we left maybe, six?

We'd decided to drive even though we anticipated hordes of humanity. On the ride down, we picked up my other son Max. Halleluiah, there was NO ONE ON THE ROAD! Fabulous, we zipped down to the parking lot under my office building at 24th and N Streets. Walked down 23rd, then east, toward the Mall. Lots of people and hardly sunrise. A buoyant and happy crowd, in spite of how cold it was. I thought we should settle near the Washington Monument and find a place in the sun and out of the wind. Jay wanted to get closer. So, we hiked down the Mall until we basically couldn't go any further because of the wall of people. We found a little spot, though we could hardly see the podium on the West portico of the Capitol building. There were jumbotrons for people to watch the inauguration and the President's speech. We were in place by nine am, and the oath of office was to take place at noon.

We stood all crammed together in the frigid air; about twenty degrees maybe, likely ten with the wind chill. The coldest I'd been since duck hunting with my Dad. But I was with my sons, and we witnessed the first African-American president being sworn into office. We were all filled with hope. The same kind of hope I felt almost fifty years before when my father took me to see John Kennedy's inaugural parade in 1961. Dad and I had seats, and it was freezing that day, too. About all I remember was the cold and watching the PT-109 float going by. It was special to be with my sons for Obama's inaugural, as I had been with Dad for JFK's. Both presidents were so inspiring. We didn't see John Kennedy's inaugural address live in 1961, but it's still one of my favorite speeches of all time. You can watch it at http://www.youtube.com/watch?v=BLmiOEk59n8; here's a short excerpt:

> *In your hands, my fellow citizens, more than in mine, will rest the final success or failure of our course. Since this country was founded, each generation of Americans has been summoned to give testimony to its national loyalty. The graves of young Americans who answered the call to service surround the globe. Now the trumpet summons us again—not as a call to bear arms, though arms we need; not as a call to battle, though embattled we are—but a call to bear the burden of a long twilight struggle, year in and year out, rejoicing in hope, patient in tribulation—a struggle against the common enemies of man: tyranny, poverty, disease, and war itself.*

The cold war was on then, but Kennedy summoned a theme of humanity's real struggles. It was scary, the tense era of stalemate with the Soviet Union. I remember the Cuban missile crisis. I remember learning to "duck and cover" under the school desks in the event of nuclear war, as if that would help. Still, Kennedy inspired us.

The economy was freefalling when Barack Obama took office, but we also held on to the hope that we could work to make our nation and the world a better place *together*.

That didn't last long.

It was scary in January 2009. It seemed we were slipping into a second great depression. I've talked about the fear gripping the financial sector in prior chapters. One of the most important pieces of legislation to keep us out of the ditch, the stimulus package, was signed into law in February 2009. The bill in the House got *not one Republican vote*. Not one. Three Republican Senators voted yes to achieve a filibuster free 60-vote tally. That was to set the stage. Republicans continued to vote in lock step against Obama's legislative agenda. After the 2010 elections, Senate Leader Mitch McConnell said the *"single most important thing we want to achieve"* was making Obama a one-term president. Not helping poor people, not bringing peace to the world, not educating the young; just beating Black Obama.

Yep, it was almost treason in my mind. The Republicans drew a line in the sand against Obama. "Or, as Ohio Senator George Voinovich summarized the strategy: 'If Obama was for it, we had to be against it.'" That was their entire mission. "The party of no," by Michael Grunwald, *Time* magazine, September 3, 2012. From the same article:

> *The stimulus debate established the pattern for the next four years. Republicans opposed the entire Obama agenda – a health care plan based on Mitt Romney's, a cap-and-trade regime that McCain had supported in 2008, financial reform after a financial meltdown. Obama squeezed his health care and Wall Street reform bills through Congress anyway, but the quest for 60 votes in the Senate forced him to cut deals that made his initiatives look ugly. And the Tea Party – which held its first rally 10 days after Obama signed the stimulus – became a powerful force opposing the Obama agenda … .*

As Fareed Zakaria wrote:

> *Today it is the Republican Part that often seems angry with America. Read the best-selling books by conservatives these days, watch Fox News or attend a Tea Party rally. They are filled with rage, often combined with a powerful nostalgia for an America that has gone away. … The Tea Partyers love America, but it's an America that is an abstraction or a memory. The nation of today – with its many immigrants, liberated women, increasingly liberated gays, myriad government programs, open trade and a Spanish-language option*

on every phone menu – seems to scare them. "The Heirs of Reagan's Optimism – Role Reversal: The Democrats are the ones celebrating America's promise," Time, September 17, 2012.

Republicans took advantage of the fears of the nation (which stemmed from 9/11 before the recession) and high unemployment, rising costs, and a changing world headed by a colored guy with a Muslim sounding name. Wayne Gilchrist, nine-term Maryland Republican House member defeated in 2008 by a conservative primary challenger, presaged the strategy with his assessment of the country's problems: "We're in this bad place as a country because of the evangelicals, the neocons, the nasty, bitter and mean … very clever ideological groups that use money, technology, fear and bigotry to lead people around. … Voting according to your knowledge and experience – that's out the window. Competence and prudence? Forget it." "Gilchrest Unloads on Know-Nothing Pols and the Rest of Us," by Marc Fisher, *Washington Post*, October 2, 2008.

Those viewpoints sowed seeds of distrust, the likes of which I don't recall in my lifetime. Remember those angry, screaming town hall meetings during the debate about the health care bill? People were *so afraid*, they became psycho angry and lost their rational minds. At one town hall meeting, "A registered nurse charged there would be 'legal genocide' if health reform passes – and questioned whether Obama was born in the United States," according to Dan Balz in an article titled, "Another Town Hall Meeting, Another Illustration of Obama's Challenge," the *Washington Post*, September 4, 2009. Balz also reported that, "When one member of the audience [at a town hall meeting] asked [Congresssman] Markey's district director, Ken Bennet, what he would report back from the meeting, he said, among other things, the sense of fear people have about changes in the system. 'It's not fear that we have,' Nancy Winters told him. 'It's mistrust. Don't tell her we're afraid. We don't trust them. There is an agenda with this Administration.'"

Of course, though, *mistrust is fear*. If one trusts, that's akin to love – no fear. If one doesn't trust, that signifies a fear that a person can't be trusted, or won't act in a way that's expected or perceived to be correct. People often don't think they're afraid, even though it's the primary cause of their emotions. In another Balz article titled, "The Grass Roots Prevail in N.Y.," in the *Washington Post* on November 1, 2009, he said, "A Republican strategist and a veteran Democratic strategist voiced their belief that anger is the most significant force in politics today and a potential threat to incumbents of both parties nest year. 'I have never seen it like this," the Republican strategist said. 'It is a breakdown of trust." There you go, again; anger based on a breakdown in trust.

Sebastian Mallaby wrote an article in the October 30, 2006 issue of the *Washington Post*, titled, "The Decline of Trust." He cited a number of books and studies showing that trust had been declining since the invasion of Iraq. He said a book called *Trust* by Francis Fukuyama in 1995 had argued that a, "society's capacity for cooperation underpins its prosperity." Mallaby says that American's trust in government went up from about 1994 through about 2003, from 21 percent saying they trusted the government to 56 percent. Corporate America was all about trust during those years, too; remember the dot.com era when decisions were pushed down so that many companies were practically democracies? But with Enron and other corporate scandals in the early 2000's, that trust evaporated, and it became more common to be auditing employees rather than trusting them. He says it would be cheaper to trust in the corporate world, rather than pay teams of lawyers and accountants to scrutinize companies for evidence of waste or fraud. Mallaby wrote:

> *Likewise, as Marc Hetherington of Vanderbilt University has demonstrated, government is constrained if nobody trusts it. The Great Society programs were possible because Americans trusted government in the 1960s; the creation of the Medicare prescription drug program arguably reflected the peaking of trust in government in 2003. But Bill Clinton's health care reform was thwarted in the low trust early 1990s, and nobody now trusts government to modernize entitlements. Meanwhile President Bush had enormous foreign policy momentum in 2002-03 because America trusted him. Thanks to the Iraq mess, Americans are now focused on holding Bush accountable, ... And the absence of trust can feed on itself.*

That's what's happening. Significant expressions of distrust – fear – increases distrust. *USAToday* reported on July 10, 2010, that "Faith in Social Security [is] Tanking." Roosevelt's signature New Deal program, the most successful government program in history, is being undercut in part just because people are afraid. The more angry people get (meaning, in particular, the Tea Party folks), the less cooperation and success we can have together in our society. Witness partisan gridlock. And the so-called "debt ceiling crisis." It seems then, it'd be better for all of us to trust one another, if only because it's a more successful strategy. We could accomplish more working and trusting ourselves and the government-that-is-us, than if we rant, rave, revile, and condemn government and its programs. I think that's common sense, but it's easier for many folks to resort to fear-based complaining and anger.

It's true, of course, that both liberals and conservatives have fears. What I'm saying, and think our NWAL review will bear this out, is that Republican *policies* are based more on fear. Underlying the conclusion that fear has been behind the Republican induced opposition to Obama is this comment by then House Minority Leader John Boehner, as reported in the *Washington Post* by Dana Milbank in an article titled, "Hard to Say He's Sorry," on September 11, 2009: "Don't underestimate the amount of emotion that people are feeling. … Americans are frustrated, they're angry, and, most importantly, they're scared to death." If being *scared to death* isn't fear, I wouldn't know what is.

You also have radio personalities fanning the flames of fear. I'm not even going to talk about Rush Limbaugh, but there were a bunch of articles on Glenn Beck about the time he led that march on the Mall in Washington in August 2010. An article in *Time* magazine on September 28, 2009 was titled, "The Agitator," with the byline "Glenn Beck is channeling the fears and anger of Americans who feel left out – but is he also stirring that anger and heightening those fears?" Duh. Of course, he was. *Fear sells.* "I'm afraid," says Beck, "You should be afraid too." According to the *Time* article:

> His fears are many … . What's this rich and talented man afraid of? He is afraid of one-world government, which will turn once proud America into another France. He is afraid that Obama 'has a deep-seated hatred for white people" – which doesn't mean, he hastens to add, that he actually thinks 'Obama doesn't like white people.' He is afraid that both Democrats and Republicans in Washington are deeply corrupt and that their corruption is spreading like the plague. He used to be afraid that hypocritical Republicans in the Bush Administration were killing capitalism and gutting liberty, but now he is afraid that all-too-sincere leftists in the Obama Administration are plotting the same.
>
> On a slow news day, Beck fears that the Rockefeller family installed communist and fascist symbols in the public artwork of Rockefeller Center. One of his Fox News Channel colleagues, Shep-

ard Smith, has jokingly called Beck's studio the *"fear chamber."* Beck countered that he preferred *"doom room."* ... Beck mines the timeless theme of the corrupt *Them* thwarting the virtuous *Us*.

Of course, he does. *That's what fear is; the perception of Them being different and separate from Us.* Ruth Marcus wrote an article titled, "The Church of Glenn Beck," in the *Washington Post* on September 1, 2009 about the rally on the Mall and quoting the reasons some of the attendees came to Washington for it:

> *"We've lost our morality. The country is headed in the wrong direction by removing God from everything," said Bob Erdt, a retired Ford engineer from Michigan. Andrea Carrasco from Colorado said she came to "ask God to restore the country. Our freedom is lost. My freedoms are lost. To be able to preach anywhere we want, to have God in our schools, to drive any kind of car we want and if I want to drive a gas guzzler, I can, if I want to eat a lot of sugar and salt, and I shouldn't be forced to buy medical care. ... To be able to burn the kind of light bulb I want. The list goes on."*

Hmm. If God were more involved in government, we could use the kind of light bulbs we want? Seems she did at least have the freedom to assemble and speak freely.

So, there's this deep-seated fear that the government is taking away rights and telling people what to do. They *are terrified* that they're loosing their individuality. But, individuality can actually grow and expand in a large, diverse, and complex society, I believe. The fact is there are just *so freaking many people*, that our collective self – our government – has to take action in the best interest of *all of us together*, not just from each individual's perspective. It's – again – oneness and interdependence versus a perception of separation – Us versus Them. Love versus fear.

Nowhere is this more clear than in the gun debate. An article titled, "A Dangerous Silence," by E.J. Dionne in the *Washington Post* on January 13, 2011, after the tragic shooting of Rep. Gabby Giffords and others in Tucson, addressed the reason we have such volatility from those against any form of gun control. Dionne said he's realized after email exchanges with "ardent gun control foes over the years, that the real passion for a let-anything-go approach to guns has little to do with culture or hunting. It is rooted in a very peculiar view of how America has maintained its freedom. Rep.

Heaven is Everywhere

Ron Paul, as is his wont, expressed it as plainly as anyone: 'The Second Amendment is not about hunting deer or keeping a pistol in your nightstand, ... It is not about protecting oneself against common criminals. *It is about preventing tyranny.*'" (Emphasis added.)

Seriously? In this great country, we have folks who go around worrying about a tyrant taking over? Sounds rather medieval, or Middle Eastern, me thinks. Paranoid, too, which obviously is fear-based. In Dionne's article, he also quotes Rep. Paul Broun (R-Ga.) at a rally in Washington:

> *Fellow patriots, we have a lot of domestic enemies of the Constitution, and they're right down the Mall, in the Congress of the United States – and right down Independence Avenue in the White House that belongs to us, ... It's not about my ability to hunt, which I love to do. It's not about the ability for me to protect my family and my property against criminals, which we have a right to do. But it's all about us protecting ourselves from a tyrannical government of the United States.*

Jimminy Crickets. Am I being naïve not to worry about tyranny here in the U.S. of A? I don't think so. But this sort of fear *is* rampant, though people don't really think of themselves as fearful. They use other descriptors, like mistrustful, angry, etc.

The debate after the Newtown shootings in 2013 highlighted this fear of tyranny meme, because the National Rifle Association and the gun-types revel in their concern about government despots here in the U.S. Check out this National Public Radio interview on April 8, 2013 with David Welna: http://m.npr.org/news/U.S./176350364. Hey, listen, we're not Syria. It's only if one has a fear that people are not good, I think, that people develop a paranoia of tyranny.

It would be nice if we the people could collectively go see my spiritual guru Avery Kanfer. That's what Eugene Robinson suggested in an article in the *Washington Post* on December 4, 2007, titled, "We the Paranoid," that the nation needs a therapist. He had attended a meeting of the American Anthropological Association and one session called, "The Insecure American."

> *[The session] turned out to be a revelation – by turns alarming, depressing and laugh-out-loud amusing – as scholar after scholar*

> *presented research showing just how unnerved this society is. Seth Low, who teaches at the City University of New York, has spent years studying the advent and increase of gated communities. People decide to sequester their families behind walls because they are afraid of crime, they feel isolated from the neighbors, and they're nostalgic for a kind of idealized Norman Rockwell past, Low reported. Nothing terribly irrational about that.*
>
> *But after extensive interviews with residents of gated communities in San Antonio and on Long Island, Low discovered that there isn't really less crime behind the walls, people don't feel more secure, and there was no greater sense of small-town closeness among neighbors. Despite the gates and guard huts, people still felt they needed to set their alarm systems. ...*
>
> *To recap; We're afraid of one another, we're afraid of the rest of the world, we're afraid of dying. Maybe if we study our insecurities, and confront them, we'll learn to keep them in check. Before we turn the whole nation into one big, paranoid gated community, maybe we'll learn that life isn't really any better behind the walls.*

I'm trying, Gene, I'm trying. I do think that the more we can address our fears, and then let them go, the better we'll be. Much better than if we celebrate our fears, which constitutes a lot of what politics is about these days. Part of the problem, though, is that we're not so smart as we think.

In an article called "5 Myths about those Civic-Minded, Deeply Informed Voters," Rick Shenkman shows that the following statements are *not* true:

- *Our voters are pretty smart;*
- *Bill O'Rielly's viewers are dumber than Jon Stewart's;*
- *If you just give American the facts, they'll be able to draw the right conclusions;*
- *Voters today are smarter than they used to be; and*
- *Young voters are paying a lot of attention to the news.*

Darn, I was hoping those would have all been true. Shenkman says:

According to an August 2006 Zogby poll, only two in five Americans know that we have three branches of government and can name them. A 2006 National Geographic poll showed that six in ten young people (aged 18 to 24) could not find Iraq on the map. The political scientists Michael Delli Carpini and Scott Keeter, surveying a wide variety of polls measuring knowledge of history, report that fewer than half of all Americans know who Karl Marx was or which war the Battle of Bunker Hill was fought in. Worse, they found that just 49 percent of Americans know that the only country ever to use a nuclear weapon in a war is their own. ...

Just before the 2003 invasion of Iraq, after months of unsubtle hinting from Bush administration officials, some 60 percent of Americans had come to believe that Iraq was behind the Sept. 11 2001, terrorist attacks, despite the absence of evidence for the claim, according to a series of surveys taken by the PIPSA/Knowledge Networks poll. A year later, after the bipartisan, independent 9/11 commission reported that Saddam Hussein had nothing to do with al-Qaeda's assaults on the World Trade Center and the Pentagon, 50 percent of Americans still insisted that he did. In other words, the public was bluntly given the data by a group of officials generally believed to be credible - and it still didn't absorb the most basic facts about the most important event of their time.

Is it possible, then, that the most important factor for a nation – and world – to learn peace might not be education of facts, but rather *psychological evolution toward an understanding of love?* If we awesome Americans are basically *stupid* when it comes to collective knowledge of factors necessary to make political decisions, wouldn't that make it all the more important to focus our attention away from fear and toward oneness and interdependence as guides? I think it's worth a try.

But I don't want to get ahead of myself. We should do our NWAL analysis of political issues. So, in the Now? Check. On to witnessing and accessing (which sometimes happen rather concurrently). Here we go:

<u>War</u>. No one would fight or kill another human being if not afraid. This is always true. Why would anyone punch someone or shoot them if not afraid in some way? Certainly, fighting and killing and murder are *not* loved-based. A good example of

collective fear leading to war is Iraq. Some saw Saddam Hussein as a *threat*, but what's a threat? A threat is being *afraid* someone or some country might attack or cause harm. Of course, when one is afraid of what someone will do, that's fear-based.

It's fair to say that those in favor of the war were primarily conservatives and Republicans. Liberals wanted to stop the war. I remember marching on the Mall to protest the war, and I guarantee you most of those folks were liberals; not many conservative protesting against the war. Conservatives see war as an Us versus Them exercise, and thus by perceiving separateness act out of fear. Liberals see more of the oneness and interdependence and sameness of all people. In other words, love. So, I believe conservatives act more out of fear when it comes to war and international conflict issues.

Taxes. What are taxes? I'd say taxes are contributions we make toward the common good. Taxes pay for the communal services that we need government to perform for us to live better, happier, healthier, and peaceful lives. If one says government is there just to provide *security*, which I believe conservatives are more likely to claim, then that's focusing on the fear side. What is a desire for security if not wanting to be safe from *attacks* by other humans? Certainly also wanting to be safe from floods, bears, and disease, but national security is all about erecting defenses because of a *fear of other people*. Liberals tend to see government more as the embodiment of us all, and that government's role is to help us in as many ways as possible; education, health care, infrastructure, housing, etc. All for one and one for all; in other words, oneness.

Regardless of the primacy of the population's need for government, however, there's no question that taxes contribute to the health and welfare of the citizenry; in other words, contribute to our own collective wellbeing. If one believes government has no right to a citizen's money, and condemns any proposals for social improvements that involve raising taxes, then that perceives the separateness, rather than the oneness. Conservatives and Republicans are definitely more likely to fight taxes than liberals and Democrats, who believe we must all pay our fair share - and the more affluent must pay more - because we're all in it together. Arguing against taxes is fear-based.

Health care. Believing that we must provide health care for all Americans (and, indeed, everyone on the planet, which would truly be *universal* health care coverage) is love-based, and contending otherwise is not. Again, liberals and Democrats tend to be in favor of such health coverage - ensuring adequate preventative and sick care for everyone, even those who can't pay - and conservatives and Republicans are not. Conservatives want everyone to fend for themselves, which clearly represents a perception of separation - Us versus Them. Shoot, the individual mandate was a *Republican* idea developed to oppose a single payer system; but now, if government (or Obama) is involved at all, they're against it.

Death Penalty. Killing anyone is not love-based. Need I say more? Seriously, I can objectively look at killing anyone for any reason and conclude unequivocally that it is *not* love-based. Imposing the death penalty is also a big-time judgment. I've heard some say the difference between abortion and the death penalty is that it's not right to take an *innocent* life; but that means someone made a *judgment* that another is not innocent, is guilty, and can no longer live. Capital punishment is fear-based, because it's theoretically justified on fear of another crime and the "ultimate justice" (to quote Governor Rick Perry), and it's totally a perception of the separateness. Democratic principles against the death penalty are love-based.

Abortion. I don't believe that anyone is really *for* abortion. But liberals and Democrats tend to view a person's right to choose to have an abortion (usually early in a pregnancy) as predominate over the right of the unborn. I'm one who actually believes *Roe v. Wade* (I've even read the decision, which is at http://caselaw.lp.findlaw.com/scripts/getcase.pl?court=US&vol=410&invol=113) was a very good and accurate judicial holding following Supreme Court precedent, which is based on an analysis of an individual's rights against compelling needs of the state. When the latter are not proportionately compelling versus the individual's rights, then the individual's rights prevail. The decision in *Roe* made the reasonable distinction that the state, i.e., government, has a compelling need to protect unborns in the last trimester, and no objective compelling need that overcame a woman's right to her own body in the first trimester.

But I'm going to surprise you and say I believe being pro-choice - the more likely position of liberals and Democrats - is fear-based, and that being against abortion is love-based. Why? Deciding not to give birth because of financial needs or because it wouldn't be convenient is not based on love and compassion; it's a decision clearly based on fear. It's the fear of being burdened by the obligations and responsibilities of having a child, and the things it would prevent one from doing. That doesn't make it right or wrong, just that it's based on fear. In a loving world, there would be care and nurturing for any child. Being pro-life is love-based.

I have a video of my comments on a number of these issues at http://www.youtube.com/watch?v=jaMeR-zETyA.

Don't the differences seem clear when you think about them? All the NWAL queries were hands down in favor of liberals, except abortion, meaning the liberal worldview is generally love-based. As my friend Peter Ainslie says, conservative thoughts are Us versus Them, and liberals are about All of Us. The perception of separation versus oneness; fear versus love. Not right or wrong. When thinking where to stand on political principles, if one wants to cast a ballot for separation, vote Republican.

And it was that fear and delusion of separateness that drove a lot of the 2012 Presidential campaign. I found the following short piece about the Republican primary in Iowa to be illustrative. Why were some god-fearing Republicans voting for former Senator Rick Santorum?

> *The Schrievers, emblematic of the evangelical conservatives who are influential in Iowa policy, said they couldn't vote for Rep. Michelle Bachmann (Minn.) because she is a woman, for Rep. Ron Paul (Tex.) because of his antiwar views, for former Massachusetts Governor Mitt Romney because he is a Mormon, for former House Speaker Newt Gingrich (Ga.) because of his past infidelity and for Texas Gov. Rick Perry because he didn't get past the surface on the issues. The Schrievers say they practice their faith through daily life and their vote for president. Washington Post, "Their faith is in Santorum," by Melina Mara, January 4, 2012.*

Well, it's a hell of a faith that is anti-woman, pro-war, and bigoted on both religious and moral grounds. That's being afraid of a lot of things. But Mitt Romney, the 2012 nominee, really let the cat outa the bag about their biggest fear; us. That's right, fear of people they think aren't hard working, industrious, and god-fearing enough for them. I'm sure you remember the 47% comment.

> *There are 47 percent who are with him [Obama], who are dependent upon government, who believe that they are victims, who believe that government has the responsibility to care for them, who believe that they are entitled to health care, to food, to housing, to you name it. ... These are people who pay no income tax. Forty-seven percent of Americans pay no income tax. ... And so my job is not to worry*

> *about those people. I'll never convince them they should take personal responsibility and care for their lives.*

It really is a remarkable statement. Do you think this ideology is "We're all in it together" or "I'm different and special"? Here's the way Lisa Miller explained it in her column in the One Faith section of the *Washington Post* on May 19, 2012, titled, "2 views of God help define Obama-Romney rivalry:"

> *Romney stands for the individualistic version of American success; Obama for the collectivist. "It's the classic American dilemma – it's liberty versus community – and it's always there," says my friend, the Boston University religion professor Stephen Prothero, who in 2004 published a book called "American Jesus." ... Romney says his religious values are rooted in his belief that God gave individual Americans the ability to conquer and withstand difficulty. ...*
>
> *Obama is saying something else entirely. Over and over again, in every recent speech, he reminds Americans that we're all in this together, that everyone must make sacrifices – including tax hikes for the rich – and that the failure of one means the failure of all. ...*
>
> *Romney is promoting the God of "I"; individual accomplishments and personal success. Obama is promoting the God of "we," in which the fates of all of us are intertwined.*

Do you think I'm being too hard on conservatives about this? Well, hear what Thomas E. Mann and Norman J. Ornstein (Norm is a pretty conservative guy, and works at the American Enterprise Institute) have to say in the *Washington Post* on April 29, 2012, in an article titled, "Admit it. The Republicans are worse."

> *We have been studying Washington politics and Congress for more than 40 years and never have we seen them this dysfunctional. In our past writings, we have criticized both parties when we believe it was warranted. Today, however, we have no choice but to acknowledge that the core of the problem lies with the Republican Party.*
>
> *The GOP has become an insurgent outlier in American politics. It is ideologically extreme, scornful of compromise; unmoved by conven-*

tional understanding of facts, evidence and science; and dismissive of the legitimacy of its political opposition. ...

But that's not really fear, you might say. What about "openness" versus "conscientiousness?" That's how science writer Chris Mooney explains it in an article titled, "Politics is all in our heads," in the *Washington Post* on April 15, 2012:

There's now a large body of evidence showing that those who opt for the political left and those who opt for the political right tend to process information in divergent ways and to differ on any number of psychological traits. Perhaps most important, liberals consistently score higher on a personality measure called "openness to experience," one of the "Big Five" personality traits, which ... means liberals tend to be the kind of people who want to try new things, including new music, books, restaurants and vacation spots, and new ideas.

"Open people everywhere tend to have more liberal values," said psychologist Robert McCrae, who conducted voluminous studies on personality while at the National Institute on Aging at the National Institutes of Health.

Conservatives, in contrast, tend to be less open – less exploratory, less in need of change – and more "conscientious," a trait that indicates they appreciate order and structure in their lives. This gels nicely with the standard definition of conservatism as resistance to change – in the famous words of William F. Buckley Jr., a desire to stand "athwart history, yelling 'Stop!'"

But, hey, it's OK. They're blameless. It's not the way I prefer to think about things. I want us all to realize our oneness; hang out and do our part. We're all here trying to figure it out; doing the best we can. Just need to do it without killing and fighting.

Generally, we do. Most people around the world are living their lives peacefully. There's the underlying fear thing going on, which manifests itself in so many ways: religion, war, politics, etc. It's important to realize the one really great thing about democracy is that it's based on – *forgiveness*. When we can *let go* of our negative feelings about others – even if we disagree with them – and not kill them and let them live their lives in peace, we are letting go and forgiving. I don't think people realize that freedom and democracy are so dependent on forgiveness. That's why it's the best of the human management styles that we have. Because when you let go of fear, then you have peace. We have a generally peaceful society, because we practice political and societal forgiveness every minute of every day. When some folks don't, that's when we get in trouble.

So, I'm going to keep on basing my life on love and forgiveness as much as I can, and making my choices as consistently with oneness as the practical circumstances of my life permit - even if people think I'm nuts. Or that my ideas about money are.

Seventeen | "Money, it's a crime…"

…. Share it fairly, but don't take a slice of my pie." Those are lyrics from the Pink Floyd song, "Money." I'd say, "Don't take a slice of my pie" captures today's major political debate. We want an egalitarian society, but don't take my money. I don't want to pay more taxes. It's MY MONEY! Pink Floyd also sang, "Money, so they say, is the root of all evil today." Does that sound harsh? Money is a crime and the root of all evil?

What about having stuff? Sure, we want money, but we also seem enamored with things. I'm in love with my iPhone, and our big back yard. We moved into this old house in 2003 without enough furniture, rugs, pictures on the wall, or plates. Now, we're bursting at the seams and I can't fit anything more in the attic. How'd we get all this stuff? Why did we want it?

John Lennon sang this:

> *Imagine no possessions*
> *I wonder if you can*
> *No need for greed or hunger*
> *A brotherhood of man*
> *Imagine all the people*
> *Sharing all the world.*

Imagine no possessions? Why is it so hard to think of things this way, in a new way? Instead, according to Eckhart Tolle in *A New Earth*:

> *The unchecked striving for more endless growth is a dysfunction and a disease. ... A large part of many people's lives is consumed by an obsessive preoccupation with things. ... Paradoxically, what keeps the so-called consumer society going is the fact that trying to find yourself through things doesn't work. The ego satisfaction is short-lived and so you keep looking for more, keep buying, keep consuming. ...*
>
> *The physical needs for food, water, shelter, clothing, and basic comforts could easily be met for all humans on the planet, were it not for the imbalance of resources created by the insane and rapacious need for more, the greed of the ego. It finds collective expression in the economic structures of this world, such as the huge corporations, which are egoic entities that compete with each other for more. Their only blind aim is profit. They pursue that aim with absolute ruthlessness. Nature, animals, people, even their own employees, are no more than digits on a balance sheet, lifeless objects to be used, then discarded.*

We talked before about how the financial system is based on fear – being afraid of not having enough. The frantic panting for more money is driven by the fear of not being secure. It's devoid of oneness. Lacking compassion. Money is pursued with a passion perhaps greater than athletes trying to win championships. Sally Jenkins in an article called, "Armchair Field Generals, Getting Sacked on Wall Street," in the *Washington Post* on February 9, 2009, writing about the financial meltdown, made this point:

> *Avarice is not the only explanation for what's happened on Wall Street. You can see that in the builds of the financial titans who testify before Congress, hard chests and muscular necks brimming out of their dark suits and crisp white shirts. You know the type: the guy who slides with his spikes high in the company softball game. Which raises the question of hyper-competitiveness, and whether it's really such a useful quality on surfaces other than dirt. ... How did so many otherwise civilized men, philanthropists, benefactors and supposed intellectuals become such unapologetic plunderers? It's easy to be cowed by the knowledge of financial people, or at least by their terms. But there is something that any of us can decipher in their behavior: Obsessive scorekeeping got in the way of responsible judgment.*

Yes, it seems something cowardly is at play in the way people seek money. I've been at fault in this. When I was about to get married the first time, I shunned a government job or working for an environmental cause because I felt I needed more money, so I took a job at the *biggest law firm on earth*. I did it for a larger salary, so I could provide for my family. I'm still at a large law firm (much less so than before), because if I quit and started working for the Potomac Conservancy or a similar organization, or just started writing full time, I'd have to give up our house.

Still, I don't understand the drive to make those million dollar bonuses. Steve Pearlstein, who has a good understanding of what makes the markets tick, said this about Wall Street traders, investment bankers, and hedge fund mangers in his article, "News Flash for Wall St. - Money Isn't Everything," in the *Washington Post* on March 10, 2010: "[W]hat's so clearly missing from their work is any sense of a 'higher purpose' other than making more money than last year. Instead, they have been encouraged to take their motivation and their sense of self-worth from their annual bonuses, which in recent years have become nothing less than an addiction. It's an addiction that demands constantly bigger doses just to keep going."

He'd been discussing the conclusion of Daniel Pink in his book, *Drive* that, "Once people achieve a reasonable level of economic comfort and security, they are likely to be less easily motivated by monetary carrots and sticks than they are by more emotional factors. And in modern workplaces, Pink argues that the most powerful emotional motivators are the desire for autonomy, the satisfaction that comes from mastering a skill or a task, and the need to serve some larger social purpose."

So, does this mean the way we've structured our financial markets and incentives doesn't really work that well? That profit is not the best motive? That the free market system actually does not allocate resources and bestow its benefits well?

Of course, the free market system isn't fair or reasonable. It's a crap shoot, where the stakes are human lives and world peace. And the free market has been awful at providing any semblance of fairness. Some people have more money than whole countries, and others hardly a nickel. This can't be good in the long run. No one is different than anyone else. We're all humans and should be able to enjoy our short chapters on earth. Sure, some people work harder, are smarter, and even luckier. We shouldn't deny people the fruits of their labor. But something is askew.

For example, in 2011, the top 25 hedge fund managers made collectively over 22 billion dollars! This from an article on April 1, 2011, by Charles Riley at CNNMoney: http://money.cnn.com/2011/04/01/news/economy/hedge_fund_pay/index.htm.

That's not making money digging ditches or farming, building houses or schools. It's basically from just doing deals. Pearlstein again, from the December 20, 2006 *Washington Post* article, "Wall Street's Season of Excess," "Nobody who is hired help and who plays with other people's money 'deserves' to earn $100 million. That's true in a moral sense. But it's also true economically." He makes the point that these folks don't have any of their own money at risk. They get paid based on a percentage of the deals. So, the rewards are outsized. According to Pearlstein:

> *Wall Street is a classic example of an oligopoly, a cozy club of competitive firms that manage somehow not to compete on price. ... My biggest problem with the rationalizations for Wall Street pay, however, has to do with the widely held misconception that top executives are somehow entitled to some fixed percentage of the profit or a percentage of the gain in a company's market value.*
>
> *This is, of course, the way we calculate waiters' tips. And it makes sense for small, closely held partnerships. But today's large, global corporations have become so big, the numbers so large, that they provide inappropriate benchmarks when calculating the compensation of a single human being. There's no limit to how big a company can get, but human beings are limited in how much they can eat, or how many homes they can occupy, or how many days they have to take vacation.*

I think that's right. Say I made three billion dollars last year. What would I do with it? It'd be nice to secure my old age. But I couldn't even spend it all on things necessary for me to live. I could buy a giant yacht, I suppose, though I don't need one. And shouldn't there be an accepted principle of giving back to society? How do oneness and interrelatedness fit into this? Where do our collective interests intersect?

In a August 1, 2010 *Washington Post* review of the book *Chasing Goldman Sachs; How the Masters of the Universe Melted Wall Street Down ... and Why They'll Take Us to the Brink Again*, by Suzanne McGee, James Ledbetter says it clearly: "The drive to maximize profits to shareholders, to improve the return on equity ... led Wall Street firms into all sorts of behavior that separated their best interests from society's."

Ah, ha! That's it! The profit motive can focus people not on what's good for society, but what's good for them. It's that Us versus Them approach again! Surely, companies providing goods and services for us to make our lives easier and better do support the public's interest. But when the drive for profits overcomes social implications, we can have counterproductive results. Nicholas Kristoff wrote an op-ed piece in the *New York Times* on August 27, 2009, titled, "Health Care Fit for Animals," about health insurers. He told the story of Wendell Potter, who used to work for an insurance company, and who helped devise strategies to block health care legislation. Then, he went to see Michael Moore's movie "Sicko" and later visited a three-day charity health care program at a county fairground where lines of people waited to be seen and treated by doctors "in stalls intended for livestock." Potter changed his mind and quit his job.

> *Mr. Potter says he liked his colleagues and bosses in the insurance industry, and respected them. They are not evil. But he adds that they are removed from the consequences of their decisions, as he was, and are obsessed with sustaining the company's stock price – which means paying fewer medical bills. One way to do that is to deny requests for expensive procedures. A second is "rescission" — seizing upon a technicality to cancel the policy of someone who has been paying premiums and finally gets cancer or some other expensive disease. A Congressional investigation into rescission found that three insurers, including Blue Cross of California, used this technique to cancel more than 20,000 policies over five years, saving the companies $300 million in claims.*
>
> *As the* Los Angeles Times *has reported, insurers encourage this approach through performance evaluations. One Blue Cross employee earned a perfect evaluation score after dropping thousands of policyholders who faced nearly $10 million in medical expenses. Mr. Potter notes that a third tactic is for insurers to raise premiums for a small business astronomically after an employee is found to have an illness that will be very expensive to treat. That forces the business to drop coverage for all its employees or go elsewhere. All this is monstrous, and it negates the entire point of insurance, which is to spread risk*

So, the main purpose of the insurers is to pay peoples' medical costs, and that is exactly what the companies *don't want* to do because it won't make them as much money. Specifically, they won't have as high profits and shareholders will make less money (for doing absolutely nothing to earn it). There is no market principle that says companies have to do any good for humans whatsoever. In the Ledbetter review of Suzanne McGee's book referenced above, he writes about the market: "Some of the system's weakness reflects human nature, but much of it, McGee concludes, is built into the fiduciary duty of public companies."

Well, how has this system, which at its core is about the fiduciary duties of directors to make profits, been at providing equitably for the people and the planet? Let me provide explanations from a few different sources:

- *Look at the latest astonishing estimates from economists Emmanuel Saez of the University of California at Berkely and Thomas Piketty of the Paris School of Economics. They find that the richest 10 percent of the [American] population received 44 percent of the pretax income in 2005. This was the highest since the 1920s and 1930s (average; 44 percent) and much higher than from 1945 to 1980 (average 32 percent). But the biggest gains occurred among the richest 1 percent. Their share of pretax income has gradually climbed from 8 percent in 1980 to 17 percent in 2005. Indeed, many others in the top 10 percent seem mainly upper middle class. For example, those in the richest 90th to 95th percentile had incomes of about $110,000.* Robert Samuelson in "The Rich and the Rest," *Washington Post* April 18, 2007.

- *The entire bottom 80 percent [of Americans] now loses a collective $743 billion each year, thanks to the cumulative effect of slow wage growth. Conversely, the top 1 percent gains $673 billion. That's a pretty close match. Basically, the money gained by the top 1 percent seems to have come almost entirely from the bottom 80 percent. ... In theory, [the market is] supposed to produce rapid economic growth that serves us all, and 30 years of free-market evangelism have convinced nearly everyone – even middle-class voters who keep getting the short end of the economic stick – that the policy preferences of the business community are good for everyone. But in practice, the benefits have gone almost entirely to the very wealthy.* Kevin Drum in "Plutocracy Now," *Mother Jones* magazine, March/April 2011.

- *Anyone wondering where all the economy's jobs are might want to look into piggy banks of the world's biggest companies. Cash is gushing into companies' coffers as they report what's shaping up to be the third-consecutive quarter of sharp earnings increases. But instead of spending on the typical things, such as expanding and hiring people, companies are mostly pocketing the money and stuffing it under their corporate mattresses.* Matt Krantz in July 28, 2010 *USA-Today*, in an article titled, "Companies are Sitting on a Pile of Cash – Enough to pay 2.4 M workers $70,000 Salaries for 5 Years – Yet They're Not Hiring."

- *Lowering the head count is the quickest way to restore profits and, from there, a company's stock price.* Robert Samuelson in *Newsweek* magazine, "The Big Hiring Freeze," August 2, 2010.

- *Americans have been watching protests against oppressive regimes that concentrate massive wealth in the hands of an elite few. Yet in our own democracy, 1 percent of the people take nearly a quarter of the nation's income - an inequality even the wealthy will come to regret. It's no use pretending that what has obviously happened has not in fact happened. The upper 1 percent of Americans are now taking in nearly a quarter of the nation's income every year. In terms of wealth rather than income, the top 1 percent control 40 percent. Their lot in life has improved considerably. Twenty-five years ago, the corresponding figures were 12 percent and 33 percent. One response might be to celebrate the ingenuity and drive that brought good fortune to these people, and to contend that a rising tide lifts all boats. That response would be misguided. While the top 1 percent have seen their incomes rise 18 percent over the past decade, those in the middle have actually seen their incomes fall. For men with only high-school degrees, the decline has been precipitous—12 percent in the last quarter-century alone. All the growth in recent decades—and more—has gone to those at the top.* Joseph Stiglitz in the May 2011 issue of *Vanity Fair* magazine in an article titled, "Of the 1%, By the 1%, For the 1%."

Stiglitz goes on to say this a huge problem, because it results in decreased opportunities for most of the population. "[A] modern economy requires 'collective action' – it needs government to invest in infrastructure, education, and technology." He says the country has benefited significantly from government action in the past, but the United States is now behind in these areas. Is there any real doubt what's feeding the dissatisfaction of the Occupy Wall Street crowd? Steve Pearlstein again in the *Washington Post* on August 6, 2010, in an article titled, "Why Sharing the Wealth Isn't Enough:"

> *As the rungs of the economic ladder grow farther apart, it's not surprisingly becoming harder to move up. Recent work by Isabel Sawhill and Ron Haskins of the Brookings Institution suggests that rising inequality in the U.S. economy is leading to lower mobility. Sawhill and Haskins found that while people born into the middle class continue to move up and down the ladder, the top and the bottom rungs are becoming much "stickier," with those born there most likely to remain there. As a result, by some measures, the United States now has less class mobility than Canada, Germany and France. "The idea that equality of opportunity is a distinctly American strength is a myth," they conclude.*
>
> *We are approaching a tipping point in America. When economic growth led to more jobs and higher incomes for wide swaths of the population, it didn't matter much that some people were smart enough or lucky enough to pull way ahead. But in recent decades, there has been a dramatic erosion in both the ideal and the reality of shared prosperity that threatens to paralyze our political system and undermine economic growth. ... Whether at an individual company or in the country at large, it is the feeling that we are all in it together that creates the basis for a truly vibrant economy and just society. Trickle-down alone won't cut it.*

E.J. Dionne also talked about this in an article in the *Washington Post* on April 17, 2011, entitled, "America's Elites Have a Duty to the Rest of Us." His premise is that an elite wealthy class in a country can be a very positive and stabilizing force. But they have an obligation "to reform and thus preserve the system that allowed them to do so well." It might be called "giving back." Sometimes the view from the top provides a broad perspective that can help determine the right path ahead. Except when the focus narrows. E.J. says:

> *An enlightened ruling class understands that it can get richer and its riches will be more secure if prosperity is broadly shared, if govern-*

> ment is investing in productive projects that lift the whole society and if social mobility allows some circulation of the elites.
> A ruling class closed to new talent doesn't remain a ruling class for long.
>
> But a funny thing happened to the American ruling class: It stopped being concerned with the health of society as a whole and became almost entirely obsessed with money.

That's not oneness; <u>it's fear</u>. The failure to see the inequality that's developed in the country can lead to a breakdown in the social system.

It's not only inequality in the United States that's problematic. There is serious health and financial inequality around the world. While the last two hundred years have seen remarkable gains, there are still significant differences between rich and poor. This is a fun video to watch, which shows the recent history of 200 hundred countries, and the continuing disparities that cause problems around the world. http://www.flixxy.com/200-countries-200-years-4-minutes.htm.

I'm not going to argue that we need to do away with capitalism, but it has some fundamental flaws that can be fixed. I'm not an economist, but here's a major issue: <u>capitalism doesn't place any value on the environment</u>. Peter Barnes wrote a book called *Capitalism 3.0*, in which according to Steven Mufson in the *Washington Post* article, "In Bill's Big Idea, Save the Climate, Share the Wealth," on April 8, 2009, he told, "how the economic system failed to protect common resources, such as air and water, because people didn't have to pay to pollute them." Barnes says, "The trouble is, markets have no appreciation for intrinsic value. They're blind and dumb and stunningly mindless; they do what they're programmed to do with ruthless aplomb." Barnes advocates putting a price on carbon emissions.

An article in the *New York Times* on December 12, 2006, "The Cost of an Overheated Planet," by Steve Lohr, said it this way:

> *Global warming can be seen as a classic "market failure," and many economists, environmental experts and policy makers agree that the single largest cause of that failure is that in most of the world, there is no price placed on spewing carbon dioxide into the atmosphere. Yet it is increasingly clear that there is a considerable cost to carbon dioxide emissions, especially to future generations, as climate specialists warn*

of declines in farm output in poor countries, fiercer hurricanes and coastal floods that could make many people refugees.

In fact, the markets don't care about nature itself. How much value would a healthy Potomac River have? The United Nations issued a report stating, "The world has vastly underestimated the economic value of nature in developing nations," according to an article in the *Washington Post* dated October 20, 2010, titled, "U.N. Report Stresses the Value of Nature to World's Economies." For example, coral reefs support some 500 million people as fisheries and for tourism, and could be valued as high as $172 billion per year. But the reefs are being threatened by climate change, and their disappearance would cause significant human hardship. Andrew Deutz of the Nature Conservancy is quoted as saying, "Our only hope to save the world's biodiversity is to help the world understand that lasting economic growth and security are wholly bound to the health and security of our natural resources."

It's clear the free market system doesn't care about people out of work or being sick, or the harmful effects of pollution, unhealthy food, or social tension caused by inequality. Wouldn't you think we could come up with a system that could help fix some of these problems? Maybe one factor could be attitude.

Paul K. Piff, a Ph.D candidate at the University of California, Berkely, and Michael Kraus published a study in the *Journal of Personality and Social Psychology*, which "found that lower-income people were more generous, charitable, trusting and helpful to others than were those with wealth. They were more attuned to the needs of others and more committed generally to the values of egalitarianism. … Empathy and compassion appeared to be the key ingredients in the greater generosity of those with lower incomes. And these two traits proved to be in increasingly short supply as people moved up the income spectrum." That was Judith Warner, writing in the *New York Times* magazine on August 22, 2010, in an article titled, "Helping Hand? Why Giving the Rich a Break Doesn't Necessarily Do anything for the Poor." She calls this, the "compassion deficit."

Maybe money clouds our vision? Rather than seeing how interdependent things are, we perceive our interests as being different or more important than others. Maybe the fear-laden essence of money acts to diminish love and oneness?

I'm going to fess up. I make charitable donations every year and give clothes to the poor. But I'm not as charitable and giving as others of my friends who probably make less money. I don't think I'm as generous to the poor as many of those less well off who live in the city or over in Southeast Washington. Maybe it's because my parents didn't seem to put an emphasis on it. My father's grandfather was a Senator from Louisiana in the 1890s. Or maybe since I grew up in Chevy Chase, went to rich kid schools like Georgetown Prep, and am a partner in a law firm, I have that *compassion deficit* Warner talks about.

Is that because I'm fearful? Do I perceive myself as separate, more deserving, or better than poor or disabled people? It may be I don't even understand oneness as fully as I'd like. They say one teaches what one needs to learn. So, perhaps I need to learn more about oneness and love. I think I do. I'm working on it.

It's not always easy to see the things I'm afraid of. If I acknowledge that poor, or fat, people are just the same as me, doesn't that make me poor or fat? No, of course not, but there are ways I feel separate and different than other people. I am concerned about money. I always try and see the oneness, but some perceptions aren't easy to overcome. I might use the excuse that I'm really busy, and I have a lot of expenses, so can't afford to give five percent of my income to charity. But I probably could if made a priority of it. I try and be nice, kind, and generous to my friends and those I interact with on a daily basis. Is that enough? All I can say, we have to do the best we can, and simply try to act in love. The more we learn about love, the better I think we can do.

I've been very critical here of the religion of my youth, and its mythical and separatist underpinnings, but I have to give some props to former Pope Benedict. I didn't agree with him on much. But in an encyclical in 2009, he called out the market system, according to an article in the *Washington Post* on July 8 of that year titled, "Pope Criticizes World Economic System, Urges Social Responsibility:"

> *Pope Benedict criticized the international economic system yesterday and called for a new global structure based on social responsibility, concern for the dignity of the worker and a respect for ethics. "Today's international economic scene, marked by grave deviations and failures, requires a profoundly new way of understanding human enterprise," Benedict wrote in his latest encyclical, which is the most authoritative document a pope can issue. "Without doubt, one of the greatest risks for business is that they are almost exclusively answerable to their investors, thereby limited in their social value."*
>
> *In the sweeping 144-page document, Benedict sketches a radically different world economy, in which access to food and water is*

Heaven is Everywhere

a universal right, wealthy nations share with poorer ones and profit is not the ultimate goal of commerce. He advocates the creation of a "world political authority" to manage the economy. He blames "badly managed and largely speculative financial dealings" for causing the meltdown. The primary capital to be safeguarded is people, he says, adding that economic systems need to be guided by charity and truth.

Benedict [also] urged that people show more respect toward the environment. [He advocated] more research into alternative energy, worldwide redistribution of energy resources and pushing more advanced countries to lower their energy consumption, either through technology or through greater "ecological sensitivity" among residents.

Steve Pearlstein also wrote about the encyclical in an article titled, "Rethinking Capitalism; How Very Enterprising," in the *Washington Post* on October 9, 2009, and quoted the Pope: "Profit is useful if it serves as a means toward an end. Once profit becomes the exclusive goal, if it is produced by improper means and without the common good as its ultimate end, it risks destroying wealth and creating poverty."

Lordy, Lordy. That's oneness thinking! Kick ASS, Benedict!

Of course, in the encyclical he also maintains that birth control, abortion, and gay marriage are immoral. I know I said abortion is fear-based, but that doesn't make it immoral or evil; it's just what is. By the way, sorry for the tangent, but what is *immoral*? Not acting or thinking in a morally good way? What does that mean? We've seen there's no right or wrong. So, how could there be moral or immoral? Would that not be judging? Sometimes we judge based on how much money people have. Or how profitable their company is. How can that be?

One person who would otherwise never agree with the Pope is Bill Maher, a great comic, author, and liberal thinker. We love watching his HBO show, "Real Time with Bill Maher." In one of Bill's articles posted on *Huffington Post* on July 23, 2009, he said:

How about this for a New Rule: Not everything in America has to make a profit. It used to be that there were some services and institutions so vital to our nation that they were exempt from market pressures. Some things we just didn't do for money. The United States always defined capitalism, but it didn't used to define us. But now it's becoming all that we are.

I agree, why does everything have to make a profit? Of course, if a company didn't make profits, it would go out of business. But how much is enough? Well, part of the issue is that the fiduciary duties of corporate directors *require* them to do whatever they can to make a profit for the company, and there's no room for social concerns. In fact, directors can be sued individually for not taking steps to maximize profits, i.e., dividends for investors. So, isn't there anything to be done about that? Bill Gates thinks so.

He wrote an article in the August 11, 2008, issue of *Time* magazine titled, "How to Fix Capitalism." Even *Bill Gates* thinks capitalism has a problem. He admits, "Capitalism has improved the lives of billions of people ... ." And I don't disagree with that.

> But it has left out billions more. ... [W]e need a more creative capitalism to stretch the reach of market forces so that more companies can benefit from doing work that makes more people better off. ... As I see it, there are two great forces of human nature: self-interest and caring for others. Capitalism harnesses self-interest in a helpful and sustainable way but only on behalf of those who can pay. Government aid and philanthropy channel our caring for those who can't pay.
>
> And the world will make lasting progress on the big inequities that remain – problems like AIDS, poverty and education – only if governments and nonprofits do their part by giving more aid and more effective aid. But the improvements will happen faster and last longer if we can channel market forces, including innovation that's tailored to the needs of the poorest, to complement what governments and nonprofits do. We need a system that draws in innovators and businessmen in a far better way than we do today.
>
> Many people assume, wrongly, that a company exists simply to make money. While this is an important result of a company's existence, we have to go deeper and find the real reasons for our being. ... People get together and exist as a company so that they are able to accomplish something collectively that they could not accomplish separately – they make a contribution to society.

Hey, Bill Gates is a wizard and rock star. I admire what he's done with all his money; starting the Bill and Melinda Gates Foundation and giving away so much to good causes. But how'd he get so much money in the first place? And I hate to argue, but *companies legally do exist to make profits* and protect against liability. I agree they shouldn't, but they do; currently, anyway.

It's interesting also the way he recognizes the duality of self-interest (i.e., perception of being on your own) versus caring for others (love). He also seems to recognize the importance of community. His plan for "creative capitalism" is essentially that companies always have a "good" goal, too - in addition to making money. Not just profits, but doing well for people. It sounds better already than the greedy market forces we've talked about.

I don't know if you're a fan of Al Gore. I am. I think he's demonstrated a lot of courage in his efforts to bring an understanding of global warming, aka climate change, to the world. His movie, 'An Inconvenient Truth" is brilliant. And his book, *Our Choice*, is a textbook of climate change issues, causes, and solutions. He said this in that book:

> *Forty years ago, Robert F. Kennedy reminded America that measures like the Dow Jones Industrial Average and gross national product fail to consider the integrity of our environment, the health of our families, and the quality of our education. As he put it, the gross national product "measures neither our wit nor our courage, neither our wisdom nor our learning, neither our compassion nor our devotion to our country. It measures everything, in short, except that which makes life worthwhile." ...*
>
> *We need a more long-term and responsible form of capitalism. We must develop sustainable capitalism.*

So, do we think *we* can, or that we can't? Develop new systems that we need, and bring peace and prosperity to the world, I mean. Isn't that the dream of every person alive? To *melt* into love and happiness.

I think so, and I believe we can. Just a few tweaks around the edges, and we're there. Listen to this perspective:

> *The New American Story puts country ahead of party. It says we can realize our dreams and we can envision how each of us can have a better tomorrow.*

"Money, it's a crime..."

And why not?

> *Once we face the truth, the good news is that there are answers to all our current problems.*

That's never happened before.

> *One look at the iconic Apollo image of Earth from space is all it takes to realize that our continuing welfare is a global proposition and each of us is responsible.*

So, it's just a matter of realization?
> *This realization leads to what might be called an "ethic of connectedness."*

Hmm, maybe true morality is about connectedness?

> *The ethic of connectedness is at the core of the New American Story.*

That sounds like us, the way we're supposed to be, doesn't it?

> *In our country's past, we rose to greater challenges: we ended slavery, won World Wars, eradicated polio, put men on the moon. The New American Story is telling us that we can, once again, make great things happen.*

You know who said that? Bill Bradley, Princeton University basketball player, who also happened to be a Senator. What if we could bring this type of thinking into the financial system? Again, I think it's just perspective. A good article about this appeared in the August 29, 2010, *Parade* magazine article by Lee Eisenberg called, "Your Money or Your Life? The Economic Crisis May Have Changed How We Define Value."

> *There are, in fact, signs that a more sober mind-set is emerging. ... [W]e're remembering what our parents taught us; that money isn't everything. In a recent Merrill Lynch survey, more than half of retired*

> respondents said that if they could do it all over again, they'd focus more on "life goals" and less "on the numbers."

I would too. Hey, wait a minute; I'm not dead yet! I can change my mind and my life (just as soon as we pay off the house, my other angel says). The article mentions a financial planner named George Kinder (nice name), who asks his client to respond to three hypothetical situations:

1. Imagine you have more money than you ever dreamed of. What would you buy? Where would you travel? At this point, Kinder says, most people go crazy: Golf memberships! A condo in Vail! A 45-foot cabin cruiser!
2. Now imagine that your doctor diagnoses a rare disease; you will die within 10 years. Suddenly, we're thinking harder about what's really important. Plenty of material goodies still wind up on the list, but it's clear that with a tighter time horizon, we're far more discerning about what really counts.
3. Now the doctor says you have only 24 hours to live. The question isn't, "What would you like to do or buy between now and tomorrow?" It's this: "What did you not get to do in life - who did you not get to be?"

> Every one of us would answer those last questions differently, of course. But Kinder says that most people answer in one of only a few ways. We wish we'd exercised our creativity more. Or given more back to the community, a church, or the planet. Or fixed a broken relationship. The telling thing; None of it costs a huge sum of money. But it does require thinking about the meaning of money – and that we open ourselves up, that we share our hopes, our dreams, our goals with others ... and that, in the end, we focus less on net worth and more on self-worth.

Or maybe more on love and community? If I only had 24 hours to live, I'd want to be with my family and friends, and be able to see outside. Look at some trees. Even if it was raining, as it is the moment I'm typing these words. I look out at the rain and a solitary tulip in front of my window, pink petals open to the sky. When thinking of life this way, as a series of perfect moments, all thoughts of money dissolve. When in the Now and feeling the oneness, there's no need or room for money. The concept of money is incompatible with the Now. Money is only about not having enough or using it to get stuff. The absence of money-consciousness when in the Now, to me, is proof of its fear roots.

Contrast these thoughts of being in the Now with the following by Colbert King in his article, "GOP to Have-Nots: Tough Luck" in the April 9, 2011, *Washington Post*. He describes the Republican plan for financial reform offered by Rep. Paul Ryan.

> *To protect and defend the American way of life, the House Republican majority would take spending back to 2008 levels, which, according to a news release from the Democrats of the House Committee on Education and Workforce, would remove 218,000 low-income children and families from Head Start; close more than 16,000 Head Start and Early Head Start classrooms; put 55,000 Head Start teachers and related staff out of work; and cause 170,000 families trying to find or keep jobs to lose child care. ...*
>
> *And to curb the government's reach, the Republican spending plan essentially removes funding from 2,400 schools that serve nearly a million low-income students. It would cut Pell Grants for all 9.4 million students eligible next year. If that's not enough, the GOP plan would kick at least 180,000 students out of the Pell Grant program altogether. To strike another blow for freedom, Ryan would replace Medicare with premium subsidies that seniors would use to pay off private health insurance companies rather than doctors and hospitals. And if the subsidies fail to cover health-care inflation in the future? Tough noogies. ...*
>
> *Oh, but Americans who are comfortable and well-off need have no fear. Tax cuts for millionaires are preserved in the GOP plan. And that nuisance 35 percent top tax rate on individuals and businesses will be taken down to 25 percent. Defense spending gets a light trim. Agricultural subsidies stay alive.*

When I read this, it makes me feel *empty* inside; sad, nervous, and concerned. Not filled with love. How can we *not* be as generous as possible to our children and young people, the poor, and elderly? Why are some people so afraid of *giving*? Do they not feel they have enough? You can't take money with you; it's ultimately meaningless.

OK, time for a quick NWAL analysis of money and our current financial system. We'll just jump right into the Now, and be as detached and objective as possible:

The market system has been very successful at creating an unbelievable society here on earth. We have remarkable goods and services to make our lives easier. Our homes, transportation, attire, electronic devices, heating and cooling systems, and basically everything we have are just amazing. We've given ourselves the possibility of fabulously comfortable and happy lives. Planes, cell phones, satellites, automobiles; just fantastic, all we've done. Money is the key to an exchange system that incents people to do better and be creative, diligent, and productive. To the extent our financial system has provided a platform enabling fulfillment of our basic needs, it promotes oneness and interdependence.

But it's still based on fear – relying on scarcity, the perceived need for more, and a concern about not having enough to ensure our security. It leads individuals to fanatical heights of excess. We spent a good amount of time earlier in the book talking about how crazed fearful people can act because of money. As I've pointed out, the market has no soul and places no value on intangibles. It cares not for the poor or sick. It's stunningly unfair, bestowing many with ungodly amounts of money, and provides others with pennies a day. It has no consciousness. It is based on self-interest and a perception of being separate and different, rather then nurturing a focus on collective conduct to lift all boats, so to speak.

Now, not everything is capable of a 100% fear or love rating. Usually, there's a mixture of fear and love. I believe the market system is primarily fear-based. While humans have advanced and enhanced their life styles, and are better off now in a physical sense than ever before, we're still leaving many people out of the game and degrading our planet at the same time. What I think is needed to counteract the fear inherent in the financial system is a dose of perspective. We need intangible goals encouraging people and businesses to do "good" for our collective populations and our planet.

I understand the free market is not the predominate system in all nations' economies. Of course, there are socialistic and autocratic regimes where the influence of the market is less strong. However, the world economy is certainly a free for all. Global economic conduct – often based on the need for security, which itself is rooted in fear – is clearly a free market system. So, how to remedy the situation is not an easy path.

One step is to accept the general principles of socialism – which are, as a practical matter, in effect in all countries. The fundamental principle of socialism is *From each according to his abilities, to each according to his needs.* That's the way Karl Marx said it. This always made sense to me. If I'm a smart and effective farmer, I can do good work and be rewarded; I can have a nice life. But if I'm more productive than someone else, who's not so smart, motivated in a different way, or producing things that have more intangible impact on life, then *why would I not be happy to contribute more?*

I don't think the basic socialist philosophy should be read to require people to give up everything they have and live in a hut. But those who have more can contribute more, to help sustain a congenial society. I pay a lot of taxes, and have even paid extra to make the point that there's nothing wrong with some contributing more than others. Shouldn't those who contribute more be proud of their extra input for the good of the social order, as opposed to feeling "I'M A TAXPAYER! It's MY money, and I shouldn't have to give any of it to the common good, because *I earned it.*" That's just so selfish, and reflects a perception of separation, for sure.

To think about it, isn't the most common socialistic unit, the *family*? One or both of the parents work to provide food, shelter, clothing, etc., and the children get what they need (at least theoretically). Why is that not considered awful by Grover Norquist? "Those children should be working to bring more into the family; why do I as a father/mother have to make all the money?" Of course, that's a ridiculous argument, but I think it's equally sane for people to claim we all simply need to do our part, and be happy to help support others who are less fortunate. That doesn't mean everyone gets exactly the same things; it means that contributing one's efforts to society, even if deemed inequitable in some ways, shouldn't make people so angry.

The more some people start acting this way – with a generous spirit of patriotic contributions to society and less anger about "what's mine" – the more other people will start acting this way. The crowd mentality is powerful. Everyone is doing it, is perhaps a better motivator than money.

An article in *Time* magazine on April 13, 2009, titled, "How Obama is Using The Science of Change," said about behavioral psychology: "[S]tudies by psychologist Robert Cialdini and other group members found that the most powerful motivator for hotel guests to reuse towels, national park visitors to stay on marked trails and citizens to vote is the suggestion that everyone is doing it. 'People want to do what they think others will do,' says Cialdini, author of the best seller *Influence*. ... 'What worked was creating a sense that we're all in this together and you're a social deviant if you don't join us,' recalls Ralph Cavanaugh of the Natural Resources Defense Council" about another study.

So, I think everyone who believes our semi-socialistic state is cool, should say that. My sister-in-law, when asked about the possibility of European-style social norms coming to the United States, said, "Bring. It. On." For sure, even increasing the amount of vacation that Americans get would have a positive impact on our feeling of community, and reduce the fear we have for one another.

There's another step we can take to transform the economic system for the better. I've mentioned fiduciary duties, and how directors of companies are legally required to make profits for the investors. What if we made it legally required that *directors consider the health and welfare of the people and the environment*?

That's right. Companies would be required to consider what's good for *all of us* when making decisions. Not just profits. Not screwing the other guy. Do what's good for us *all*.

Could they do it? Could modern *rich-as-shit* humans actually decide to do something that wasn't so influenced by money? It might be hard to do.

So, we'll require *by law* all directors of corporations to consider oneness when they make decisions about what the company should do. We'll tell them that company profits are good, but that company profits AND doing good for the planet are better and complementary. We'd all win in the end. Our children would also win.

There's precedent for this. My home state of Maryland was the first to pass a law in 2010 allowing the establishment of 'benefit corporations" in which profits are not the only goal. A benefit corporation is required to operate with the purpose of creating, "a material positive impact on society and the environment, as measured by a third-party standard, through activities that promote a combination of specific public benefits." Directors of a benefit corporation can base their decisions not just on what's good for the company, but on goals such as preserving the environment, improving human health, advancing the arts, or providing benefit to society or the environment.

Sounds great, right? Actually, to me, it sounds the way corporations *should* operate. An article in the *Washington Post* by Danielle Douglas, titled, "'Benefit Corporations' Sign Up," on January 24, 2011 said:

> At its core, benefit corporations blend the altruism of non-profits with

> the business sensibilities of for-profit companies. These hybrid entities pay taxes and can have shareholders, without the risk of being sued for not maximizing profits. Companies can consider the needs of customers, workers, the community or environment and be well within their legal right.

That's what I'm talking about! In fact, my hope is that one day soon, *all* businesses will be benefit corporations. I think it'd be a different world, if that happened. I don't consider it unrealistic to believe there could be significant changes in our global market system. An article on the *Huffington Post* published April 8, 2011, titled, "U.S. Support for 'Free Market' Capitalism Drops Below China, Brazil, Poll Finds," reported on a survey by the company GlobalScan concluding that popular support for free market capitalism has dropped significantly in the last decade. You'd expect that those in the United States would be strongly supportive of the free market as the best system, but only 59 percent think so, which is less than 67 percent in China and 68 percent in Brazil. And about a quarter of Americans don't trust the market *at all*.

This is not bad. This is good. It's important to think outside the box, and to have a healthy skepticism of the way things *have been*. Another *Huffington Post* article, "Poll: Americans See Clash Between Christianity, Capitalism," by Nicole Neroulias, on April 20, 2011, said:

> A new poll released Thursday ... [by Public Religion Research Institute] found that more Americans (44 percent) see the free market system at odds with Christian values than those who don't (36 percent), whether they are white evangelicals, mainline Protestants, Catholics or minority Christians. But in other demographic breakdowns, several categories lean the other way: Republicans and Tea Party members, college graduates and members of high-income households view the systems as more compatible than not. ... In other findings:
>
> - Half of women believe that capitalism and Christian values are at odds, compared to 37 percent of men.
> - A majority (53 percent) of Democrats believe capitalism and Christian values are at odds, compared to 37 percent of Republicans and 41 percent of independents. A majority (56 percent)

> *of Tea Party members say capitalism is consistent with Christian values.*
> - *Nearly half (46 percent) of Americans with household incomes of $100,000 a year or more believe that capitalism is consistent with Christian values, compared to just 23 percent of those with household incomes of $30,000 a year or less.*
> - *Most Americans (61 percent) disagree that businesses would act ethically on their own without regulation from the government. White evangelicals (44 percent) are more likely than Catholics (36 percent), white mainline (33 percent) or minority Christians (34 percent) to say unregulated businesses would act ethically.*

This says to me that religious folks tend to believe in the capitalistic system, more so than others. It's just another instance of fear-based thought finding consistency with other fear-based principles.

The worldwide economic system would be much less fear-based if some of the improvements we talked about could be made.

But we've got other windmills to tilt at.

Eighteen | NWAL Becomes Intuitive

It was about 7th grade. Somehow, I heard there was going be a rumble. Well, maybe not like *West Side Story*, but a fight up at school. I wasn't sure who the other guys were. Had no idea what it was about. But we went up to the Blessed Sacrament gym.

Sure enough, there were some black kids up there. We were all just hanging around. One of them came over to me. He said, "Hey, man, wanna fight?" I said, "Sure." So, I put up my fists and he punched me in the mouth.

I never expected that. I never thought he'd hit me. I rubbed my jaw and said, "OK, that's enough for me." I could shoot at imaginary Krauts and Japs in the alley behind the school, but I didn't really want to fight a real person. Never have.

My tooth was tender for years. I was lucky. Here's what happened to Bobby Tillman in Douglasville, Georgia, according to *The Press Democrat* newspaper on November 9, 2010 in an article titled, "Partygoers Beat Random Stranger to Death," by Greg Bluestein:

> *It began with a brawl outside a house party. A woman hit a man, and the man refused to strike back, saying he wouldn't hit a girl. Instead, he vowed to attack the next male who walked by, even if that person was a random stranger.*
>
> *That's when 18-year-old Bobby Tillman happened to approach a group of four partygoers, who swiftly stomped, kicked and punched him to death while dozens of bystanders watched.*

Right around the Beltway in Prince George's County, Maryland, three teenagers killed a 56-year-old man who had denied their request for a cigarette. "3 Teens Admit Beating Man to Death," by Ruben Castenada, *Washington Post*, January 6, 2009. The teenagers were part of a gang called the "Skull Crushers."

You're with me if I say this type of conduct isn't love-based, right? Not much question about that. It's fear-based, because they think it shows they're better, more powerful, or cooler than the victim. They think it's actually OK to hit, kick, punch, or smash someone, because that's *some other guy*. Us versus Them.

But Eckhart Tolle says, "It's always the case that both victim and perpetrator suffer the consequences of any acts of violence, oppression, or brutality. For what you do to others, you do to yourself."

That sounds like something Jesus would say, doesn't it? But it's not about Jesus. It's about the fact that there's no separation at all between us, and when we perceive we're separate, it can lead to lots of awful things. When we do something to another person, we really do it to ourselves, because we're the same. When people think they have the right to do just as they want, no matter the impact on others, that's not oneness. I don't think we even need to NWAL this type of conduct, do we? It goes without saying that acting this way is not conducive to love.

Here's another example of similar behavior, from an article titled, "'Off-Road Rage' Climbs as Trails Get More Crowded" in the *Washington Post* on August 12, 2008, by Karl Vick:

> *As more and more Americans light out for backcountry trails, officials are seeing a parallel rise in episodes of "off-road rage": unpleasant, even violent encounters between drivers of all-terrain vehicles and hikers, mountain bikers and others. ... "I hate these things. They're loud. They're obnoxious," said Bill Burgund, 61, an amputee with one leg who was walking on a Bitterroot National Forest trail in Montana last year when an ATV careened around a corner, snagging his crutch, wrenching his shoulder and knocking him to the ground. ...*
>
> *Critics point out that ATV riders account for 10 percent of visitors to public land, at most. Yet their impact -- whining engine noise, dust clouds visible for miles and nuisance driving, especially by young operators -- profoundly affects the other 90 percent. ... "It's totally about culture," said Bethanie Walder of Wildlands CPR, which opposes off-roading. "I think that's where the problem derives. They prefer to*

> ride off-trail. They want to blaze their own trail. The culture's one of 'I can do whatever I want.' ... In Idaho, a Payette taxidermist was found guilty in 2005 of punching and threatening the life of a hunter who confronted him for driving an ATV on a national forest trail closed to motors.
>
> "Put a bullet in her head," a man called out in Darby, Mont., in January, as a woman spoke against ATVs at a profanity-laced public meeting the Forest Service had convened.

I wanna do whatever I want, and if someone tries to stop me, I'm gonna shoot the bastard! Not so loving, is it? The ATV's are loud and obnoxious. Most of the people who go to our fabulous national parks want to enjoy the remarkable scenery and the natural world in its pristine state. But some folks think they don't have to worry about that; they can make as much noise as they like.

I was riding my bicycle along MacArthur Boulevard near our house a couple years ago, and a motorcycle came by, putting out about 100 decibels, way in excess of what the local noise ordinance allows. So loud, I literally cringed. As we pulled up to a stoplight, I turned to him and mouthed, "You're too loud." He shouted, "Fuck you," and roared off.

If a person thinks they can do whatever they want, be as loud as they want, drive as fast as they can, or pollute and degrade the earth as much as want, they're *not* acting in harmony with others on our planet. They perceive they're different, separate, entitled, and deserving, but they're not thinking we're all in this together.

It now seems easy to see the difference between love and fear induced behavior. Don't you think it's easy? We've been focusing on these concepts for a lotta pages. Sure, we can go through the full NWAL analysis, but it's almost intuitive at this point.

These off-roaders are like the commuters who race down Canal Road, also called Clara Barton Parkway, near our home. It's a crazy situation. The Parkway is a National Park Service road, which is intended (seriously) to provide a nice and leisurely drive through the park. It's a two lane road from Georgetown almost out to the Beltway, except for a half-mile stretch right near our neighborhood. There, it opens up into two lanes in each direction. The speed limit is 35 mph, so the cars routinely go 45 mph or more. But many race to pass two or three cars before the road goes back to one lane. Invariably, heading south, they end up stopping just a mile later at the light at Chain Bridge.

The nutty thing is, right in the middle of the four-lane stretch is a *pedestrian crosswalk*. Here is where people cross to get to the C&O Canal, where they walk with their kids or dogs, go running, or fishing. There's a special feeder canal that George Washington's company actually built two hundred years ago, where kayakers train for the Olympics because the water there is deep, fast moving, and doesn't freeze up in the winter. That's the place we spread Eileen's ashes (if you remember I told you about that).

So, all these nature-loving folks cross the Parkway, at the exact spot where the automobiles are accelerating to sometimes 60+ mph to pass – *right at the crosswalk*! They don't usually stop for pedestrians in the crosswalk, and sometimes yell at the people crossing. Kids carry kayaks on their heads across the Parkway, and have to *literally* dodge speeding cars. Yet, the cars continue to race through the crosswalk. It's insane and dangerous. About ten years ago, then Congressman Richard Armey quashed a Park Police effort to install speed cameras there, because he considered it an invasion of privacy. Privacy to do what; speed and kill old people, kids, and dogs?

This is insane behavior. It annoys me to no end, I have to admit. I made a video of the insanity down there on the Parkway. You can watch it here: http://www.youtube.com/watch?v=f9EYVSuYKX8

Why should people slow down and be careful when they drive? Well, for one, "Traffic injuries are the leading cause of death in people ages 10 to 24 around the world – a huge overlooked and largely preventable public health problem, the World Health Organization said," as reported in the *Washington Post* on April 20, 2007, in an article titled, "Traffic Deaths a Global Scourge, Health Agency Says," by David Brown. Traffic fatalities for that age group kill more people that AIDS, violence, and war. There are about 400,000 traffic deaths each year worldwide. "In Mozambique, about 65 percent of road injuries and deaths involve pedestrians. In Cambodia, 75 percent were motorcyclists." These accidents can kill breadwinners and destroy families. Over 30,000 people were killed in 2009 in traffic incidents in the United States. http://www-fars.nhtsa.dot.gov/Main/index.aspx

These might seem like just statistics, but each one is a human life, like you and me. And for each of those people, getting killed was a *significant personal event*. I think that's enough reason to be careful when driving close to kids with kayaks on their heads. A perspective of care and concern for one another is reflective of love and oneness.

If there were an instrument that killed a whole bunch of kids, one would think we as a group of adults in the wealthiest nation in the universe would take collective action to do whatever was necessary to halt the carnage, right? *No.*

> *In 2004, more preschoolers than law enforcement officers were killed by firearms, according to the Children's Defense Fund. The number of children killed by guns in the United States each year is about three times greater than the number of servicemen and women killed annually in Iraq and Afghanistan. In fact, more children – children – have been killed by guns in the past 25 years than the total number of American fatalities in all wars of the past five decades. Washington Post,* April 22, 2007, in an article by Jonathan Safran Foer, titled, "Some People Love Guns. Why Should the Rest of Us Be Targets?"

All of those were tragic, but how about this: a 4-year-old boy in Indianapolis grabbed a .45 caliber handgun lying on the kitchen table in his home and shot a 3-year-old to death. Actually, *every day* in the United States, 8 children and teenagers die from gun violence. Each year, according to an article in *Time* magazine on January 24, 2011, about 3,000 teens and children are killed by gun violence. The *Time* article was titled: "15 Seconds to Fire the Glock, 31 Bullets in One Clip, 19 Victims with Six Killed, 1 Madman and a Gun." Every year, some 17,000 people kill themselves with guns, and a total of over 100,000 people in American are shot in murders, assaults, suicides, accidents, or police interventions.

We discussed before how having guns is based on fear; if one wasn't afraid, there'd be no need for a gun to use in self-defense. But someone is convincing people they *need* guns to be safe. Of course, that's a canard. Do you know, however, what percent of gun deaths involve self-defense? How about one percent? Yep, that'd be right, according to the *Time* article.

You'd think that'd rally us to some reasonable course of action to protect ourselves from guns, right? *Wrong.*

After the shooting of Rep. Gabriel Giffords in January 2011, over 2,400 Americans were shot and killed in the next two months! About 400,000 people have been shot and *killed* by guns since 1968 when Martin Luther King, Jr. and Robert Kennedy were shot. And legal-handgun carriers stopped *not one* of the 18 mass shootings since May 2007, according to an article in *Newsweek* dated March 21, 2011 and titled, "2405 Shot Dead Since Tucson," by Andrew Romano and Pat Wingert.

And know what? As of May 31, 2013, the number of gun deaths in the United States since the December 14, 2012 mass shooting of children in Newtown, Connecticut exceeded the number of deaths in the Iraq war! Yep, according to the National Journal, 4,499 Americans were killed by guns during that period compared to 4,409 U.S. armed forces during the Iraq war. "Gun Deaths Since Newtown Now Surpass Number of Americans Killed in Iraq," by Matt Vasilogambros. Truthfully, that's just sick. Fareed Zakaria wrote in the *Washington Post* on December 20, 2012, in an article titled, "A clear solution," that:

> *What we should be trying to understand is not one single event but why we have so many of them. The number of deaths by firearms in the United States was 32,000 last year. Around 11,000 were gun homicides.*
>
> *To understand how staggeringly high this number is, compare it to the rate in other rich countries. England and Wales have about 50 gun homicides a year – 3 percent of our rate per 100,000 people. Many people believe that America is simply a more violent, individualistic society. But again, the data clarify. For most crimes – theft, burglary, robbery, assault – the United States is within the range of other advanced countries. The category in which the U.S. rate is magnitudes higher is gun homicides.*
>
> *The U.S. gun homicide rate is <u>30 times</u> that of France or Australia, according to the U.N. Office on Drugs and Crime, and 12 times higher than the average for other developed countries. …*
>
> *The data in social sciences are rarely this clear. They strongly suggest that we have so much more gun violence than other countries because we have far more permissive laws than others regarding the sale and possession of guns. With 5 percent of the world's population, the United States has 50 percent of the guns.*

And who do you think has all those guns? Maybe the afeared people? Harold Meyerson wrote in the *Washington Post* on that same day in an article titled, "The party of the packing," that, "There's a name for those gun buyers: Republicans. As the FiveThirtyEight blog noted Tuesday, the 2010 General Social Survey showed that 50 percent of adult Republicans owned guns, while only 22 percent of adult Democrats did."

And why is that? Because they're hepped up on fear. According to Michael

McNulty in the *Washington Post* on July 29, 2012, "The NRA's real interest," "If you search YouTube for 'The NRA's Circus of Fear,' you'll find a collection of [NRA President Wayne] LaPierre's reasons why he lives in fear, and his arguments as to why you should, too. Personally, I think that 'living in fear' is inconsistent with being an American, and I'm not going to play." Here's the video: http://www.youtube.com/watch?v=Hd73D44L6cM.

I agree with that. I don't buy into the fear, but maybe that's because I don't see evil-doers behind every tree. In an article titled "Colorado House passes gun-control measures," by Ivan Moreno, in the *Washington Post* on February 19, 2013, there's this quote from a Colorado legislator, "This bill will never keep evil people from doing evil things," said Rep. Jerry Sonnenberg (R). And what do we recognize that as now? The perception of separation that constitutes fear.

It's gone so totally off the chain that a new series of laws is being pushed that celebrates, extols, and rewards fear – the so-called stand your ground laws. These laws came to my attention after the Trayvon Martin killing in Florida in February 2012. George Zimmermann was able to claim he was afraid for his life, and if that's the case, Florida law lets him off even though he's the shooter. Several states have also passed similar laws, and they really do seek to protect fearful people owning guns who shoot other people if they're afraid. I've had Facebook discussions with friends and relatives who said having a gun makes you strong, completely missing the point that *no one needs a gun unless they're afraid*. And being afraid is not being strong.

And these new stand your ground laws – brought to us by the people who think 47 percent of Americans are out to get them somehow – have led to an increase in "justifiable homicides," according to an article titled, "Justifiable killings up as self-defense is redefined," in the *Washington Post* on April 8, 2012. And a *Washington Post* Editorial on July 21, 2013, titled "Give Ground" said, "Stand-your-ground states saw more homicides than their peers – about 600 more over the period they studied. One possible explanation is that stand-your-ground laws encourage people to escalate conflicts rather than withdraw." Do ya think? President Obama was quoted in that editorial as saying, "If we're sending a message as a society in our communities that someone who is armed potentially has the right to use those firearms even if there's a way for them to exit from a situation, is that really going to be contributing to the kind of peace and order we'd like to see?" Ah, Mr. President, no.

So, why don't we have some laws to stop this carnage? Well, "the National Rifle Association has done a brilliant job persuading some gun owners and many politicians that even modest restrictions represent ominous steps toward tyranny," according to the January 24, 2011, *Time* magazine article. There's that pesky fear of tyranny again. Clearly, the use of guns in domestic situations is based on fear, and efforts to stop reasonable gun control are based on fear. We could do a NWAL on guns, but it's just so obvious. Can we please stop being so afraid, fellas?

These are all domestic situations. Huge amounts of resources are poured into much bigger weapons around the world. If one has a gun for personal protection because of being afraid, doesn't that apply to countries, as well? In 2010, according to the Stockholm International Peace Research Institute, the United States spent 687 billion dollars on defense! That was almost *seven times* more than China, which only spent 114 billion. France was next with 61 billion, then the United Kingdom with 57 billion, and Russia with only 52 billion.

Doesn't this make us the *scarediest* humans on the planet; that we need to protect ourselves so much? I understand we've had this sort of policeman role around the world, where we act like the Lone Ranger and fight for peace, justice, and the American way. I don't deny that someone has to have a big enough stick to keep the peace. But seriously, this seems crazy. The U.S. also spent over a trillion dollars on the Iraq and Afghanistan wars, and the costs of caring for all the soldiers who were injured for the rest of their lives is staggering.

According to a piece titled, "Multiple Wars Weigh on Continent's Economics," in the *Washington Post* on October 11, 2007, "Africa's nearly two dozen wars in recent decades have robbed the continent of about $18 billion a year that could have gone to helping one of the worlds poorest regions build stronger economies, ..." Everyone is *afraid* of everyone else and what they might do; the Hutus and Tutsis, Arabs and Israelis, etc., etc., etc.

So, the actual cost of fear to humans is *ginormous*. What if we spent the amount of defense spending around the world on education, health care, and green energy? Wouldn't that make a ton more sense?

I don't remember ever being around a gun that wasn't for hunting, meaning when I was with my Dad or some friends who hunted. No, wait; I take that back. I went to a party one time where I was told one of the guys was packing heat. It was Superfly's party.

That was in the summer of 1974. I had a job working construction, making a union wage of $7.23 per hour, which was serious good money for a college student. A friend of mine named Tyrone, black dude about 20-something, was nicknamed Superfly. I never knew why; maybe I was dumb. Anyway, he was always real nice to me, even though I was the proverbial rich white kid. He had invited me to a party at his house, musta been a Friday night. I'd heard about a happy hour at the Capitol Hill Club, which is a Republican social club near the capitol. I went to that event first, and boy the drinks were flowing. That was back in the days of three martini lunches that President Carter criticized when he was in office. I met this woman at the happy hour; she was a few years older than me, but I got her phone number, wrote it down, and put the paper in my wallet. Then, I drove off to find Superfly's party.

Problem was, I'd been drinking and needed to take a leak. I got out of my car on some residential street downtown, and found a tree. Unfortunately, a police officer saw me and took me to the police station nearby. I sat there for a while, and finally they released me without charge, but wouldn't take me back to my car. So, I took off walking through this unknown neighborhood. Two black guys came up to me, and asked what I was doing. I said I was trying to find my car, but then I was going to Superfly's party, if they wanted to go. They said sure, we found my car, and drove out of the city to Superfly's house.

I was the only white guy there, but was dancing with all the women and having a good old time. I think we smoked some weed, too. When I was leaving, Superfly took me aside, told me his friend had a gun, and said he'd escort me to my car. He said not to take the two other black dudes with me. I didn't realize why, but said OK. When we got outside, the other two guys showed up and said, "You're not gonna leave us out here, are you?" I felt guilty, so said I'd give them a ride home. Superfly's friend with the gun in his pocket reluctantly allowed them to get in the car with me.

I followed their directions, but they were leading me all over the place. Finally I said, "Listen, it's like three in the morning. I gotta drop you off."

The dude in the front seat stuck a finger in his coat pocket toward me, and said, "Give us your wallet." I'm shocked, tired, and saying, "But I took you guys to a party." They grabbed my wallet and jumped out of the car. Damn.

The wallet didn't have too much money, but had a few things I cherished, like the scores from my high school football team (we were city champions), my Turtle Club card, and that girl's phone number. Somehow, I was able to eventually track her down and explain why I hadn't called her sooner. I remember inviting her to go cross-country skiing at a place in the mountain with some friends, but we never hooked up. Memorable things about that trip were: I wrote her a song I called, "Arrested for Taking a Pee by the DCPD," my brother who went with us didn't have skis so had to run along behind us, and we shot a coconut with a rifle to make a bong.

And the point of the story was … ? Well, drinking was probably a lot more dangerous for me than smoking pot, but pot is illegal and booze is not. And they robbed me! First, let's talk about marijuana.

In my mind, alcohol is much worse than weed. I do like to drink, I have to say. Maybe too much. But it's legal. Pot is such a chill out. It's fun and engenders great appreciation for – *everything*. I know some people get all paranoid, and I think that may be inner fears on display. So, why is pot illegal? Why not regulate it and tax the sales to help with government deficits?

Remember that old movie, "Reefer Madness," from back in the 1930's? What a great college flick. It showed how cannabis could make you kill people and then go insane. It was initially financed by a church group as a morality play. Whereas alcohol is accepted (my Dad made it almost a religion), people think marijuana is the first step to hell. We go through this big charade with "medical" marijuana being permitted under some state laws. Is it bad for you or not??

Still, people are afraid society will go down the toilet if we let everyone get zoned out on blow. Now, you noticed that I said people are *afraid* of what it will do. Fear-based, once again.

I did cocaine a few times in my twenties, but not anything much stronger. Marijuana itself is not addictive. Guns are much worse for society. Apart from the Second Amendment to the Constitution, which doesn't at all prohibit appropriate regulation and licensing of guns, there's no reason to allow every scared citizen and his brother to have a gun. But potheads don't kill hundreds of thousands of people. Doesn't make any sense. I believe arguments against legalizing drugs are fear-based. The *Economist* magazine published a position piece on this subject on March 7, 2009, titled, "How to Stop the Drug Wars:"

> *The United States alone spends some $40 billion each year on trying to eliminate the supply of drugs. It arrests 1.5m of its citizens each year for drug offences, locking up half a million of them; tougher drug laws are the main reason why one in five black American men spend some time behind bars. In the developing world, blood is being shed at an astonishing rate. In Mexico more than 800 policemen and soldiers have been killed since December 2006 (and the annual overall death toll is running at over 6000). …*
>
> *According to the UN's perhaps inflated estimate, the illegal drug*

> industry is worth some $320 billion a year. In the West it makes criminals out of otherwise law-abiding citizens. ... Legalization would not only drive away the gangsters; it would transform drugs from a law-and-order problem into a public-health problem, which is how they ought to be treated. Governments would tax and regulate the drug trade, and use the funds raised (and the billions saved on law-enforcement) to educate the public about the risks of drug-taking and to treat addiction. ...
>
> Plenty of American parents might accept that legalization would be the right answer for the people of Latin America, Asia and Africa; they might even see its usefulness in the fight against terrorism. But their immediate fear would be for their own children. That fear is based in large part on the presumption that more people would take drugs under a legal regime. That presumption may be wrong. There is no correlation between the harshness of drug laws and the incidence of drug taking ... .

Just like my Dad said about booze; prohibition made criminals out of everyone. "From 1965 through the election of Barack Obama, our government arrested 20 million people for possession of marijuana. ... In 2006 alone, the last year for which statistics are available – 829,625 people were arrested on marijuana-related charges, according to the FBI ... ." That was Dan Sweeney writing in the *Huffington Post* on March 25, 2009 in an article titled, "Taking the Pro-Pot Position (Because Somebody Has to)." After Olympic swimmer Michael Phelps was caught on camera taking a bong hit, Kathleen Parker wrote an article, "One Toke Over the Line," in the *Washington Post* on February 4, 2009, and said, "Our marijuana laws have been ludicrous for as long as most of us have been alive."

Another way fear rears its ugly head in the "war on drugs" is the violence engendered by enforcement efforts. They exacerbate fear and pit the police versus criminals in an insane fight in which no one wins, except the cartels get rich, don't pay taxes, and everyone has plenty of drugs anyway. Johann Hari wrote in the *Huffington Post* on February 10, 2009, titled, "Obama Must end the War on Drugs – Or Mexico and Afghanistan Will Collapse:"

> *In the past decade, the U.S. has spent a fortune spraying carcinogenic*

> chemicals over Colombia's coca-growing areas, so the drug trade has simply shifted to Mexico. It's known as the "balloon effect": press down in one place, and the air rushes to another. When I was last there in 2006, I saw the drug violence taking off and warned that the murder rate was going to skyrocket- - but I didn't imagine it would reach this scale. In 2007, more than 2,000 people were killed. In 2008, it was more than 5,400 people. The victims range from a pregnant woman washing her car to a four year-old child to a family in the "wrong" house watching television. Today, 70 percent of Mexicans say they are frightened to go out because of the cartels. ...
>
> It's true that where drugs are decriminalized, like the Netherlands, levels of addiction are much lower than in the U.S. It's true that when several U.S. states decriminalized marijuana in the seventies, there was no increase in use. ...
>
> Legalization would also be the single biggest blow for civil rights in the U.S. since Lyndon Johnson. Today, 13 percent of American drug users are black, yet they make up 74 percent of the drug offenders in prison. A whole generation of black men has been destroyed by prohibition: Barack Obama could easily have become one of them if the police had walked into the wrong party at the wrong time.

So, our efforts to stop people from getting high have led to the crumbling of societies. That's not oneness; that's paranoia gone *wild*. Misha Glenny wrote in the *Washington Post* on August 19, 2007, in an article titled, "The Lost War: We've Spent 36 Years and [hundreds of] Billions of Dollars Fighting It, but the Drug Trade Keeps Growing:"

> The trade in illegal narcotics begets violence, poverty and tragedy. And wherever I went around the world, gangsters, cops, victims, academics and politicians delivered the same message: The war on drugs is the underlying cause of the misery. Everywhere, that is, except Washington, where a powerful bipartisan consensus has turned the issue into a political third rail. ... Now the drug war is undermining Western security throughout the world.

Can you believe how counter-productive this war on drugs is? Funny thing, some drugs could have a fabulously positive effect on people. An Associated Press article written by Malcom Ritter appeared in *The Globe and Mail* from Canada on July 11, 2006, titled, "Hallucinogens Lead to Mystical Experiences, Study Finds – Subjects Speak to Spiritual Awakening After Ingesting Mushroom-based Drug."

> *People who took an illegal drug made from mushrooms reported profound mystical experiences that led to behavioral changes lasting for weeks – all part of a U.S. experiment that recalls the psychedelic Sixties. Many of the 36 volunteers rated their reaction to a single dose of the drug, called psilocybin, as one of the most meaningful or spiritually significant experiences of their lives. Some compared it to the birth of a child or the death of a parent.*
>
> *Psilocybin has been used for centuries in religious practices, and its ability to produce a mystical experience is no surprise. But the new work demonstrates it more clearly than before, Dr. Roland Griffiths said. Even two months after taking the drug, most of the volunteers said the experience had changed them in beneficial ways, such as making them more compassionate, loving, optimistic and patient. Family members and friends said they noticed a difference, too.*

Instead of welcoming the potential benefits of such recreational drugs – I swear marijuana has made me a more insightful and understanding person – we throw users in jail; a lot of users. *The Washington Post* reported on December 1, 2006, in an article titled, "New High in U.S. Prison Numbers," by N.C Aizenman, that the United States set a prison population record with 7 million people behind bars, on probation, or on parole. About 7 percent of them are women. "Misguided policies that create harsher sentences for nonviolent drug offences are disproportionately responsible for the increasing rates of women in prisons and jails," said Marc Maurer, executive director of the Sentencing Project, a Washington-based group that supports criminal justice reform. Those with drug convictions count for almost 50 percent of inmates. Plus, there are significant racial disparities, with blacks and Hispanics more likely to be in prison.

You'll be happy to know that we set another new record for prison population in 2007 and at the time led the world in *both the numbers and percentages* of citizens in jail, according to a *Washington Post* article, "Growth Attributed to More Stringent Sentencing Laws," by N.C Aizenman, on February 29, 2008.

Again, Fareed to the rescue, from an article titled, "Incarceration Nation," in *Time* magazine on April 2, 2012:

> The U.S. has 760 prisoners per 100,000 citizens. That's not just many more than in most other developed countries, but seven to 10 times as many. Japan has 63 per 100,000, Germany has 90, France has 96, South Korea has 97, and Britain - with a rate among the highest - has 153. Even developing countries that are well know for their crime problems have a third of U.S. numbers. Mexico has 208 prisoners per 100,000 citizens, and Brazil has 242.
>
> [The reason] is the war on drugs. Drug convictions went from 15 inmates per 100,000 adults in 1980 to 148 in 1996, an almost tenfold increase. More than half of America's federal inmates today are in prison on drug convictions.

I see this as another "Us versus Them" situation. The bad guys go to jail to keep us safe. It's just that a lot of the "bad" guys are only doing drugs, which isn't really any worse than alcohol. We should be rehabilitating them, not jailing them. People make mistakes (actually, I don't believe in mistakes, but that's another discussion). Really, we should be forgiving them.

Archbishop Desmond Tutu led South Africa's Truth and Reconciliation Commission in the mid-1990's. I read some quotes by him in a book by the Dalai Lama and Victor Chan called *The Wisdom of Forgiveness*. Tutu said:

> Forgiveness is not cheap. And reconciliation is not easy. But with forgiveness, we open the door for someone. Someone who might have been shackled to the past, to break loose from the shackles, walks back through the door and into a new future. ...
>
> In our country we speak of something called ubuuntu. When I want to praise you, the highest praise I can give you is to say, you have ubuuntu - this person has what it takes to be a human being. This is a person who recognizes that he exists only because others exist; a person is a person through other persons. When we say you have ubuuntu, we mean you are gentle, you are compassionate, you are hospitable, you want to share, and you care about the welfare of

others. This is because my humanity is caught up in your humanity. So when I dehumanize others, whether I like it or not, inexorably, I dehumanize myself. For we can only be human, we can only be free, together. To forgive is actually the best form of self-interest.

Yea, Desmond, that's right. When we throw so many people into prison, we degrade our society. We're basically "doing unto others." Sure, there are hardened criminals; we need to restrict some folks. But I believe that anyone can be rehabilitated; anyone can learn about love. Maybe we should give the prisoners mushrooms.

Another way the war on drugs injures us more than we realize is the high cost of incarceration. According to that February 29, 2008 article in the Post, some states spend *more money on corrections than on education*! How's that for getting your priorities straight?

Thankfully, there are signs we might be moving in the right direction. States have been looking for ways to cut the enormous expenses of having more and more people in prison. They are reducing probation terms, trying alternative sentencing ideas, and reducing the number of people sent to prison for technical violations. A Republican Governor, Nathan Deal of Georgia, declared when taking office that putting so many drug addicts in jail was placing an "unsustainable financial and civic burden" on the states. *Huffington Post* by Nick Wing in an article titled, "Nathan Deal: Jailing Drug Addicts is 'Draining to Our State,'" on January 10, 2011. I don't care if a desire to cut government spending is the reason we might begin to change our bad habit of tossing people into jail. Even the Texas prison population was reduced by 6 percent in 2009, according to an article titled, "States Seek Less Costly Substitutes for Prison," by Keith Richburg, in the *Washington Post* on July 13, 2009. That article also said:

> *A powerful motivator for alternative sentencing is recidivism. For nearly 20 years, national recidivism rates have remained the same, with half of all freed inmates returning to prison within three years. But evidence shows that those who get treatment for drug and alcohol abuse have a far lower rate of returning to prison. For example, about 70 percent of people enrolled in drug courts complete the program, and 75 percent of them have not been arrested again, ...*

How about that? Treating the inmates so they don't go back to jail; what will they think of next?

Hey, remember we were talking about the time I got robbed after Superfly's party? So, why does anyone want to rob or steal from someone else? When someone steals, doesn't that mean they're *afraid* they won't have enough, or won't get what they want, without stealing? If a person had everything he or she needed, and totally accepted his or her current state in the now, why would they want to steal from anyone? I don't think they would. It's only if you're *afraid* you can't get enough honestly that you'd steal.

There aren't that many thieves among us, though, are there? Well, ever think about how much security is needed to keep people from stealing out of retail stores? Almost everything in our lives is impacted by the fact that people steal, a lot. From those impossible to open plastic cases that products come in, to all the passwords one has to remember to sign on to computers or the internet, to locks on our doors, *we live in a world afraid of theft*. While we think our proud fellow 'Merican citizens are upstanding, honest, and truthful, there may be some dispute about that.

I participated in a crime one time - we called it the "McDonald's Heist." We were in high school and a bunch of my friends had been working out getting ready for football season. After the workout, we were hungry, but didn't have enough money for our lunch. One of us (I will not reveal who) came up with the idea of ripping off McDonald's for some burgers. The idea was one friend would get in line, order a bunch of burgers, tell the cashier the guys behind him were paying, and head out of the store. I was one of those guys. When Charlie got his burgers and took off, we got to the front of the line and pretended we had no idea what he was talking about. My friend John was really good at making it seem we didn't know who Charlie was. We bought a couple burgers, Charlie handed the big bag off to another guy, and we all met in the woods and had a feast.

I'm not proud of that. I was just a teenager; what can I say? But I'm not the only one. An article in the *Washington Post* on December 1, 2008 by David Crary was titled, "Survey Finds Growing Deceit Among Teens."

> *In the past year, 30 percent of U.S. high school students have stolen from a store and 64 percent have cheated on a test, according to a new, large-scale survey suggesting that Americans are apathetic about ethical standards. ... Other findings from the survey:*
> - *Cheating in schools is rampant and getting worse. Sixty-four per-*

cent of students cheated on a test in the past year and 38 percent did so two or more times, up from 60 percent and 35 percent in a 2006 survey.

- *Thirty-six percent said they used the Internet to plagiarize an assignment, up from 33 percent in 2004.*
- *Forty-two percent said they sometimes lie to save money – 49 percent of the boys and 36 percent of the girls.*

Despite such responses, 93 percent of the students said they were satisfied with their personal ethics and character, and 77 percent affirmed that "when it comes to doing what is right, I am better than most people I know."

There's that blasted perception of separation again. When it comes to lying, here's how I look at it. No one would ever lie if they weren't *afraid* of the truth, or to own up to the truth. So, no question, a quick NWAL-like inquiry demonstrates cheating and lying are fear-based. But it's not just kids – *oh, no.*

According to the National White Collar Crime Center, it is estimated that employee theft and embezzlement cost businesses and organizations $400 billion per year. According to the Federal Bureau of Investigation's Uniform Crime Reports, a motor vehicle is stolen in the United States every 26.4 seconds. Twenty-two major retailers lost more than $6 billion to shoplifters and dishonest employees in 2008, according to the 21st Annual Retail Theft Survey conducted by Jack Hayes International, a loss-prevention consulting firm. Here are some more statistics complied by the Schulman Center for Compulsive Theft, Spending and Hoarding at http://www.theshulman-center.com:

- *There are nearly 300 million shoplifting incidents per year.*
  - 2010 Jack Hayes International, Inc.
- *Shoplifting is estimated to account for between 30-40% of retailer's lost profit*
  - 2010, Univ. of Florida & Jack Hayes International, Inc.
- *In 2009, retailers lost an estimated $12.58 billion per year to shoplifting*
  - 2010, Jack Hayes International, Inc.

- In 2009, 1,014,817 shoplifters were apprehended from 24 large retail companies, up 16.8% from 2008.
  - 2010, Jack Hayes International, Inc.
- For every $1.00 recovered from retail theft, $31.74 is lost; only 3.15% is recovered total.
  - 2010, Jack Hayes International, Inc.
- Did you know that 60% of inventory losses are caused by employees?
  - National Retail Federation – 2010
- U.S. retailers and small businesses lost $33 billion in revenue last year due to theft
  - National Retail Federation – 2010
- Employee theft is on the rise according to national statistics and company surveys
  - National Retail Federation – 2010
- Businesses lose 20% of every dollar to employee theft!
  - American Society of Employers
- In 2009, one in every 28.4 employees was apprehended for theft from their employer
  - 2010 Jack Hayes International, Inc. survey
- 75% of employees steal from work and most do so repeatedly
  - US Chamber of Commerce

Pretty depressing. An editorial titled, "Let's Face It: We're All Cheaters," in the *Wausau Daily Herald* on December 7, 2008, said it well:

> Fact is, though, there's no evidence that more people cheat today than they ever have. Cavemen probably stole one another's clubs and grabbed more than their share of mastodon ribs from their cavemates. Today, kids buy pre-written term papers on the Internet, and athletes score Mexican steroids at the gym.
>
> And you – yes, you – probably are guilty too. Ever file a fudged expense report at work or declare a tax deduction that wouldn't pass an auditor's scrutiny? Some of us who never would think of stealing something from a store or hitting someone on the head and grabbing

his wallet probably feel OK about gaming the IRS. ... Eventually, most of us land in a place where we can live with ourselves – not as honest as we possibly could be, but by and large pretty trustworthy.

I think people are basically trustworthy. It's the fear of failing or not having enough that drives so much questionable behavior around the globe. Just like the reduction in recidivist rates from alternative sentences and treatment, I'd say we can also handle some of these problems with stealing and robbing if we can teach more about love and oneness. The environment is one subject that could use a lot of love education.

Nineteen | Nothin' but High and Low Water

His name was Clemmy Chesledine. He was a friend of Joe and Bernie Wise, the caretakers of the Jefferson Island Club starting in 1946 for several decades. I admired the "Wise Men" when I was growing up. Dad used to take us to the Island so often Joe and Bernie were father figures. We didn't know at the time the locals called them "Fuzzy" and "Stump." I stayed at Joe's house near the Island a week every summer for several years when I was maybe twelve.

I remember getting up god-awful early, having a hearty breakfast of eggs and bacon cooked up by his wife Rose, and then going out with Joe to fish the crab pots. Even though they managed the Island, they'd grown up around the water. Joe had a bunch of pots, which were chicken wire boxes with dead fish as bait. The crabs could get in, but not out. I'd go on the water with Joe 'bout sunrise. He'd drive the boat, pull up the pot, empty the crabs and re-bait, clean seaweed off the pot with a wire brush, and do that about a hundred times. I'd sit at the front of the 12-foot wooden skiff and make sure no stray crabs got up under the seat.

Life was pretty good then, in a lot of ways. There were plenty of crabs. Joe used to take bushels to Cap't Sam's Crabhouse. They'd put 'em in a walk-up refrigerator. Bushel baskets sitting there in rows four or five high, stuffed with Chesapeake Bay Blue Crabs.

For meals at the Island, the cook would make us a bounty for dinner. We'd have fresh rockfish, crab cakes, trout, summer beans, ripe local tomatoes, and sweet corn. The cook would be there for breakfast, and make us any kind of eggs we wanted; I remember having soft-boiled eggs in little green eggcups.

Since I've been going to the Island for over fifty years, and lived on the Island, I became the ersatz historian. I published a second edition of the Club history, *My Love Affair with an Island*, in 2006. I've been giving the occasional talk about the Club's history, and one time some local folks told me I should go see Clemmy. He was in his 90s, with a bit of dementia. I sat down and interviewed him for about an hour a year or two back.

I didn't tape record him, but am going to use his voice to tell the story. These weren't his exact words, but you get the drift.

> *Was born in 1915 right down the road here. Went to school for a while, but quit when I was 16 and never went back. Worked the water pretty much ever' day after that til just recent. All seasons. Sometimes took off Christmas, but was out there most days.*
>
> *I'd do everthin' on the water. We'd fish a lot; string up long nets and catch maybe 200 boxes a fish sometime. Couldn't always sell 'em all. Lots of perch, spot, rock, all sorts of fish. Crabbed, too, 'deed we did. Lord, there was lots of crabs. Caught 'em ever way you could. We'd use trot lines, push nets through the grass to catch the soft shells, nets off the pier, and crab pots. I hauled a lot of pots outa the water in my day.*
>
> *Got a lot a ducks and geese, too. There was quite a variety of ducks; mallards, canvas back, redheads, wood ducks - you name it, we killed it. Swans, too. Those were some big birds. Used to be so many geese the clouds'd turn black when they'd leave the fields to go out overnight on the river.*
>
> *You heard about the sneak boats? They's these little small boats, real low to the water. Man had to sort a lie down in 'em. Had a great large gun on it; long and big barrel. You'd go out at night when the wind and moon was right. Had these small little sneak paddles to move through the water; often had 'em tied together with a leather strap. Would go quiet through the water and sneak up on a big rafta ducks. Get the gun lined up and – BLAM! Could take out hunerd or more ducks at a time. Then sell 'em up at market. That's why they'd called 'em 'market gunners.'*
>
> *In the winter, also could dip terrapin outa the water. On a cold day, the water'd be clear as glass, and could see all the wayta the bot-*

tom. See this outline shape of a turtle, and it was easy pickens, just jabbin' that net down and snatch him up. Turtle soup was mighty good. The missus could really cook up that stuff, too. Never was gonna starve round here back then.

Did oysterin' in the winter. Man, they'd been lots a oysters. They'd be big reefs of 'em; oysters layered up and boats could get even wreck on them reefs. We'd mostly tong 'em. Had these long wooden handles, 'bout 12 or maybe 16 feet, with wire baskets at the end. You'd scoop 'em up off the bottom. Let them handles down, move the tongs back and forth, grab a bunch and then close the tongs and shimmy 'em right up and bring 'em on the boat. Water was cold, but I didn't care much for gloves.

Those oysters was big, Lord, cut some of 'em like a steak with a knife. That ole Stump was handy with the nipper tongs. He could send that tong down there with the little couple of nippers to the one oyster that he could see he wanted, and he'd pull that sucker right up.

We all used to work the water and it was good to us. Took a lot out of it though. I remember Stump's father said to us one day, "Well, boys, I might not see it, and you might not neither, but one of these days, there's not gonna be nothin' in that river but high and low water."

Or, maybe - one a these days, ain't gonna be nothin' on this here planet but clouds and dust?

The oysters in the Chesapeake Bay used to filter the water of the entire Bay in three days. Nowadays, the few remaining oysters *maybe* could do it in a year. According to an article in the *Washington Post* on September 2, 2011, titled, "Halt oyster harvest, study says," by Darryl Fears, " … nearly 100 percent of the oyster population has been lost since its peak in the early 1800s and more than 90 percent has been lost since 1980." There also used to be huge sturgeon in the Chesapeake Bay. Capt. John Smith when he explored the Bay in the 1600s proclaimed that the James River, "had more sturgeon than could be devoured by dog and man." They can grow to several hundred pounds; the largest on record was over 800 pounds. Sturgeon can be up to 14 feet long and live to age sixty. But they're endangered because of overfishing and pollution, and very rare now. http://www.bayjournal.com/article.cfm?article=3960.

"[T]he Potomac will almost certainly lose its oldest, largest and most distinctive fish." "A Potomac with no sturgeon," by Darryl Fears, *Washington Post*, January 28, 2013.

We've been decimating the seas for a long time:

> *Human activities are affecting every square mile of the world's oceans, according to a study by a team of American, British, and Canadian researchers who mapped the severity of the effects from pole to pole. The analysis of 17 global data sets, led by Benjamin S. Halpren of the National Center for Ecological Analysis and Synthesis in Santa Barbara, Calif., details how humans are reshaping the seas through overfishing, air and water pollution, commercial shipping and other activities. ...*
>
> *The team of scientists analyzed factors that included warming ocean temperatures because of greenhouse gas emissions, nutrient runoff and fishing. ... Almost half of all coral reefs, they wrote, "experience medium to very high impact" from humans. Overall, rising ocean temperatures represent the biggest threat to marine systems.* "Study Finds Humans' Effect on Oceans Comprehensive," by Juliet Eilperin, *Washington Post*, February 15, 2008.

As another example, it's been reported that, "Sharks are being fished at a rate that is 30 to 60 percent higher than they can sustain, ... Sharks are primarily targeted for their fins, which are used in the Asian delicacy sharks fin soup, though they are also caught accidentally by vessels seeking tuna, swordfish and other species." "New study estimates that commercial fishing kills 100 million sharks annually," by Juliet Eilperin, *Washington Post*, March 5, 2013.

According to another article titled, "Obama's Chance for a Blue Legacy," by Vikki Spruill, in the *Washington Post* on January 6, 2009:

> *Covering 71 percent of the planet, the oceans are our life support system. They provide most of the oxygen that we breathe and much of the food that we eat. As the engine that drives our climate, the oceans are the front line of the global climate challenge, absorbing half of the carbon dioxide we've pumped into the atmosphere and more excess heat from greenhouse gases than all the rain forests combined.*

> *Indeed, the oceans are the unsung hero in the climate change battle – but they are also the most vulnerable victims.*

The oceans are also where we get a lot of fish to eat – a lot of fish. Just the same way that our love of Chesapeake oysters and crabs has caused their population to plunge, it's happening all over the world.

> *Over the past 100 years, some two-thirds of the large predator fish in the ocean have been caught and consumed by humans, and in the decades ahead, the rest are likely to perish, too. In their place, small fish such as sardines and anchovies are flourishing in the absence of the tuna, grouper and cod that traditionally feed on them, creating an ecological imbalance that experts say will forever change the oceans. ...*
>
> *This grim reckoning was presented at the American Association for the Advancement of Science's annual meeting ... during a panel that asked the question: "2050: Will there be fish in the ocean?"*
>
> *The panel predicted that while there would be fish decades from now, they will primarily be smaller varieties currently used as fish oil, fish meal for farmed fish and only infrequently as fish for humans. People the experts said, will have to develop a taste for anchovies, capelins and other smaller species. One startling conclusion: More than 54 percent of the decrease in large predator fish has taken place over the past 40 years.* "Experts sound alarm over decline of predator fish," by Marc Kaufman, *Washington Post*, February 21, 2011.

This makes me sad. Sometimes I wonder if my grandchildren will even know what fish were. Is that possible? How can things be so bleak? *Can man be at fault?*

Rose told me every spring, after the farmers put pesticides on the fields, rain would wash the chemicals off into the Potomac and all the sea grass around the shore would die. At one time, the grass was so thick they had to cut paths *through the seaweed in the water* so the boats could get to and from the dock on the Island. The grass where crabs would molt - shedding their skin, turning soft, then hardening up – is mostly gone now.

I also remember how Bernie Wise used to have a shedding float in the water. He'd pick out the crabs about to change into softshells, and put them into a half-submerged wooden box floating in the water. The River water flowed through the boxes, which were meant to allow the crabs to be in moving water. Bernie could tell which crabs were about to shed just by looking at them. They called those crabs "peelers," because they were about to peel off their hard outer shell. Bernie'd go down to his dock a few times a day and check for soft crabs. He'd dip them out with a net, then sell them. Soft shell crabs sold for more than hard shells.

But, know what? The water got so bad in recent years that the watermen couldn't leave the crabs to shed in the water; they'd *die*. The watermen had to build their shedding apparatus on the land and pump filtered water for the crabs to live in till they turned soft. Can't even shed out crabs in the River anymore; it's pathetic. The reason? Pollution from me, my friends, family, neighbors, and everyone who lives in the Chesapeake watershed, specifically including farmers. Yep, humans are the main cause of this mess. We're changing the environment in multiple ways.

As another example, "Over in the Chesapeake Bay region, 523 million chickens generate 42 million cubic feet of waste each year, the Pew Environmental Group reports, enough to fill the Capitol dome 50 times over. The result is a recurring dead zone that spans from Baltimore Harbor to south of the Potomac River." *Mother Jones* magazine, September/October, "The Fishy Truth About Your Meat," by Tom Philpott. How about this; people shit is killing coral in the Caribbean. "The elkhorn coral, named for its resemblance to elk antlers and known for providing valuable marine habitat, was once the Caribbean's most abundant reef builder. But the 'redwood of the coral forest has declined by 90 percent over the past decade, due in large part to a highly contagious white pox disease, … . Now, researchers say they have found the reason for this decline: human excrement." *Washington Post*, August 23, 2011, "Study blames bug in human excrement for decline in Caribbean's elkhorn coral," by Gisela Telis, *ScienceNow*.

For example, "About a third of the 186 invasive species in the Great Lakes are thought to have entered on ocean-going ships in the ballast water they take on for stabilization when carrying little or no cargo. … Zebra and quagga mussels from the Black Sea clog intake water systems and power plants. The mussels also gobble plankton so voraciously that little is left for other organisms." "Major Shipping Route Fosters a Plague of Sea Life," by Kari Lydersen, *Washington Post*, August 31, 2009.

And who can forget how we're blowing out whale eardrums? The use of sonar by navy ships and submarines posed a serious threat to whales and other marine mammals by disrupting their navigation and communication abilities. http://www.nrdc.org/wildlife/marine/sonar.asp. Of course, in the Bush administration, the White House exempted the Navy from environmental laws because use of the sonar in training was "essential to national security." "Navy Wins Exemption from Bush to Continue Sonar Exercises in Calif." by Mark Kaufman, *Washington Post*, January 17, 2008. Meaning we need the whales to be deaf so we can protect our homeland?

All sorts of strange things are happening because of human activity. Scotland's Soay sheep are shrinking because of global warming. "In 25 years, Soay sheep have gotten 5% smaller, on average, according to Tim Coulson of Imperial College London. It's not that evolution has been repealed in Scotland; rather global warming has simply made it easier for smaller, less fit Soay sheep to survive. And plenty of other species are quickly adapting to the changing climate in similar ways. It seems global warming, which by one forecast could threaten up to one-third of the world's species if left unchecked, is emerging as Darwin's new enforcer." *Time* magazine, "Why Are Scotland's Sheep Shrinking," by Bryan Walsh, July 30, 2009.

Did you also know that frogs, toads, and salamanders are vanishing from our country at an alarming rate. Researchers think the causes could be invasive species (which we probably introduced), pesticides, and climate change. Some folks are calling for a ban on that old high school routine of dissecting frogs. Good idea. "U.S. study finds steep decline in amphibian numbers," by Darryl Fears, *Washington Post*, May 23, 2013. And how many types of birds do you think are going extinct in the Amazon region? Well, 1,331 types of birds are at risk on the Red List of Threatened Species published by the International Union for Conservation of Nature, but only 100 species in the Amazon region are on the critically endangered category. Whoo, hooo! "100 bird species in Amazon are said to risk extinction," by Jenny Barchfield, *Washington Post*, June 8, 2012.

Introduction of a nonnative species of brown tree snakes by man into Guam fifty years ago has led to a near extermination of Guam's birds and is changing the way its forests grow, as well as leading to a huge increase in the spider population. "Without birds, which eat the seeds of certain trees and then spread them in their droppings, those trees are losing out to others that do not depend as much on bird middlemen. The seeds of the trees that relied on birds are now falling mostly near the trunks of the parent tree, where they are more likely to be spoiled by fungus and less likely to grow into healthy trees. ... Birds typically make up a small part of the life of a forest, but they are important not only for spreading seeds but also for pollinating flowers and controlling some insects that feed on plants." "Snakes Impact on Guam Appears to Extend to Flora," by Marc Kaufman, *Washington Post*, August 11, 2008.

An article on http://inhabitat.com, dated May 12, 2011, that I saw on a friend's post on Facebook was titled, "More Scientific Studies Indicate that Cell Phones are Killing Bees." Apparently recent studies have led researchers to conclude that cell phone signals confuse bees and can result in their death. We know that bees are critical to pollination of plants and flowers and, if we're killing them with text messages, that would truthfully suck.

Although other studies have indicated that certain pesticides – called neonicotinoids – may be wiping out the bees. Ironic, huh, that bees are critical for pollinating plants, so to grow the plants we are now killing the bees. One thing's for sure, they wouldn't be dying off without us involved. "Research links bees' decline to class of pesticides," by Mark Kaufman, *Washington Post*, March 30, 2012.

Of course, hunting also kills many animals. I used to go hunting with my Dad at the Island, as you know. Later, when I was manager of the Island, I was the proverbial chief cook, bottle washer, and *hunting guide*. I'd wake up early and set the decoys out, like Joe and Bernie did. I'd drive the hunters out to the blinds in the boat, then sit on the shore watching for downed ducks. I'd get in the boat and grab them out of the water when they were shot. I even learned to clean ducks, something I never conceived I'd do. The cook taught me how to take the feathers off, including those damned pinfeathers you had to pluck out sometimes one at a time. I was proud of myself.

One day, two hunters came down and, after a morning of hunting, killed two ducks. When they came out of the blind, I asked if they wanted me to clean the birds they'd shot. They said, "Oh, no, we don't care about those. You can have them." I was dumbfounded. They just killed them - to kill something. I decided I didn't like hunting any more after that.

Has man been as irresponsible with the animal population as a whole? According to a short piece in *Time* magazine on October 20, 2008, "Hunting and Habitat Loss Threaten More Than 20% of the World's Mammals with Extinction." Holy smokes! *Twenty percent of all the mammals in the world?* Sounds like a mass extinction, and that'd be right. A later article in *Time* dated April 13, 2009, titled, "The New Age of Extinction," by Bryan Walsh said:

> There have been five extinction waves in the planet's history – including the Permian extinction 250 million years ago, when an estimated

70% of all terrestrial animals and 96% of all marine creatures vanished, and, most recently the Cretaceous event 65 million years ago, which ended the reign of the dinosaurs. Though scientists have directly assessed the viability of fewer than 3% of the world's described species, the sample polling of animal populations so far suggests that we may have entered what will be the planet's sixth great extinction wave. And this time the cause isn't an errant asteroid or mega-volcanoes. It's us.

Through our growing numbers, our thirst for natural resources and, most of all, climate change – which, by one reckoning, could help carry off 20% to 30% of all species before the end of the century – we're shaping an Earth that will be biologically impoverished. ... Why does the loss of a few species among millions matter?

For one thing, we're animals too, dependent on this planet like every other form of life. The more species living in an ecosystem, the healthier and more productive it is, which matters for us .... The worst-case scenarios of habitat loss and climate change – and that's the pathway we seem to be on – show the planet losing hundreds of thousands to millions of species, many of which we haven't even discovered yet.

Michael Novacek, provost of the American Museum of Natural History, said it this way on January 13, 2008 in an article titled, "It Happened to Him. It's Happening to You," in the *Washington Post*:

Assuming that we survive the current mass extinction event, won't we be okay? The disappearance of more than a few species is regrettable, but we can't compromise an ever-expanding population and a global economy whose collapse would leave billions to starve. This dismissal, however, ignores an essential fact about all those species: They live together in tightly networked ecosystems responsible for providing habitats in which even we humans thrive. Pollination of flowers by diverse species of wild bees, wasps, butterflies and other insects, not just managed honeybees, accounts for more than 30 percent of all food production that humans depend upon. [I]t is the double whammy of climate change combined with fragmented, degraded natural habitats – not climate change alone – that is the real threat to many populations, species and ecosystems, including human populations marginalized and displaced by those combined forces.

The scientists keep referring to climate change like it's definitely happening, in spite of some who think it isn't and there's nothing to worry about. I'd analogize this to the old argument made by religious types: "What if you're wrong?" Their point being that, if I'm wrong about God and Jesus, or Allah or whoever, then I'm screwed (more accurately, damned) when the rapture comes about. On the climate issue, though, if they're wrong, and we didn't act when we could have, *we all are really screwed.*

Why not pull together and do the best we can to minimize the potential causes of climate change? Oh, right, *jobs*. There's a funny scene in Al Gore's movie, "An Inconvenient Truth," where he mentions a Bush White House briefing book with a picture of a scale balancing gold bars and the planet earth. "Hmm," he says, "A few gold bars, or the *entire planet earth*?" Not much of a choice, in my mind.

I don't think there's any denying climate change. I've noticed it in my life. The winters are different, more extreme; the summers are hotter – much hotter. According to a *Washington Post* article on January 12, 2011, the last decade had nine of the ten hottest years on record! The year 2010 tied 2005 for hottest year ever. And, seriously, anyone who's ever just stood on an asphalt parking lot in Washington, DC in July and felt that blast furnace where grass and trees *used to be*, has got to understand that we're changing the weather around us. For goodness sake, it's warmer in the city than in the suburbs. Doesn't *that* say something?

The editorial page of the *Washington Post* on May 16, 2011 proclaimed:

> *"Climate change is occurring, is very likely cause by human activities, and poses significant risks for a broad range of human and natural systems."* So says – in response to a request from Congress – the National Research Council of the National Academy of Sciences, the country's preeminent institution chartered to provide scientific advice to lawmakers. [The editorial also went on to excoriate Republicans and climate-change deniers for being "willfully ignorant, lost in wishful thinking, cynical or some combination of the three." I'll

only say again that those who perceive separation are generally consistent across the board.]

Adrian Higgins in the *Washington Post* on January 29, 2009, in his article "Study Calls on 'Citizen Scientists' To Tap Their Inner Thoreau," said that scientists have been studying the dates when flowers and plants bloom back to the time of Henry David Thoreau, who wrote *Walden's Pond*. Higgins says the research shows: "They determined that temperatures have risen nearly 4 degrees in the past 150 years, that flowers now open a week earlier on average and that some plants are far better equipped to deal with this phenomenon than others. More than 60 percent of the plants Thoreau tracked either have gone extinct locally or are on the brink of disappearance." Higgins reports that similar findings have been made in the Washington, DC area, where out of about 100 plants, 90 of them are flowering an average of five days sooner than only ten years ago.

In another report in an article titled, "Report: Climate Change is Hurting Crops," by Brian Vastag and Juliet Eilperin, in the *Washington Post*, on May 6, 2011, scientists have found that corn and wheat production was down throughout much of the world due to rising temperatures. "Extreme rainstorms and snowfalls have grown substantially stronger, two studies suggest, with scientists for the first time finding the telltale fingerprints of man-made global warming on downpours that often cause deadly flooding," as reported in the *Huffington Post* on February 16, 2011 in an article titled, "Scientists Connect Global Warming to Extreme Rain," based on an article in *Nature* magazine. Scientists have done computer modeling that essentially proves current weather disruptions, floods, blizzards, etc. are caused by greenhouse gases, according to the article.

For those who think a warming planet can't cause lots of snow and cold weather, please read the article titled, "A Warming Planet Can Cause a Brutal, Chilling Winter," by Ben Harder in the *Washington Post* on February 8, 2011. Basically, the cold air over the North Pole gets warmer and the pressure increases, causing the cold air to slide down the globe. The main problem is melting sea ice in the Arctic Ocean. Recently, there's been more ice melting than freezing back. It reminds me of the Island, and the stories Joe and Bernie used to tell about how the River would freeze solid most of the way across. Nowadays – it never does.

In his book, *Hot, Flat and Crowded*, Tom Friedman demonstrates no doubt about climate change, or global weirding, as he calls it.

> *At the beginning of the Industrial Revolution, and particularly in the last fifty years, the amount of CO2 in the earth's atmosphere shot up from 280 ppm to 384 ppm, where it has probably never been for twenty million years, and at a speed of increase that took the sun thousands of years in each cycle to produce. And we are on a track to add 100 or more ppm of CO2 to the atmosphere in the next fifty years. This extra CO2 is not coming from the oceans. It is coming from humans burning fossil fuels and from deforestation ...(which accounts for some 20 percent of all CO2 emissions).*
>
> *In the same twenty minutes that will see some unique species vanish forever, Conservation International notes, 1,200 acres of forests will be burned and cleared for development. The CO2 emissions from deforestation are greater than the emissions from the world's entire transportation sector – all the cars, trucks, planes, trains, and ships combined. ...*
>
> *Imagine a world without forests. Imagine a world without coral. Imagine a world without fish. Imagine a world where rivers run only in the rainy season. Not only is that possible in more and more places, it is possible in our lifetime ... .*

In fact, the levels of CO2 in the atmosphere are now at an all time high. http://thinkprogress.org/romm/2011/11/21/374141/heat-trapping-co2-new-high-growth-methane-levels-are-rising-again

This makes it all the more critical that we humans understand how much things on this earth are *interconnected*. For example, did you know dust from deserts blowing across the globe acts to accelerate the melting of glaciers in the Himalayas by making the ice pack darker and heating more in the sun? "Dust From Far-off Deserts May Speed Up the Melting of Glaciers in the Himalayas," *Washington Post*, January 11, 2011. Some seem to think one thing that happens over here, can't affect what happens there. That's a failure to recognize our interdependence – our oneness. In fact, there's the theory that the earth is really just *one big living organism*.

That's what James Lovelock talks about in his book, *Gaia, The Practical Science of Planetary Medicine*. The theory is that our planet is not a dumb lump of dirt and stone moronically circling around the sun. Rather:

> *Gaia is the name of the Earth seen as a single physiological system, an entity that is alive at least to the extent that, like the other living organisms, its chemistry and temperature are self-regulated at a state favourable for its inhabitants. ... Gaia is an evolving system, a system made up from all living things and their surface environment, the oceans, atmosphere, and crustal rocks, the two parts tightly coupled and indivisible. ... Make no mistake, to understand the physiology of the Earth, how Gaia works, requires a top-down view of the Earth as a whole system; if you like, as something alive....*

Did you ever think of the earth as alive? We talked about the Hubble images; in a funny way, don't those galaxies and star clusters look *alive*? For that matter, doesn't the earth look like it's a living entity in those iconic photos from space? That's what Lovelock says:

> *Even if in the end Gaia should turn out to be no more than metaphor, it would still have been worth thinking of the Earth as a living system. ... All I do ask is that you consider Gaia theory as an alternative to the conventional wisdom that sees the Earth as a dead planet made of inanimate rocks, ocean, and atmosphere and merely inhabited by life. Consider it as a real system, comprising all of life and all of its environment tightly coupled so as to form a self-regulating entity.*

Lovelock's book is chock-a-block full of scientific facts, theories, and principles. It's not easy to understand for a non-scientist like me. But we've spent a lot of time talking about allegory and myth. A major emphasis of my inquiry here has been to develop a theory of life, my own personal allegory of what I believe. As Tom Harpur said earlier in this book, the only way we can understand the divine is through story and allegory. So, in this regard, is it more healthy to think of the earth as dead, *or alive*? Who could prove it one way or another?

We've also seen that – at the smallest level of being, from a sub-molecular, atomic level – we're all connected and thus everything is *alive* in some sense. I've told you I believe everything's alive. Isn't the Gaia principle just a reasonable corollary to that? *Why not* think of our planet as being alive? As a true Mother Earth?

Trees are a main focus for Lovelock. He says the forests are critical to our lives as humans. "We are failing to recognize the true value of the forest as a self-regulating subsystem that keeps the climate of the region, and to some extent the Earth, comfortable for life. Without the trees there is no rain, and without the rain, there are no trees."

Is that right? Are trees *necessary* for rain? Lovelock believes this is true, and we are sowing the seeds of our self-destruction by cutting down so many trees.

> *[B]y far the most dangerous malady afflicting the Earth is that of exfoliation – destruction of its living skin. In human medicine, the loss of skin from whatever cause is a serious threat to life; the loss of more than 70 percent of the skin by burning is usually fatal. To denude the Earth of its forests and other natural ecosystems and of its soils is like burning the skin of a human. And we shall soon have destroyed or replaced with inefficient farmlands 70 percent of the Earth's natural land surface cover. ...*
>
> *Perhaps the best known example of the pathology of forest loss is Harrapan in Western Pakistan. The region was once abundantly forested and enjoyed an adequate rainfall during the monsoon season. It was a fine example of a self-sustaining forest ecosystem. The forest was gradually cleared by peasant farmers who kept cattle and goats that grazed on the scrub and grass that replaced the forest trees. The rainfall was sustained over the region until rather more than half of the forests had been cleared. But after that the region became arid and the remaining forest decayed. The region is now so dry that as a semi-desert it can support only a fraction of the people and other organisms that were once there.*

Wow, that reminds me of Jared Diamond's book *Collapse*. I had no idea forests could turn into deserts because the trees were cut down. That's what Diamond said happened on Easter Island in the Pacific.

> *[F]or hundreds of thousands of years before human arrival and still during the early days of human settlement, Easter was not at all a barren wasteland but a subtropical forest of tall trees and woody bushes. ... But the Easter Island seen by ... European visitors had*

> very few trees, all of them small and less than 10 feet tall; the most nearly treeless island in all of Polynesia. ...
>
> Deforestation must have begun some time after human arrival by A.D. 900, and must have been completed by 1722, when [explorers] arrived and saw no trees over 10 feet tall. ... The overall pictures for Easter is the most extreme example of forest destruction in the Pacific, and among the most extreme in the world; the whole forest gone, and all of its tree species extinct. Immediate consequences for the Islanders were losses of raw materials, losses of wild-caught foods, and decreased crop yields. ... The further consequences start with starvation, a population crash, and a descent into cannibalism. ... Easter's isolation makes it the clearest example of a society that destroyed itself by overexploiting its own resources.

Question: What in the world caused a lush Easter Island to become a barren wasteland?

Answer: They cut down the trees to make ropes and slides to move big huge statues to various gods all around the Island. Seriously. Jared continues:

> Easter's land surface was divided into about a dozen (either 11 or 12) territories, each belonging to one clan or lineage group ... .
> Each territory had its own chief and its own major ceremonial platforms supporting statues. The clans competed peacefully by seeking to outdo each other in building platforms and statues, but eventually their competition took the form of ferocious fighting. ... The "average" erected statue was 13 feet tall and weighed about 10 tons. The tallest ever erected successfully, known as Paro, was 32 feet tall but was slender and weighed "only" 75 tons ... . The increase in statue size with time suggests competition between rival chiefs commissioning the statutes to outdo each other. ... I cannot resist the thought that they were produced as a show of one-upsmanship. They seem to proclaim: "All right, so you can erect a statue 30 feet high, but look at me: I can put this 12-ton pukao on top of my statue; you try and top that, you wimp!"

Seemed pretty silly, right, erecting medieval skyscrapers to see who could build the biggest one? Even sillier, though, was that they cut down all the trees on the Island to move the stones they carved. How'd they do it? Well, the natives in the area had used "canoe ladders," that is, pieces of parallel wooden rails over which the large canoes were dragged to the water. The versions employed for statues used huge numbers of logs, and rope made of vine and bark, to move the heavy stone statues. According to one study, "50 to 70 people working five hours per day and dragging a sled five yards at each pull, could transport an average-sized 12-ton statue nine miles in a week." The people synchronized their pulling efforts to move the huge stones. With more people, they could haul bigger statues.

Diamond gives other examples of situations in which humans made individual decisions that didn't seem harmful out of context, but that collectively amounted to terrible and self-destructive behavior. It's not unlike all of us in the Chesapeake Bay region eating all the fish, crabs, ducks, and geese, and polluting the Bay, from which we get our life-giving water. Basically, destroying our own homes and local food supply. Like a cancer; destroying the host. This situation can be played out in a thousand ways, involving countless species and resources we need. Here's what Diamond says:

> *The parallels between Easter Island and the whole modern world are chillingly obvious. Thanks to globalization, international trade, jet planes, and the Internet, all countries on Earth today share resources and affect each other, just as did Easter's dozen clans. Polynesian Easter Island was as isolated in the Pacific Ocean as the Earth is today in space. When the Easter Islanders got into difficulties, there was no-where to which they could flee, to which they could turn for help; nor shall we modern Earthlings have recourse elsewhere if our troubles increase. Those are the reasons why people see the collapse of Easter Island society as a metaphor, a worst-case scenario, for what may lie ahead of us in our own future.*

Why? Lovelock responds:

> *Humans on the Earth behave in some ways like a pathogenic microorganism. We have grown in numbers and in disturbance to Gaia, to the point where our presence is perceptibly disabling, like a disease. ... As a species, we live neither as free and independent individuals, nor*

as completely integrated social organisms like the bees. Rather, we live tribally; and our tribal behavior is all too often far below the standard of the best among us. Intelligent we may be as individuals; but as social collectives we behave churlishly and with ignorance.

Yea, we produce skads of plastic bags and other non-biodegradable stuff. Watched the move *Bag It!* recently. http://www.bagitmovie.com. It's depressing, how we're putting so much awful detritus into the environment, and particularly into the waters of this world.

Plastic bags, which only really came into common use when I was in high school, are now ubiquitous. And they're not going away. Plastic lasts – well, basically, much longer than you or me. Lots longer. Relatively speaking, *forever*.

Can you think of many things today that are *not* packaged in plastic? Cheese, TV dinners, toys, CD's, and particularly all those hard-plastic product cases that are not possible to open unless you have super-powers. (My Mom would have no chance opening some of those plastic packages.) From http://ecologicfoodservice.com/video/garbage-patch:

> *Most of our waste today is comprised of plastic. Plastic, which is made from petroleum, is a material that the Earth cannot digest. Every bit of plastic that has ever been created still exists, except for a small amount that has been incinerated, releasing toxic chemicals.*
>
> *In the ocean, plastic waste accumulates in swirling seas of debris, where plastic to sea life ratios are 6:1; where birds and mammals are dying of starvation and dehydration with bellies full of plastics; where fish are ingesting toxins at such a rate that soon they will no longer be safe to eat.*
>
> *The largest of these garbage swills is known as the Pacific Gyre, or the Greatest Garbage Patch. It is roughly the size of Texas, containing approximately 3.5 million tons of trash. Shoes, toys, bags, pacifiers, wrappers, toothbrushes, and bottles too numerous to count are only part of what can be found in this accidental dump floating midway between Hawaii and San Francisco.*

The ocean currents swirl around and trap all the plastic that one way or another winds its way down to the sea. None of us mean it to get into the ocean; when we toss a water bottle out, we don't want it to go out into the Atlantic Ocean and create an ocean-borne trash dump. But it does.

In the '50's, when I was fishing off the old wooden boat the Club owned, the Seahawk, after you drank a soda, you'd just pitch the can over the side. Or the Coke bottle, if that's what you had. Yep, right over the side. The river was just too immense to be affected by our puny waste, right? Lovelock, again:

> *None of the environmental agonies now confronting us – the destruction of the tropical forests; the degrading of land and seas; the looming threat of global warming; ozone depletion and acid rain – would be a perceptible problem at a global population of 50 million. Even at a billion people, these pollutions would probably be containable. But at our present numbers – more than six billion - and present way of living, they are unsupportable. If unchecked, they will kill a great many of us and other species, and change the planet irreversibly. ... Gaia is suffering from Disseminated Primatemia, a plague of people. ... The statement, "There is no pollution but people" carries an awful truth.*

According to http://www.worldometers.info, our world population as I write this is now over seven billion people! We produce about 4 billion tons of food, wipe out 2 million hectares of forest and 2 million hectares of arable land, put 13 billion tons of carbon dioxide into the air, and release 4 million tons of toxic chemicals into the environment – each year. And we've consumed so much oil that we've only got about 15,000 days worth of oil left on the planet, at our current consumption levels. We's got a big problem.

Here are some more fun facts from Kid's Post in the *Washington Post* on November 12, 2008:

- In the United States, $43 billion in household food is wasted every year.
- In France, each person eats an average of 57 pounds of cheese annually.
- In Australia, 3.9 million kangaroos were killed in 2003 for meat and skins.
- In Japan, each person eats an average of 146 pounds of fish each year.
- While in Mali, 91 percent of the population lives on less than $2 per day.
- In Greenland, 20 percent of the population eats seal four times a week.

- In China, only 12% of rural families have a refrigerator.
- Americans each consume 157 pounds of sugar and sweeteners annually.

So, we're not even doing a very good job of raping the planet on an egalitarian basis. There are too many of us, and we're crazy to boot. Eckhart Tolle proves the point in *The Power of Now*:

> How is it possible that humans killed in excess of one hundred million fellow humans in the twentieth century alone? Humans inflicting pain of such magnitude on one another is beyond anything you can imagine. ... Do they act this way because they are in touch with their natural state, the joy of life within? Of course not. Only people who are in a deeply negative state, who feel very bad indeed, would create such a reality as a reflection of how they feel. Now they are engaged in destroying nature and the planet that sustains them. Unbelievable but true. Humans are a dangerously insane and very sick species. That's not a judgment. It's a fact. It is also a fact that the sanity is there underneath the madness. Healing and redemption are available right now.

I believe that, Eckhart. I do. I have faith in us. But it's not easy. It's almost easier to believe in Jesus than in us. It can be depressing watching the news. Tolle again, from *A New Earth*:

> We only need to watch the daily news on television to realize that the madness has not abated, that it is continuing in the twenty-first century. Another aspect of the collective dysfunction of the human mind is the unprecedented violence that humans are inflicting on other life-forms and the planet itself – the destruction of oxygen producing forests and other plant and animal life; ill-treatment of animals in factory farms; and poisoning of rivers, oceans and air. Driven by greed, ignorant of their connectedness to the whole, humans persist in behavior that, if continued unchecked, can only result in their own destruction.

I believe one can see truth in minor actions, so check this out. An article in the *Wash-*

ington Post titled, "Overnight at Ocean City, a Degraded Beach Undergoes a Restoration," by John Kelly on August 18, 2009, told the story of a maintenance worker in Ocean City, Maryland, who cleans the beaches each night in the summer before the people go out and play. His name is Dustin.

> "I think it's just laziness, to be honest with you. ... There are spots [on the beach] where you see there was a camp-out of people, a little circle, and their cups and cans and bottles are just sitting right there where they never picked them up. And five feet away is a trash can. That's what's amazed me: the general nature of people, how people just don't really care, it seems like. ..."
>
> "Anything you find at the dump you can find out here. Food, chicken bones, dirty diapers, clothes. I've found complete outfits, including underwear, and nobody around." Condoms, keys, cellphones, necklaces, rings, boogie boards, flotation noodles, cameras, memory cards, sunglasses, dead horseshoe crabs. "Anything."

Yep, those my peeps. However, Lovelock says:

> I think by far the greatest damage we do to the Earth, and thus by far the greatest threat to our own survival, comes from agriculture. We shall soon have taken away more than two thirds of Gaia's natural terrestrial ecosystem and replaced them with agricultural systems. When we replace natural forests with food crops or cattle farms we diminish the ability of the land and surface to control its own climate and chemistry. ... When we farm, unless we do so very sensitively, we are evading our contractual obligation to Gaia – and most farming, especially agribusiness, is grossly insensitive. ...
>
> As polluters, we alter the atmosphere, waters and soils of the Earth and so increase the stress to which the natural ecosystems are subject. But as farmers, we do still greater harm, by clearing the land and so reducing the capacity of the whole system to deal with stress. ... Take away the trees, and the rain goes with them. ... If we delay our decision to stop felling trees until 70 percent are gone, it might be too late; the rest would die anyway. If we let deforestation continue

> *we may soon reach the day when at least a billion people are living in those once-forested regions, but in a hostile climate, hot and arid – an unprecedented human political problem as threatening as a major thermonuclear war.*

That's serious stuff. After I read that about agriculture, another book called out to me, *Eating Animals*, by Jonathan Safran Foer. He started looking into how food is produced, because he and his wife were about to have kids. He wanted to research the best dietary choices for his offspring; omnivore or vegetarian. What he found is startling. He says, "Animal agriculture makes a 40% greater contribution to global warming than all transportation in the world combined; it is the number one cause of climate change." Yikes, so it's true what Lovelock is saying! And with this huge impact on the planet just from our food, it's ironic to note that, "Americans choose to eat less that .25% of the known edible foods on the planet."

That doesn't sound like we're taking full advantage of the diversity of sustenance available to us. Growing up, we ate the typical stuff; hot dogs, hamburgers, steaks, fish (particularly on Friday's when the Catholic church said we couldn't eat meat), Campbell soups, spaghetti, mashed potatoes, all sorts of sandwiches, cereal, white bread, etc. Never had whole wheat bread, cheese other than the yellow stuff, sun-dried tomatoes, or any real Asian or Mexican food, for example. It was just pretty limited (except when we went to the Island). Back in those days, however, most of the meat came from regular farms, not the factory farms of today on land and sea. That's a big difference. Foer says:

> *For every ten tuna, sharks, and other large predatory fish that were in our oceans fifty to a hundred years ago, only one is left. Many scientists predict the total collapse of all fished species in less than fifty years … . As my experience with the world of animal agriculture deepened, I saw that the radical transformation fishing has undergone in the fifty years are representative of something much larger. We have waged war, or rather let a war be waged, against all of the animals we eat. This war is new and has a name; factory farming.*
>
> *Globally, roughly 450 billion land animals are factory farmed every year. (There is no tally of fish.) Ninety-nine percent of all land animals eaten or used to produce milk and eggs in the United States*

> *are factory farmed. ... Once the picture of industrial fishing is filled in – the 1.4 billion hooks deployed annually on longlines (on each of which is a chunk of fish, squid, or dolphin flesh used as bait) the 1,200 nets, each one thirty miles in length, used by only one fleet to catch only one species; the ability of a single vessel to haul in fifty tons of sea animals in a few minutes – it becomes easier to think of contemporary fishers as factory farmers rather than fisherman.*

That really puts Clemmy Cheseldine and the others of his day in perspective. Almost puny. The scale of the current system that produces our food is mind-boggling. But what really gives me pause is the way the animals are treated. There is a disregard for them that's shameful. Foer again:

> *[T]wenty of the roughly thirty-five classified species of sea horse worldwide are threatened with extinction because they are killed "unintentionally" in seafood production ... (sea horses are one of the more than one hundred sea animal species killed as "bycatch" in the modern tuna industry [and] ... shrimp trawling devastates sea horse population more than any other activity). ... [This] in the name of affordability treats the animals [we] eat with cruelty so extreme it would be illegal if inflicted on a dog.*

He explains that there are not just sea horses in the bycatch. Tuna fishing regularly kills 145 other species in its bycatch, including gulls, dolphins, whales, and all sorts of seabirds and turtles. He suggests an appropriate label for a can of trawled shrimp from Indonesia might read: 26 POUNDS OF OTHER SEA ANIMALS WERE KILLED AND TOSSED BACK INTO THE OCEAN FOR EVERY 1 POUND OF THIS SHRIMP.

The cruelty of the land animal production is also difficult to fathom, but it's mostly kept secret from us. Nobody wants to know where the tasty grilled chicken or steak comes from. But it's pretty awful. Since 99 percent of the land animals produced through the factory farm system are birds, we should learn a little about them. There are two types of chickens now, broilers that we eat and layers that lay eggs. Let's listen to Foer tell us a bit about that:

> [B]roilers ... are lucky: they tend to get close to a single square foot of space. ... Chickens once had a life expectancy of fifteen to twenty years, but the modern broiler is typically killed at around six weeks. Their daily growth rate has increased roughly 400 percent. ... [Male layers] serve no function. Which is why all male layers – half of all the layer chickens born in the United States, more than 250 million chicks a year – are destroyed. ... Most male layers are destroyed by being sucked through a series of pipes onto an electrified plate. Other layer chicks are destroyed in other ways, and it's impossible to call those animals more or less fortunate. ... [Some] are sent fully conscious through macerators (picture a wood chipper filled with chicks).
>
> [C]hicken factory farms – well run or poorly run, "cage-free" or not – are basically the same; all birds come from similar Frankenstein-like genetic stock; all are confined; none enjoy the breeze or warmth of sunlight; none are able to fulfill all (or usually any) of their species-specific behaviors like nesting, perching, exploring their environment, and forming stable social units; illness is always rampant; suffering is always the rule; the animals are always only a unit, a weight; death is invariably cruel.
>
> Needless to say, jamming deformed, drugged, overstressed birds together in a filthy, waste-coated room is not very healthy. Beyond deformities, eye damage, blindness, bacterial infection of bones, slipped vertebrae, paralysis, internal bleeding, anemia, slipped tendons, twisted lower legs and necks, respiratory diseases, and weakened immune systems are frequent and long-standing problems on factory farms. Scientific studies and government records suggest that virtually all (upwards of 95 percent of) chickens become infected with E. coli (an indicator of fecal contamination) and between 39 and 75 percent of chickens in retail stores are still infected. ...
>
> Every year fifty billion birds are made to live and die like this. It cannot be overstated how revolutionary and relatively new this reality is – the number of factory-farmed birds was zero before ... 1923 ... .

He's basically saying that most chickens we eat, even if "cage-free," are produced by the factory farms in the same cruel way. The chickens can't really move around and act like chickens. They're produced in filthy, stinking, awful henhouses, fed hormones to keep them alive, and then we eat them - ingesting that pleasant and mellow experience of life they've had. If you want to learn more about this, check out *Eating Animals*.

An excellent visual summary of our food production system is the movie *Food, Inc.* The movie tells how it's changed so much that, back around the turn of the last century, the average farmer could feed six or eight people; now the average American farmer can feed 128 people. The movie shows how a handful of companies have changed how we eat. "McDonald's brought factory work to the back of the restaurant and changed how restaurants operated. They changed the way meat was produced. The companies redesigned the way chickens are produced; they redesigned chicken breasts. The chickens they grow never see sunlight. Antibiotics are put in the feed and pass through so they don't work anymore."

Another great book on the topic is Michael Pollan's *An Omnivore's Dilemma*. He refers to the industrial food producers in our country as Concentrated Animal Feeding Operations – CAFOs. They're so different from the farms and ranches of yesteryear that a new term was needed. Pollan describes the incredible changes that have taken place across our nation. "The new animal and human landscapes were both products of government policy. The postwar suburbs would never have been built if not for the interstate highway systems, as well as the G.I. Bill and federally subsidized mortgages. The urbanization of America's animal population would never have taken place if not for the advent of cheap, federally subsidized corn."

That's a theme that's emphasized in the book. So many things are made from corn; it's like all we do is drink and eat corn, particularly corn syrup. And we feed huge amounts of it to cattle, *that don't normally eat corn*! They're grass eaters, but we feed them corn and animal parts because it's cheaper, even though it makes them sick. Then, all the antibiotics and hormones end up in our bodies and in our children, and in our waterways now producing *intersex fish*. Yay, us!

According to Pollan, we've robbed animals of their humanity, so to speak. We don't allow them to be individuals with personalities and do the things chickens, pigs, and cows are supposed to; we don't let them do anything they're born to do. The more I read these books, the more disgusted I get with the system we've created for ourselves. Both he and Foer describe some pretty awful things about the animal harvesting situation, and here's an example from Pollan:

Egg operations are the worst from everything I've read; I haven't managed to actually get into one of these places since journalists are unwelcome there. Beef cattle in America at least still live outdoors, albeit standing ankle-deep in their own waste and eating a diet that makes them sick. And broiler chickens, although they are bred for such a swift and breast-heavy growth they can barely walk, at least don't spend their lives in cages too small to even stretch a wing.

That fate is reserved for the American laying hen, who spends her brief span of days piled together with a half-dozen other hens in a wire cage the floor of which four pages of this book could carpet wall to wall. Every natural instinct of this hen is thwarted, leading to a range of behavioral "vices" that can include cannibalizing her cage mates and rubbing her breast against the wire mesh until it is completely bald and bleeding. ... Whatever you want to call what goes on in those cages, the 10 percent or so of hens that can't endure it and simply dies is built into the cost of production. And when the output of the survivors begins to ebb, the hens will be "force-molted" – starved of food and water and light for several days in order to stimulate a final bout of egg laying before their life's work is done. ...

This is another example of the cultural contradictions of capitalism – the tendency over time for the economic impulse to erode the moral underpinnings of society. Mercy toward the animals in our care is one such casualty. The industrial animal factory offers a nightmarish glimpse of what capitalism is capable of in the absence of any moral or regulatory constraint whatsoever. ...

No other country raises and slaughters its food animals quite as intensely or as brutally as we do. No other people in history have lived at quite so great a remove from the animals they eat. Were the walls of our meat industry to become transparent, literally or even figuratively, we would not long continue to raise, kill, and eat animals they way we do. Tail docking and sow crates and beak clipping would disappear overnight, and the days of slaughtering four hundred head of cattle an hour would promptly come to an end – for who could stand the sight? Yes, meat would get more expensive. We'd probably eat a lot less of it, too, but maybe when we did eat animals we'd eat them with the consciousness, ceremony, and respect they deserve.

We don't need hotels, bars, and restaurants on every river and lake on the planet. We have to realize that taking too much only hurts us all ultimately. Putting out too much of our trash will crush the Earth's ability to sustain us. In *Our Choice*, Al Gore says:

Human civilization and the earth's ecological system are colliding and the climate crisis is the most prominent, destructive, and threatening manifestation of this collision. It is often lumped together with other ecological crises, such as the destruction of the ocean fisheries and coral reefs; the growing shortages of freshwater; the depletion of topsoil in many prime agricultural areas; the cutting and burning of ancient forests; the extinction crisis; the introduction of long-lived toxic pollutants into the biosphere and the accumulation of toxic waste from chemical processing, mining, and other industrial activities; air pollution; and water pollution.

These manifestations of the violent impact human civilization has on the earth's ecosystem add up to a worldwide ecological crisis that affects and threatens the habitability of the earth. But the deterioration of our atmosphere is by far the most serious manifestation of this crisis. It is inherently global and affects every part of the earth; it is a contributing and causative factor in most of the other crises; and if not quickly addressed, it has the potential to end human civilization as we know it.

For all its complexity, however, its causes are breathtakingly simple and easy to understand. All around the world, we humans are putting into the atmosphere extraordinary amounts of six different kinds of air pollution that trap heat and raise the temperature of the air, the oceans, and the surface of the earth.

I don't know why that's so hard for some folks to understand. Guys like Frosty E. Hardiman, a 43-year-old computer consultant and evangelical Christian. As reported in an article titled, "Gore Film Sparks Parents' Anger," by Blaine Harden, in the *Washington Post* on January 25, 2007, he believes the warming planet is "one of the signs" of Jesus Christ's imminent return for Judgment Day. You see, Mr. Hardiman, father of seven, wouldn't let one of his daughters watch Gore's movie *An Inconvenient Truth* in her 7th grade class. "No, you will not teach or show that propagandist Al Gore video to my child, blaming our nation – the greatest nation ever to exist on this planet – for global warming," he wrote in an email to the school board.

I feel sorry for this dude in oh-so-many ways. First, why is he so defensive about our country? What's he afraid of? We have to believe we're the best or we're not any good? And, really, global warming is a sign that Jesus is coming? Plus, what's wrong with his daughter seeing the movie and making a judgment *for herself*?

Of course, following the extreme paranoia evoked by 9/11 and then the presidency of black Barack Obama, fewer Americans have been believing in global warming, according to a *Washington Post* article titled "Fewer Americans Believe in Global Warming, Poll Shows," by Juliet Eilperion, on November 25, 2009. According to the poll cited in the article, almost 90 percent of Democrats say global warming has been happening, compared to about 50% of Republicans, and more college graduates believe it's happening versus non-college graduates. The perception of separation lives on in popular climatology. (But please note that even one of the main global warming skeptics, after formally studying the scientific facts, now agrees with the prevailing science. "Robert Muller, Global Warming Skeptic, Now Agrees Climate Change is Real," by Seth Borenstien, October 30, 2011, in the *Huffington Post*.)

It's OK, though; it's just a human thing. It might even be an American thing. We are the world's leaders, right? The land of the free and the home of the brave? *We are exceptional!* We are the best!! Except that, according to a *Washington Post* article titled, "It's Natural to Behave Irrationally," by David Farenthold, on December 8, 2009:

> *"We are collectively irrational, in the sense that we should really care about the long-term well-being of the planet but when we get up in the morning it's very hard to motivate ourselves," said Dan Ariely, a professor of behavioral economics at Duke University. ... Psychologists studying the issue say that the now familiar warnings about climate change kick at emotional dead spots in American brains. Researchers have only theories to explain why people in the United States have done less than those in such places as Europe and Japan. Some think Americans are culturally leery of programs the government might develop to target climate change, trusting instead that the free market will solve major problems.*

That sort of explains things. The American spirit that we all celebrate and extol is afraid to solve problems if government is involved. It gets back to this fear of government - of anarchy - that is – excuse me – *fucked up*. If you go around all the time looking for the bogeyman, you *will* find him. Here, we're just talking about trying to protect the planet. What's so scary about that? Why can't we work on it together through our communal institutions? The article continues:

> *Another problem with climate change is called, more obscurely, "system justification." This refers to humankind's deep-seated love for the status quo and willingness to defend it.*

Hey, that's conservatism! Defending the status quo. Dadgone it, life is good, and we're happy and making money, so why concern ourselves with possible implications to the planet. It'll be OK. From the article again:

> *A third problem is that psychologists say humans can fret about only so many things at once – the technical term is the "finite pool of worry."*

So, if you're worried about damnation, money, attacks, and all sorts of mean and nasty things, it's just hard to get jazzed about real threats to the species.

> *Psychological researchers say one possible way to overcome all these obstacles needed to curb carbon emissions as "saving" the American way of life, instead of changing it. Another is to pair warnings about climate change with concrete suggestions about what to do, so people can act instead of just stewing in worry.*

In other words, we can dedicate ourselves to ourselves, instead of to fear. Because, if we all get caught up in fear, we're doomed. In other words, if we can't solve our problems together, could be a heap 'o trouble. For example, according to a *Time* article, Land of Hope," by Alex Perry, on December 13, 2010:

> *Africa's deserts, in short, are all kinds of dangerous. And climate change means the badlands are growing. The U.N Food and Agricul-*

> ture Organization says that on the southern edge of the Sahara, an
> area the size of Somalia has become desert in the past 50 years. The
> U.N. Environment Programme(UNEP) says 14 African countries
> currently experience water scarcity or stress, a number that will rise
> to 25 by 2025. In a May 2008 report, the U.N. Convention to Combat
> Desertification (UNCCD) said 46% of Africa is threatened by land
> degradation. ...
>
> That's the gloomy consensus: Africa, beset by conflict, hunger and
> disease, is being hit by a new disaster that combines them all – though
> Africa produces just 2% of global emissions. For some, that's a call to
> action: African governments will make desertification a key issue at the
> annual global Climate Change Conference,...

Do you think anyone will listen to what *Africa* wants, or needs? But can't we see that such environmental and economic stress in various parts of the world will lead to conflict? Here's another example. I didn't know Peru had 70 percent of the world's tropical glaciers that are crucial for human use of water for drinking, plant production, and even electric power. Rising temperatures and changes to the water supply over the last half-century, however, have hurt crop production and the supply of fish. "Without international help to build reservoirs and dams and improve irrigation, the South American nation could become a case study in how climate change can destabilize a strategically important region, according to Peruvian, U.S., and other officials." "Melting Glaciers Threaten Peru on Many Fronts," *Washington Post*, by Heather Somerville, January 17, 2011. What would happen if millions more Africans and South Americans were starving to death; how would that affect chances for global peace?

> Global warming could lead to warfare in three different ways. ... The
> first is conflict arising from scarcity. ... The second cause of climate
> wars is the flip side of scarcity; the problem of an increase in abun-
> dance. ... [W]hat happens if some tempting new [oil] field pops up
> in international waters contested by two great powers? Or if smaller
> countries with murky boarders start arguing over newly arable land?
> ... Finally, we should also worry about new conflicts over issues of
> sovereignty that we didn't need to deal with in our older, colder world.

> *Consider the Northwest Passage, which is turning into an ice-free corridor from Europe to Asia during the summer months. ... [W]e need to get our heads around the idea that global warming is one of the most serious long-term threats to our national and personal security.* "Global Warming Is Just the Tip of the Iceberg," by James Lee, *Washington Post*, January 4, 2009.

Ban Ki-moon, the Secretary-General of the United Nations, wrote an article in *Time* magazine on April 28, 2008 titled, "The Right War ... why a greener planet would be a more peaceful one." Interesting concept, huh, that focusing on environmental solutions can help bring peace. Here's what he said:

> *Many of the challenges we face, from poverty to armed conflict, are linked to the effects of global warming. Finding a solution to climate change can bring benefits in other areas. A greener planet will be a more peaceful and prosperous one too. ... The basic building block of peace and security for all peoples is economic and social security, anchored in sustainable development. It is a key to all problems. Why? Because it allows us to address all the great issues – poverty, climate, environment and political stability – as parts of a whole.*
>
> *Consider Darfur. [Its] violence began with the onset of a decades-long drought. Farmers and herders came into conflict over land and water. If this root problem is not addressed – if the challenges of poverty alleviation, environmental stewardship and the control of climate change are not tied together – any solutions we propose in Darfur will at best be a temporary Band-Aid. ... No place is immune, neither the arid Sahel of Africa nor the grain exporting regions of Australia nor the drought-prone Southwest of the U.S. ...*
>
> *[W]e might recall the historic importance of American leadership in this fight. In 1963, President John F. Kennedy told the U.N. General Assembly, "The effort to improve the conditions of man ... is not a task for the few. It is the task of all – acting along, acting in groups, acting in the United Nations. For plague and pestilence, plunder and pollution, the hazards of nature and the hunger of children are the foes of every nation. The earth, the sea and the air are the concern of*

every nation. And science, technology and education can be the ally of every nation."

Of course, they can. By working together, we can achieve anything. Not by working against each other, or working just for money, but by combining our forces and resources for the greater good, we can do whatever we want as a species. Could we reach a goal of 100% renewable energy in 20 years? According to an article titled, "100% Renewable Energy Achievable by 2030," by Joanna Zelman, on January 25, 2011 in the *Huffington Post*:

> *New research says we can. A report published in the journal Energy Policy claims that by 2030, the world can achieve 100 percent renewable energy if the proper measures are taken. What exactly are these measures? According to PhysOrg, over 80 percent of our world's energy supply currently comes from fossil fuels. We would need to build approximately four million wind turbines, nearly 2 billion solar photovoltaic systems, and about 90,000 solar power plants. The 5 MW wind turbines needed are up to three times the capacity of most of our current wind turbines.*

Sure it's daunting, but don't you think humans could do it, if they/we just put their/our minds to it? How about save the oceans; could we do that?

> *Amid the collapse of once-rich fisheries around the world, policymakers, fisherman and environmentalists have been debating a controversial question; Can a fishery be saved by giving those who harvest the sea a guaranteed share of its bounty, rather than have them compete to see who can extract the most the fastest? A study published ... in the journal Science, conducted by two economists and an ecologist, suggests that the answer is yes. The authors – two from the University of California at Santa Barbara and one from the University of Hawaii – surveyed 121 fisheries worldwide where individuals receive a predetermined portion of a fishery's catch limit and found that they were half as likely to have collapsed as those without "catch share" system.*
>
> *In addition, the researchers found that when a fishery that had*

> *relied on traditional methods – such as seasonal limits or overall catch restrictions – was converted to using catch shares, the change did not just slow the fishery's decline; it stopped it. Once people are given a fixed share in a fishery, said lead author Christopher Costello, they are less likely to overfish, because they have a financial interest in having the species thrive.* "Study Suggests Sharing the Catch Could Save Fisheries," by Juliet Eilperin, *Washington Post*, September 22, 2008.

Isn't that amazing? If all the fishermen are in it together, they can share the catch and maintain the fish. If it's dog-eat-dog, so to speak, the fisheries collapse. In other words, if the individuals have the perspective of maintaining the bounty collectively, they can; but if it's a free market approach – *so long, Charlie*. I'd say the former is oneness thinking, and the latter clearly not. If everyone is out just for themselves, they're afraid they won't get enough, so they take as much as they can. This highlights as well as any example the great potential with a new way of thinking that throws out former destructive concepts.

This also is a perfect juncture to do a little NWAL. We'll assume we're in the *now*. Let's *witness* the different perspectives.

I see many who treat the planet simply as a shopping mall and personal trash dump. They ruthlessly take the resources out of the land and the sea, and expel waste indiscriminately. Money and jobs are more important than the impact on the environment. Eating well and inexpensively is more important than harvesting food morally. Like the hunters at the Island who didn't even want to eat the ducks they killed, there's no conscious awareness of the importance of oneness. I see people using those heinous plastic bags all over town. I hear the climate change deniers poo-poo scientists.

Then, I see people who treasure the bounty and biodiversity of the planet. They try to use recycled materials, they don't use plastic bags or water bottles. A lot of my friends are that way. There are so many people who try to live simply, so that others may simply live.

Now, *access* the nature of these different approaches; which are based in fear and which in love? The thought process that says the earth's bounty – all of its oil, plants, animals, minerals, and fish – are there for the taking – are *mine*? The political position that says no need to worry about deforestation or preserving the fish in the sea, because *we need the jobs*? The view that ignores the suffering and cruelty to animals, and at the same time supports an industry that makes foods that are not even good for us?

Or, the principle that says we have to maintain the fullness and diversity of our environment, and our clean air, water, and land – we can't take more than we need? That we need to look at the impact on the entire system? It's here for all of us; *we are a part of it?* That animals are part of it too, and we must respect them and all living things?

It doesn't take much analysis to determine which principles are about oneness, and which are only seeing separation. The political position that favors jobs over the environment screams: *us against them, mine not yours,* and *drill baby drill*, as opposed to the famous Iroquois saying:

> *In our every deliberation, we must consider the impact of our decisions on the next seven generations.*

When we have run our NWAL analysis in prior chapters, one thing we haven't focused on is the "L" for Love – proceed to act in a *loving* way. When talking about religion, I think NWAL would suggest that we forsake doctrine and just try to act and think in accordance with love and oneness. When we try and figure out what side we are on politically, NWAL suggests that liberalism is preferable, because it is much more based in oneness. We can decide to vote for liberals. When thinking about money and the financial system, we can simply try and not to be too influenced by the greedy desire to have more money at all costs. Making and having more money is totally cool, but obsessing about it to the detriment of our oneness ultimately perceives separateness.

But what do we do about the environment? If conservation of the environment and promoting/protecting our biodiversity is love-based, how do we do that? Can we do as much as we need to overcome humans' impact on the planet?

It's a tall order. Not one of us can solve the problem entirely. However, we can each do our part to help protect the planet. It's difficult to live without having some effect on the planet. Here in Bethesda, we have electricity, heat our houses, flush our toilets, use plastic, and drive our cars. All those things have an effect – constitute our true footprint on the planet.

So, to me, the best we can do is – *the best we can do*. I try and recycle paper, plastic, cans, etc. We put in a solar water heater, added insulation to our attic, installed thick blinds, and even painted the west side of our roof white to cut down on the heat absorbed by the planet. My wife drives a hybrid; I couldn't afford one, yet. We want to install geothermal heat to minimize our energy usage and reduce costs. I'm sorta vegetarian, not eating much meat any more, but fish, vegetables, eggs, and cheese. I should go the whole way, but every little bit helps. I'm working on it.

But it's hard to do as much as I want. I understand we're not all saints and can't live on ethereal vibes. I will think, act, and vote to improve environmental policies and human conduct. That is a positive step.

We have such a huge effect on the planet, just based on *how we think* about things. And we do think about sex a lot.

Twenty | Bigger Love

Let me be honest. With apologies to my wife and kids, I have to say I think about sex a lot. When I drive home in the summer through downtown D.C., I can't help but check out women walking on the street. It's a distraction worse than texting.

I remember my friend Ernie years ago saying the same thing. I thought it was bold to admit. He was sitting with my ex-wife and me at the Island, and said: "I think about sex constantly. Sitting here, looking at you [meaning, her], I'm thinking about sex."

My wife's sister says most guys think they don't get enough. I don't know if all men are like this; but, my guess is, pretty much. In fact, researchers conducting a study at the University of Montreal failed to find a single man who had never watched porn. "All Men Watch Porn, Scientists Say," by Jonathan Liew, *The Telegraph*," December 2, 2009. Professor Simon Louis Lajeunesse said that the men interviewed regularly watched Internet porn, but maybe it's just part of our make-up; you know, in our jeans? Here's the answer to a question about this on the web by Gunborg Palme, certified psychologist:

> *According to evolutionary scientists, natural selection has favoured men who want to have sex with as many women as possible. Men with such a genetic code have bigger chances to spread their genes. On the other hand, a man knows that his own wife would be furious and heartbroken if he cheated on her. Some married men argue that they have too little sex or have it too seldom. Those men can calm down their genes by looking at pornography, thus avoiding the risks of infidelity. ... However, men's interests in pornographic material*

often hurt women's feelings. Some women feel that they are not good enough and that their men dirty their love with repulsive pictures. It is difficult to find a solution for such a problem. A man who cannot manage to give up his porno watching must be very discrete in his hobby so that his wife will not discover it. Men, who both cheat on their wives and watch pornographic material, usually say that it is their wives they love, the other ones are just sexual objects. Women find it very difficult to accept. http://web4health.info/en/answers/sex-porn-men.htm

It's interesting that watching people having sex can be thought of as so bad, when everybody does it, it's the reason for being here, and what's keeping us here. Women can be sorta judgmental about men in that regard.

However, a report from the Czech Republic, "suggests that making pornography more accessible could lead to a drop in child sex abuse. Rates of abuse in that country declined after a ban on all sexually explicit material, including child pornography, was lifted by the new democratic government in 1989. Scientists think the reason may be that potential offenders use porn as a stand-in for sex crimes." *Time* magazine, December 13, 2010.

So, could porn be a good thing? I know many women feel it's degrading to females generally. I agree, for some of it. But for the everyday schmos posting on www.youporn.com, I'd say it's just fun. What's better than having sex? And, if nothing else, porn is educational and can free people from their inhibitions and repressed desires.

But many humans have serious issues with sex. I described earlier how fears of men killing children led to significant physiological changes in women. Flip side, men's fears give rise to their murderous ways in the first place. I wonder if this inherent need for men to spread their seed is related to the seemingly pathological need for power, such as national power? Men seem to work out their deepest fears by having wars and killing other people.

Remember my Mom said, the "Lord made the sex urge a little too strong." How strong? It drives some men crazy and leads them to do heinous stuff. But could this drive be used to benefit society? I think it's fair to say, given the choice, most men would take a blowjob rather than kill someone. Don't you? "Here, man, you can shoot this dude in the head for any reason, or you can have this foxy young woman just go crazy on your dick." I'm thinking the dude is picking fellatio; *every time*. Let's not deny that aspect of men, and try to work *with* it. Maybe we can figure out a way for sexuality in society to work in our favor and help lead to a more peaceful world? I think we should be open to potential solutions.

For one thing, I don't think we should be so weirded out by sex. For example, many people are fraught with anxiety about being *nude*. We look at naked animals all the time, but seeing a naked person brings up tension for lots of folks. Here's a response from "Ask Amy" in the July 14, 2007, *Washington Post*, to a question from a guy who liked to do house and yard work in the nude:

> *I understand that as a naturist you are attached to your clothing-optional way of life, but if there's any activity that I think of as being incompatible with wearing clothing, preferable many layers of it, it would be yard work. Good god, man, watch out for that hedge trimmer!*
>
> *The website for the Naturist Society (www.naturistsociety.com) offers this very practical and sensitive advice for you: "Typically, women are more wary than men of clothing-optional venues. But everyone, male and female, has 'body issues.' For some, the idea of being seen nude – and seeing others nude – is filled with psychological tension."*

Yikes! Body issues? *Psychological tension?* Are we really so squeamish about our bodies? And isn't that a fear? For women, a fear of being assaulted, and maybe for men that we wouldn't look as well-endowed as we'd like?

Funny thing, my teacher Avery Kanfer told me that being comfortable with the nakedness of a partner is *love*. That's right. If you don't have fear of someone seeing you in the buff, that's love.

But here's another example of our collective prudishness. Lisa de Moraes of the *Washington Post* in an article, "Some Local Stations Cautious in Gauguin Painting Coverage," on April 6, 2011, reported on the television coverage of an attack on Paul Gaugin's $80 million masterpiece by a woman who said the painting's semi-nudity was evil. OK, that's f'ed up in the first place, but how did the TV stations cover the event?

> *Well, if you're Fox-owned-and operated station WTTG, you blur out the nipples on the two semi-clad Tahitian women portrayed in the*

famous late-19th-century painting. Only then, it kinda appears that Gaugin painted the women to be anatomically incorrect. ...

And if you're Allbritton-owned ABC affiliate, WJLA, you go with the Bouncing Banner. That is, you push up that banner of type that usually runs at the bottom of the screen, so it serves a dual purpose: conveying the salient point "Gaugin Painting Attacked," while also modestly covering up the native women's breasts. ...

TV stations are understandably very wary of women's breasts ever since Janet Jackson showed us hers, by way of breaking up the monotony of CBS's broadcast of Super Bowl XXXVIII in 2004. Remember, it was the CBS stations that the Federal Communications Commission slapped with those hefty fines when the aging popstar experienced her wardrobe malfunction.

Isn't it something that people get so freaked seeing (or having their children see) nakedness? More than that, many Americans flip about sex education. An article in *Time* magazine on March 30, 2009, titled, "How to End the War Over Sex Ed," by Amy Sullivan, called teen-pregnancy and birth rates in the United States an "epidemic compared with those in other Western countries." One-third of states don't require sex ed, but the teen birth rate in the United States is the *highest* in the developed world. In Texas and Mississippi, the teen birth rates per 1,000 in 2006 were 63.1 and 68.4, respectively, the U.S. average was 42.5, nearly double the next highest U.K. at 26.7, and those godless nations like Switzerland and the Netherlands were 4.5 and 3.8.

In the *Washington Post* on April 14, 2007, an article titled, "Study Casts Doubt on Abstinence-Only Programs," by Laura Stepp, said: "A long-awaited national study has concluded that abstinence-only sex education, a cornerstone of the Bush administration's social agenda, does not keep teenagers from having sex." Another article in the *Washington Post* on December 29, 2008, titled "Premarital Abstinence Pledges Ineffective, Study Finds," by Rob Stein, said: "Teenagers who pledge to remain virgins until marriage are just as likely to have premarital sex as those who do not promise abstinence and are significantly less likely to use condoms and other forms of birth control when they do, according to a study released today."

Still, the Catholic Church, "teaches that because contraception suppresses the possibilities of procreation, and therefore violates the natural law, it is always wrong." *Washington Post*, "Bishops Stand firm on Birth Control," by Tim Townsend,

November 25, 2006. So, when I got the snip-snip, that means *every time since* then was a sin? I'm just not gonna feel guilty about that. It's insane that we don't teach our children more about one of the most important aspects of being human. It's like we're afraid of life.

Many of the problems we have with sex clearly come from religion. Christianity ends up with priests and brothers abusing children because they're horny. And Islam is probably worse than Christianity. Women are routinely abused in Muslim countries; no secret about that. Religion lets men get away with a lot of crap when it comes to how women are treated.

> *A major obstacle to recognizing abuse, experts say, can be Islam itself. The religion prizes female modesty and fidelity while allowing men to divorce at will and have several wives at once. Many Muslim men believe they have the right to beat their wives. An often-quoted verse in the Koran says a husband may chastise a disobedient wife, ... A closely related problem is nervousness about the prospect of [women] leaving home. In many Muslim societies, women are protected and housed by their fathers and then by their husbands; if they date or live alone, they risk being tarred as prostitutes. ... [E]ven U.S.-educated women can be browbeaten into enduring abuse for fear of shaming their families or facing cruel gossip at the mosque.*

Interesting thing, this article was written about abuse of Muslim women in the Washington, D.C. area. "For Some Muslim Wives, Abuse Knows No Borders," by Pamela Constable, *Washington Post*, May 8, 2007. This abuse by men in the name of religion leads women in some countries to try and escape their tribulations by – *setting themselves on fire*. Shayma Amini is the head nurse at Herat Regional Hospital's burn unit in western Afghanistan. According to an article in *Mother Jones* magazine, January/February 2011, "Trial by Fire," by J. Malcolm Garcia:

> *Amini guesses she's seen at least a thousand self-immolation cases since she started at the hospital 13 years ago – almost all of them young women seeking to escape abusive marriages or the prospect of being turned out into the street by men who no longer want them. Women have been beaten – or starved, as a controversial 2009 law*

> *allows some angry husbands to do. ... The UN Development Fund for Women has found that 70 to 80 percent of Afghan females are forced into marriage – most before the legal age of 16, more than 60 percent into physically abusive households.*

I'd say there's *not* a whole lotta love at the procreational foundation of these societies. Sure, culture is an issue, and religion is at the heart of the problem, but so is sex. These are men wanting sex exactly as they want it, with no regard for the women.

Men are a huge problem when it comes to striving for world peace and more fulfilling societal relationships. We talked before about the issues with sex trafficking, aka slavery. We've got to figure out this situation for the better. But remember who we're dealing with – males who used to kill human children if fathered by a different guy, to the point women changed physiologically to try and avoid the murderous cads. To further illustrate the point about men's idiocy, here's an excerpt from a piece in the *Washington Post* on December 6, 2006, about a dude in Allentown, Pa.

> *A man testified that he repeatedly sexually assaulted two girls because his wife spent too much time playing bingo. Floyd Kinney Jr., 49, said he began molesting children as a way to lash out at his wife, who "was never home."*

Really, Dude. That's sick. Yet around the world, as we saw earlier, cultures allow men to do it with whoever they want, except the women they might actually marry. It's like we men have a fear of women that causes us to put them down. It's the men that desperately need some educatin' to help our world mature.

For example, according to the Health & Science column in *Time* magazine on July 25, 2011, roads would be safer with more women behind the wheel, because from 1998 to 2009, "77.3% of fatally injured drivers were male because men often take risks like driving drunk and speeding." Also, the Los Angeles Police Department, "paid out $66.3 million for police-brutality in the 1990s, 96% for acts by male officers." And, "A 2009 study showed that 47% of male investors held on to stocks too long, compared with 35% of women. From 2000 to 2009, hedge funds run by women averaged annual returns of 9%, compared with the 5.82% achieved by men."

It's also fair to say men's aggressiveness can be blamed for most economic problems. Debora Spar wrote in the *Washington Post* on January 4, 2009, in an article titled, "One Gender's Crash:"

> [A]s the financial crisis unfolds, I can't help but notice that all the perpetrators of the greatest economic mess in eight decades are, well, men. Specifically, they are rich, white, middle-aged guys, same as the ones who brought us the Watergate scandal in the 1970s, the Teapot Dome Scandal in the 1920s and, presumably, the fall of Rome. ... A Catalyst Research study last year found that women make up almost 60 percent of the workforce in Fortune 500 finance and insurance companies but account for only 17.9 percent of corporate officer positions and none of the chief executive officers. ... It may be that women perceive and act on risk in subtly difference ways; that they don't, as a general rule, embrace the kind of massively aggressive behavior that brought us a Dow of 14,000 and then, seemingly overnight, a crash of epic proportions.

I also don't think humans down through the centuries have done so swell with men at the helm. Just think of all the awful wars and genocide; you know it's the dudes at fault. We need to teach men how to chill out. We don't need to have them fighting all the time. We need them not to be so mad, angry, and horny. And figure out how to use their incredible aggression and sex drive for positive results. We must teach men *love* – that's what I'm saying.

You know what? I'm trying. This whole effort is to help folks see reality in a different light. I have faith in men, and in women. We should collectively work on educating everyone to be more loving humans. I think we can do it. One thing the experts do agree on, to have a better world, is that we must improve the lot of women.

> This is the tantalizing idea for activists concerned with poverty, with disease, with the rise of violent extremism: if you want to change the world, invest in girls. ... Across the developing world, by the time she is 12, a girl is tending house, cooking, cleaning. She eats what's left after the men and boys have eaten; she is less likely to be vaccinated, to see a doctor, to attend school. In sub-Saharan Africa, fewer than 1 in 5 girls make it to secondary school. Nearly half are married by the

> time they are 18; 1 in 7 across the developing world marries before she
> is 15. Then she get's pregnant. The leading cause of death for girls 15
> to 19 worldwide is not accident or violence of disease; it is complica-
> tions from pregnancy.
>
> There are countless reasons rescuing girls is the right thing to do.
> It's also the smart thing to do. Consider the virtuous cycle: An extra
> year of primary school boosts girls' eventual wages by 10% to 20%.
> An extra year of secondary school adds 15% to 25%. Girls who stay
> in school for seven or more years typically marry four years later and
> have two fewer children than girls who drop out. Fewer dependents
> per worker allow for greater economic growth. And the World Food
> Programme has found that when girls and women earn income, they
> reinvest 90% of it in their families. They buy more books, medicine,
> bed nets. For men, that figure is more like 30% to 50%. Nancy Gibbs,
> "The Best Investment," *Time* magazine, February 14, 2011.

In his book, *Our Choice*, Al Gore says that the four most important factors to stabilize population, and therefore reduce the stress on Gaia, are:

1. *The widespread education of girls.*
2. *The social and political empowerment of women to participate in the decisions of their families, communities, and nations,*
3. *High child-survival rates, leading parents to feel confident that most or all of their children will survive into adulthood.*
4. *The ability of women to determine the number and spacing of their children.*

It sounds like such a good idea, and things are improving, but progress is slow in many areas. Here's what a UNICEF release said on April 18, 2005:

> More children than ever are going to school, in part because more
> girls are going to school: That's the good news from UNICEF's latest
> *Progress for Children* report, focusing on gender parity in primary
> school attendance. However, millions of girls are still denied a basic
> education. While the gender gap in primary school attendance is

shrinking globally, in many parts of the world it still yawns wide. The barriers keeping girls out of school in the developing world not only rob them of future opportunity, but impact their very health and survival.

"Education is about more than just learning. In many countries it's a life-saver, especially where girls are concerned," said UNICEF Executive Director Carol Bellamy, at the launch of the report. "A girl out of school is more likely to fall prey to HIV/AIDS and less able to raise a healthy family."

The world has made impressive gains towards getting equal numbers of girls into primary schools as boys. Some 125 out of 180 countries for which data were available are on course to reach gender parity by 2005 - a target set by the UN as part of the Millennium Development Goals. Yet the global average masks huge pockets of inequity. Three regions -- Middle East/North Africa, South Asia and West/Central Africa -- will not meet the gender parity goal.

Gender parity is a prerequisite if the world is to achieve universal primary education by 2015, the target date set by the UN for a key Millennium Development Goal. The shrinking gender gap has helped reduce the total number of children denied a primary education. According to projections, fewer than 100 million children may be out of primary school by 2005, down from an estimated 115 million in 2001. Whatever the exact figure, it is clear that far too many are still shut out of the classroom, and at the present rate of increased school attendance, the goal of universal primary education by 2015 won't be met.

I'm all for education of girls, but let me ask one question. Though it seems there's more education provided for boys and young men, why are men *still* fighting and killing? Do you think we need a different kind of education? And isn't a big problem globally that many men won't let girls and women be educated? If men don't know how to devote the resources to help their families, why isn't that a top priority? Aren't men the ones who continue this rapacious need for young girls around the world that leads to sex trafficking? And it's not just a problem over there.

Contrary to the belief that the sexual exploitation of children only

> takes place in Third World countries, it occurs daily in the United States. A well-known trafficking highway leads down the Eastern seaboard, through Miami, Charlotte, Atlanta, Phoenix, Las Vegas, Los Angeles, Portland, Seattle, Hawaii, Minneapolis and more. No city - regardless of its size – is immune.
>
> While child prostitution is an equal opportunity crime, Caucasian girls bring top dollar on the streets - $60 to $300 for 15-minute to one-hour increments. Blondes with blue eyes known as "swans" command even more and in Minnesota, Native American girls are in high demand. Mickey Goodman, "Sex Trafficking in the United States; Children Across America are Unseen Victims," *Huffington Post*, January 23, 2011.

Seriously, what's up with this? We talked about the problems in many other countries, particularly Muslim nations, where affluent men will only marry virgins so put their sexual energy prior to that on prostitutes. But here in America, men – presumably some guys we might even know – like to *fuck children*?

When I hear about all this, though I understand about being a man, it's clear that *men have to change*. I mean even Al Gore was alleged to have groped a Portland massage therapist at an upscale hotel in 2006. Then, on June 1, 2011, he and Tipper announced they were separating. *Washington Post*, "Police in Ore. Reopen Investigation of Groping Allegations Against Al Gore," by Nigel Duara, July 1, 2010. Don't you think he just hit the age where he wanted something different for a change?

A female friend of mine said to me recently, "I don't understand men." But I don't think it's hard. Let me try to explain.

Men want love and want to be loved. No difference from women on that score. That's the basic "need," if you will. Men want to be respected and admired, which is being loved. Same with women. But from there, the way men attempt to achieve love seems to me different than women.

Men want to be seen as strong, potent, and powerful, and want history to reflect that. Their maleness compels them to demonstrate that strength. They want to create a lineage of might. They don't want to be weak and forgotten. They want to be Kings! If they can't – or when they're *afraid* they won't - succeed in being loved, they too often accommodate those fears by trying to become richer and more important than others, by beating others – at games or physically, or even killing others. That's to show they're not less than the other men. There's a physical aspect to men working out their fears, which is not generally conducive to a fair, just, and loving society.

On a very elemental level, men want to spread their seed far and wide, to the furthest corners of the earth. They want to conquer time and place, and defeat death. They want to send their sperm out – their *strong and magnificent sperm* – to be consumed and admired by the women. Not just one woman; many women. Different women. Men see themselves in heroic positions depositing their seed among gorgeous women capable of producing kings. Women who want them, who have saved themselves for their super man! *Seventy-two virgins might not be enough!*

Men are obsessed with procreating. Logically, we know we can't achieve what our genes drive us to do. *But we try*. Whether men prefer women or men, this immutable force carries them along. We all want to be the *big swinging dick* in the locker room. Admired for our prodigious lingams. Achieving orgasms is just the successful conclusion of these deep desires. Having as many as possible, in many different ways with lots of different women, is a need akin to breathing. I know men's obsession with sex seems just way out of control, and it is. But, it is what it is. I think we've got to come up with some new options to allow men to fulfill their sexual desires, but channel it in a better way for society. Any ideas?

This need also is physiologically affected by fatherhood – temporarily. Recent studies show that testosterone levels in men decrease when they become fathers and become nurturing caregivers of their children. "Fatherhood, Testosterone Decline Linked," *Washington Post*, September 13, 2011. Just as women have evolved based on fears, men have fears of not successfully procreating. In other words, men chill out a bit when they finally have kids. What the study didn't show, but what I believe is true - when the kids are out of the house, the testosterone goes back up. Maybe I'm wrong, but it seems like that happened to me.

On the other hand, what's up with women? I think women need to love and be loved in different ways than men. (This is treacherous territory for a man to bring up; but here goes.) Women want to be lovely. They want to be like the quintessential Blessed Virgin Mary. Above debauchery. Beautiful, without sin, perfect. They want to be the wise grandmother; the matriarch. They want to be stylish, dressed *so fine*, and smell fabulous. Some like to look sexy; others not.

Women want to be secure. Security goes hand in hand with love for women. They don't want to be alone. A female friend whose husband passed says she just doesn't like to be alone, and she has five kids. I don't think men like being alone either, but there's something different about it.

As I've noted before, it seems women have more of a tendency toward *low self-esteem*, being afraid (a fear, right?) that they're not adequate from sort of a spousal perspective. There's even an acronym for that – *LSE*. I learned that in the book, *How to Love Like a Hot Chick*, by Jodi Lipper and Cerina Vincent. Yea, LSE –

> *Stands for low self-esteem. It is a disease that infects everyone from time to time, but Hot Chicks try really hard to cure themselves of this plague. ... LSE is not hot, ladies, and recovering from this deadly infection is the first and biggest step to truly being a Hot Chick.*
>
> *If you're not sure what we're talking about, here's an example:*
> *I am smart, I am sexy, I am funny, and I totally deserve a hot, loving relationship. I am secure with myself, I am not jealous of other girls, and I have the power to get exactly what I want. I think dating can be fun, and I want to make it fun. I will not let my LSE run my life ..."*

Many women don't want men to treat them as sex objects; many feel demeaned by that, meaning, afraid they're not worthy as humans. They don't like porn, and feel degraded by it, as I said, like it's all the men ever want. Maybe; sometimes. But in a funny way, not really degrading to women, but celebratory of women, I think.

Women aren't as into winning, killing, and beating other people. That's a real plus, in my mind. They can be bitchy, with raging hormones all over the place. Guys tend often to be focused in one direction; women have to deal with rainbows of emotions. Where some men could do it on the hour; women by the month or year sometimes. Women don't understand that "dick" thing. Women want to love and nurture their children; men want to have lots of offspring, but not nurture them so much. But women do get pretty hot about one thing – unfaithful men. Lordy, lordy.

Happens all the time. It's always the lecherous men who can't keep their flies up. Much difficulty arises around this pattern. Just the other day I read an obituary in my trusty *Washington Post*, August 26, 2011, about Shirley Eskapa. Don't know her?

> *Shirley Eskapa, 77, a South African writer [] attracted international attention for her 1984 book on marital infidelity, "Woman Versus*

Woman," which explored the tensions among libidinous husbands, betrayed wives, and the "other women," ... .

Mrs. Eskapa observed how two of her friends separately dealt with cheating husbands. One woman handled the situation by telling her husband she would wait for him to leave his mistress and return to her. He did. The other friend and her husband, both physicians, were "great intellectuals" who spoke four languages and had children who could play piano "to concert standard," ... . Mrs. Eskapa said she watched as her friend behaved, "like a shrew, all her intelligence deserted her," and "the marriage ended in catastrophe." What interested Mrs. Eskapa most, however, was how the wives directed their anger not at their spouses but toward the "other women."

Mrs. Eskapa interviewed hundreds of divorced husbands and wives and former mistresses to understand the conflicts between the women. ... She wrote about how some wives discovered their husbands' indiscretions. One hint was a husband whose tie was knotted one way at breakfast and another way at dinner.

She found that once women learned of their husbands' cheating, many wives blamed themselves and their perceived imperfections. Mrs. Eskapa, however, argued that the women weren't necessarily at fault; she said men were hard-wired for promiscuity. ... Her advice for women was to wait for the husbands' "crisis of ecstasy" to run its course. A little trickery, teamwork, or reverse seduction, of course, couldn't hurt. In her book, Mrs. Eskapa recounted how one woman took up yoga to learn how to contort her body into flirtatious positions.

Mrs. Eskapa "certainly felt strongly that instead of being enemies that they needed to see they were in a similar boat and that the man was not being honest and fair to either one of them, said Washington-based marital psychologist Audry Chapman. "She said 'the two of you need to team up.' That was unique to say at the time."

A few things here. That pattern I was talking about – male strays – repeats over and over. Sure, some women leave their husbands after having an affair. Some women are crazed by sex more than men. The basic sitch, though, is the men have a hard time not wanting to do it with more women. As Mrs. Eskapa says, men are "hard-wired for promiscuity." As another female friend says, "I hear men just want to jump you." Maybe less glamorous language than I'd prefer (as a man), but not altogether incorrect.

You also have the "hell hath no fury like a woman scorned" theme in what Mrs. Eskapa addressed. It's a deep fury, no doubt. If you take away a women's security, *she's pissed*. But the interesting aspect studied by Mrs. Eskapa is that the women took it out on the other women, apparently more so than on the man! What causes *that*? Maybe women get used to and expect men to be the way they are – from centuries of experience – but it just makes them so angry that *another woman* would contribute to their aloneness?

Given all this, is it fair to argue that marriage (as the legal structure of monogamy) may *not be* working? Well, according to an article in *Time* magazine from November 29, 2010, titled, "Marriage, What's It Good For?" by Belinda Luscombe, as reported in the Pew Research Center poll, only 28% said marriage was becoming obsolete in 1978, but 62% say that today. And 69% say there is *not* one true love for each person (though 79% of the 28 percent who said there is believed they'd found theirs). The article also noted that "Fewer U.S. adults are married, more are living alone, and more kids are born to unmarried women."

Obviously, change is happening. The statistics are different around the globe, of course, so it's risky to generalize. But here's a quote in the article from Andrew Cherlin, a sociologist from Johns Hopkins University, who wrote a book entitled, *The Marriage-Go-Round: the State of Marriage and the Family in America Today*: "One statistic I saw when writing my book that floored me was that a child living with unmarried parents in Sweden has a lower chance that his family will disrupt than does a child living with married parents in the U.S."

Ladies and gentlemen, again, does this mean *marriage is failing us*? Al Gore, Bill Clinton, Elliot Spitzer, Newt Gingrich, my father, my brother, me, all of us guys, don't really want to do just one woman forever. *It's hard*. And much more fun to share intimacy with many women; I would think that's the same for women.

In fact, despite what I've been generalizing about, that men are sex-crazed, women in some respects are more so. An interesting book is *Sex at Dawn*, by Christopher Ryan and Cacilda Jetha. They basically say that humans do not have a monogamous history, and simply aren't meant to be in exclusive relationships, like bonobos.

> *[F]emale chimps and bonobos go wild regularly and shamelessly. Females often mate with every male they can find, copulating far*

more than is necessary for reproduction. [Jane] Goodall reported seeing one female at Gombe who mated fifty times in a single day. ... [Mary Jane] Sherfey [says]: "The sexual hunger of the female and her capacity for copulation completely exceeds that of any male," and To all intents and purposes, the human female is sexually insatiable. ..."

So, maybe the conventional wisdom about sex isn't accurate in some ways? Maybe we deemphasize its importance because sex is kind of a taboo subject. Are we way off base? Here's an excerpt from a review of journalist Daniel Bergner's book, *What Women Want: Adventures in the Science of Female Desire*, positing the theory that we've sublimated how things really are when it comes to sex:

In fact, Bergner found that female sexuality is everything we tell ourselves about male sexuality - that it's base, ravenous and animalistic - is true of female sexuality. Certain qualities society has traditionally attributed to women - that they are inherently and biologically better suited to monogamy, that women's desires are based in romantic love - are 'scarcely more than a fairy tale', writes Bergner. From the website: http://www.capitalbay.com/latest-news/346692-women-s-sexual-desire-is-animalistic-and-ravenous-claims-new-book-that-turns-female-sexual-stereotype-on-their-head.html

How can we misunderstand all this? Again, from *Sex at Dawn*:

For most men and many women, sexual monogamy leads inexorably to monotony. It's important to understand this process has nothing to do with the attractiveness of the long-term partner or the depth and sincerity of the love felt for him or her. Indeed, quoting [Donald] Symons, "A man's sexual desire for a woman to whom he is not married is largely the result of her not being his wife." Novelty itself is the attraction. ... Maybe this is why some twenty million American marriages can be categorized as no-sex or low-sex due to the man's loss of sexual interest. ...

If it's true that most men are constituted, by millions of years of evolution, to need occasional novel partners to maintain and active

> and vital sexuality throughout their lives, then what are we saying to men when we demand lifetime sexual monogamy? Must they choose between familial love and long-term sexual fulfillment? Most men don't fully appreciate the conflict between demands of society and those of their own biology until they've been married for years – plenty of time for life to have grown very complicated, with children, joint property, mutual friends, and the sort of love and friendship only shared history can bring.

It's not just men though. According to an article in the *New York Times Magazine* on May 26, 2013, titled, "There May Be a Pill for That," by Daniel Bergner, women often develop a lack of lust that is technically called hypoactive sexual-desire disorder, or HSDD. It could amount to between 15 to 30 percent of women.

> [F]or many women, the cause of their sexual malaise appears to be monogamy itself. It is women much more than men who have HSDD, who don't feel heat for their steady partners. … [F]or women who've been with their partners between one and four years, a dive begins – and continues, leaving male desire far higher. (Within this plunge, there is a notable pattern: over time, women who don't live with their partners retain their desire much more than women who do.)
>
> Lesbian couples seem to fare no better, and maybe worse, in keeping their sexual ardor for each other. The term "lesbian bed death," coined by the University of Washington sociologist Pepper Schwartz in the '80's has been critiqued as overstatement but not quite as fiction. "In the lesbian community, the monogamy problem is being aired more and more," Lisa Diamond, a professor of psychology and gender studies at the University of Utah told me. "For years, gay men have been making open arrangements for sex outside the couple. Now, increasingly, gay women are doing it."

Why aren't straight folks doing that? Well, they are, just forced to keep it under wraps. The article also asks whether HSDD is really about boredom. A friend of mine said, in several conversations with women friends, that even though they are married to great and successful guys, they go out to dinner with their spouses and are bored out of their minds. Again, from the *New York Times Magazine* article: "Every woman raised a mix of possible reasons. There were demands of work, medical issues, men who weren't always as kind or nearly engaged as they could be. But at bottom there seemed to be one common cause: they had all grown tired of sex with their long-term partners."

Maybe that's the reason matrimony is not working. The divorce rate in America is 50% percent for first marriages, 67% for second, and 74% for third marriages, according to Jennifer Baker of the Forest Institute of Professional Psychology in Springfield, Missouri. www.divorcerate.org. Yet, society puts us men in a box and tells us to chill. Stay together, even though your soul leads you elsewhere. It can even happen that happily married men can fall for other men, partly because of that search for novelty "thing." http://www.huffingtonpost.com/rex-oso/extramarital-affair_b_976380.html.

For a more peaceful and loving society in the long run, I'm coming to think we need to change our relationship system. I understand that a stable familial situation is important to support kids. I don't think the specific format of the family is crucial, just that the adults love their children and try and help them, stand with them, and teach them love.

Remember my dear Bible-thumping cousin? The one who thought spiders didn't have souls? We had another conversation electronically about gay marriage. I must have asked her what the problem would be if gay humans got married.

She replied, "Well, if a man and a man could get married, or a woman and a woman, then why couldn't two men get married to a woman, or two men marry three women?"

I was *flabbergasted*. That had never occurred to me. Why would several people want to be married to one another? I didn't think anyone was suggesting that. But, in recent years, I've thought about it more. Could marriage – or more accurately, committed relationships - involving more than two people be a good thing? Could people be *happier* with several spouses? Well, one thing is you could have sex with more than just one person, like the bonobos. That doesn't sound so bad. What's the reason for monogamy, anyway?

Jared Diamond in *Why is Sex Fun?* tells us human history has seen many different mating systems: monogamy, harems, and promiscuity. In many Muslim countries today, men can have multiple wives. In the United States, we have a monogamous mating system ostensibly based on love and God's law; but that law was once interpreted so the men owned the women like cattle, which doesn't give it much credence. Relying on the Bible or Koran for relationship training ain't working, IMHO. Marriage hasn't always, and isn't now, only based on the union of one man and one woman. There also apparently used to be sanctioned same sex marriages in Greece, Rome, and the early Church. http://rationalreasons.blogspot.com/2005/05/brief-history-of-marriage.html.

The history of marriage shows very diverse civil approaches to regulating access to women and that our current system isn't the only option. http://en.wikipedia.org/wiki/Marriage. There's no one right way to arrange a society, just like there's no right way to live or die.

Animal societies, as we have seen, also demonstrate systems that could be described as monogamous, patriarchal or matriarchal harems, or promiscuous. In fact, most animals don't seem monogamous. I frankly now don't think that humans are meant to be monogamous. Men certainly have a hard time with it; some women, too.

Here's another consideration; isn't *fear* inherent in marriage? The fear of being left and alone. Being joined in matrimony helps reduce the fear of being alone. If he's married to her, he's not supposed to go off and boink some other chick and leave her. Marriage is protective in that the parents also can have more certainty about paternity, so there's no fear of another man being the father. The mother *theoretically* doesn't have to worry about the husband murdering her, or her children. Husband and wife also don't have to be concerned about sexually transmitted diseases. But how many marriages can sustain that emotion we call love?

Being afraid that your partner will go off with someone else. Jealousy is certainly fear based, right? The fear that someone will take your lover. Well, according to the book, *Opening Up*, by Tristan Taormino:

> *Jealousy is really an umbrella term for a constellation of feelings including envy, competitiveness, insecurity, inadequacy, possessiveness, fear of abandonment, and feeling left out. ... Insecurity or low self-esteem is intertwined with envy and is at the heart of most jealous feelings. It may take the form of self-doubt, self-judgment, constant comparisons to others (especially to a partner's other partners), or feeling not good enough. ...*
>
> *Behind jealousy, and the insecurity that fuels it, is fear. Usually the fear concerns a change in the relationship or its end. Fear of the unknown is a very powerful emotion. You fear that your partner will fall in love with someone else ... . You fear that your lover will leave you for another person. ... You fear that you'll be a failure because the relationship failed. You may even fear that you will be alone forever.*

Whole lotta fear going on there. Here's a letter to Ask Amy Dickinson in the *Washington Post* on August 10, 2012, which demonstrates a severe case of the jealous fears in a marriage:

> *My husband and I have been together for 10 years and we have four beautiful children together – twin 8-year-olds, and 7- and 3- year-olds. I recently discovered my husband sending nude pictures over the Internet to an unknown e-mail address and asking the recipient for a no-strings-attached sexual encounter. I was completely shocked and heartbroken! To me, this is cheating!*
>
> *He said he was seeking attention from other women because our sexual relationship hasn't been up to par in his mind. He says he has never cheated on me. He claimed he was just looking to see if he still "had it."*
>
> *He works a lot out of town, and now my recent findings have my mind racing about what he is doing every minute of every day or whom he is doing it with. I can't seem to get this e-mail or image out of my mind. It's all I think about from the time I wake up to the time I go to bed! It absolutely disgusts me! I question him all the time about where he is going or what he is doing. I find I am constantly checking his emails and his phone for some indication that he might be having conversations with other women and, of course, he gets mad at me and says if I can't trust him, then why be with him?*

I don't think that relationship will last long, with such a core of fear. Too bad, because it sounds like they have a nice life. But he's just doing what he is biologically designed to do, no matter what you think of it. About.com says: "Infidelity leads to divorce and is probably the single most damaging thing that can happen to a marriage. Unfortunately, it is also one of the most common problems a married couple will face. Statistics vary on this subject, but it's widely reported that 60% of men and 40% of women will participate in an extramarital affair at some point during their married life." So, lots of good marriages break up, and people's lives change, because one of the partners wants to have more or different sex. Seems kinda dumb for lives to shatter just based on wanting some nookie.

Love can definitely decline in the face of fear. In fact, lots of marriages end up devoid of love. My parents' infatuation with one another didn't last. Mom says my Dad initially was "her god." The last half of their marriage, they were angry with one another, but by that time were stuck.

Marriage also prevents intimacy with others. It's strange to think one's position could be that, "if you don't love me to the exclusion of all others, I can't love you or be with you." Doesn't marriage *limit* love in that respect? It doesn't allow us to experience intimacy or sexuality with another person other than one's spouse. And to reiterate a point made earlier, according to religious tenets, if you're not married, you shouldn't be doing it at all!

Marriage can be good – very good – don't get me wrong. I'm not criticizing anyone in a happy marriage. But people do evolve and preferences change in many ways. Sexually, of course. Menopause affects women and typically slows down the sex drive. In the meantime, guys may feel they only have a short time left to fling those little guys, and to experience different intimate situations. Just like Mrs. Eskapa said, it feels to me that *I am* hard-wired for promiscuity. Is that the best way to describe it?

In researching this topic (which is the most difficult of my book here), I was led to youtube. Rachel Rabbit White has a number of videos, and in this one, she opines that men are inherently *polyamorous*. http://www.youtube.com/watch?v=Xdzb-lbnmHM. What's that mean? Do other people know what that means? Wikipedia says polyamory, "is the practice, desire, or acceptance of having more than one intimate relationship at a time with the knowledge and consent of everyone involved. Polyamory, often abbreviated as *poly*, is often described as *consensual, ethical, or responsible non-monogamy*." http://en.wikipedia.org/wiki/Polyamory. Huh, ethical non-monogamy? Interesting, so I looked into it further.

In *Opening Up*, Taormino says: "I would define polyamory as the desire for or the practice of maintaining multiple significant intimate relationships simultaneously. These relationships may encompass many elements, including love, friendship, closeness, emotional intimacy, recurring contact, commitment, affection, flirting, romance, desire, erotic contact, sex and a spiritual connection." What is wrong with that?

This is from an article in the *Washington Post* on February 13, 2008, titled, "Pairs with Spares," by Monica Hesse.

> *Polyamory isn't about sex, polys tell you. It is about love. It is about loving your primary partner enough to love that they have a new secondary partner, even when their New Relationship Energy with that*

person leaves you, briefly, out in the cold. It's about loving yourself enough to acknowledge that your needs cannot be met by one loving person. It's about loving love enough to embrace it in unexpected form – like maybe in the form of your primary's new secondary! – in which case you may all form a triad and live happily together. ...

"Many of us tried to make monogamy work," [Anita] Wagner says. But monogamy, she says, often seemed to throw the baby out with the bath water, so to speak. Its practitioners would break off "perfectly good relationships" just because of intellectual incompatibility, for example, or because one partner liked ballet and the other liked bowling. Doesn't it make more sense, polys ask, to keep the good parts of the relationship, and find another boyfriend who likes "Swan Lake"? ...

Thought: Maybe you can have it all. You just can't have it with one person.

It's the thought that illustrates a paradox in polyamory: Its practitioners have astonishing optimism for humans' endless capacity to love, to share, to forgive, to grow, to explore. But that optimism is rooted in a cynical belief that the monogamous are stuck in a myth, one that leads to cheating, unhappiness or divorce court. They believe, as do some evolutionary biologists, that most humans do not have endless capacity to be faithful to just one person.

Or maybe that's *not* cynical? Maybe practical? Just a realization there's little precedent in nature for monogamous relationships and lots of evidence that the one-man/one-woman model just doesn't work well. Marriage is based on the concept that, if a man has sex with a different person than his spouse, that's a breach of trust and acts to break up the first relationship; if you consider this from a 10,000 foot perspective, it's ridiculous. It's entirely fear-based. You love a man, unless he puts his penis in some other woman? Then you don't love him anymore. The man may still love his wife, and they may be able to live a long time *both* happy and fulfilled. So, what changed? Trust? Or fear realized?

Thinking further, though, if the sex occurred with another person who was *committed* to both the man and wife, and there was no fear or jealousy, wouldn't that be *wonderful*? Wouldn't that be much better than the typical dysfunctional situation that Mrs. Eskapa wrote about?

Then aren't we really talking about sharing one another? Since you get what you give, isn't it better to share love and openness with more people? There's a quote in *Sex at Dawn* by anthropologists William and Jean Crocker about their studies of the Canela people in the Brazilian Amazon:

> *It is difficult for members of a modern individualistic society to imagine the extent to which the Canela saw the group and the tribe as more important than the individual. Generosity and sharing was the ideal, while withholding was a social evil. Sharing possessions brought esteem. Sharing one's body was a direct corollary. Desiring control over one's goods and self was a form of stinginess. In this context, it is easy to understand why women chose to please men and why men chose to please women who express strong sexual needs. No one was so self-important that satisfying a fellow tribesman was less gratifying than personal gain.*

Since a lot of relationship problems arise from one or the other spouse, usually the man, being intrigued by another person, wouldn't it be a better system if the feelings were about love, forgiveness, and sharing, rather than exclusivity? Do you think humans have the capacity for that *bigger love*? Here's an article by Jessica Bennet from *Newsweek* magazine on July 29, 2009, "Only You. And You. And You."

> *Researchers are just beginning to study the [polyamory] phenomenon, but the few who do estimate that openly polyamorous families in the United States number more than half a million, with thriving contingents in nearly every major city. Over the past year, books like* Open, *by journalist Jenny Block;* Opening Up, *by sex columnist Tristan Taormino; and an updated version of* The Ethical Slut — *widely considered the modern "poly" Bible — have helped publicize the concept. Today there are poly blogs and podcasts, local get-togethers, and an online polyamory magazine called* Loving More *with 15,000 regular readers.*
>
> *Celebrities like actress Tilda Swinton and Carla Bruni, the first*

Bigger Love

> lady of France, have voiced support for nonmonogamy, while [Terisa] Greenan herself has become somewhat of an unofficial spokesperson, as the creator of a comic Web series about the practice—called "<u>Family</u>"—that's loosely based on her life. "There have always been some loud-mouthed ironclads talking about the labors of monogamy and multiple-partner relationships," says Ken Haslam, a retired anesthesiologist who curates a polyamory library at the Indiana University-based Kinsey Institute for Research in Sex, Gender and Reproduction. "But finally, with the Internet, the thing has really come about."

One of the things that's intriguing about polyamory is the need for openness, almost like oneness. In this video, the panelists stress how important open communication is in polyamorous relationships, more so than in traditional marriage because there are more than two people involved. <u>http://www.youtube.com/watch?v=jmodZ351SC4&feature=related</u>. When there's honest communication, *that is love*. Think about it; if you had to manage intimate relationships among a number of people, you'd have to be very good at communicating, and even better at loving and forgiving. It may take a higher level of consciousness to fully experience consensual nonmonogamy. More honesty and respect; more trust. Some polys believe the call is to evolve to a more compassionate, nurturing, and sensual society. I'm down with that.

I'd like to go back to one thing Rachel Rabbit White said in the video I noted above. She says women are *hypergamous*. That means they basically stick with one person, unless they can move up to someone who has more money or who can treat them better. <u>http://en.wikipedia.org/wiki/Hypergamy</u>. Do you think that's right? This is actually consistent with what Jared Diamond said back in chapter 8. Men tend to be more interested in sexual variety, casual sex, and brief relationships, which is certainly not monogamous, but could be polyamorous. Women tend to search for a lasting relationship with a man best able to provide for her. What do you think of that?

On the other hand, aren't *both* men and women really promiscuous, i.e., nonmonogamous? Men and women have affairs. Aren't we going against our nature with this monogamy stuff? I mean, how many sexual partners have you had? Julie said she had 14 (I was surprised to hear), and when I counted, my number was about the same. Maybe we're kidding ourselves when we say we have a monogamous society. Cheating seems to be a more understood concept than openly having more than one partner. Many folks would be freaked by the concept of polyamory, just like polygamy. But

I don't think there's anything wrong with *adults* having pretty much any kind of relationships they want. Gay, straight, bi-sexual, polyamorous, whatever. The problem arises in, for example, the stereotypical polygamous relationships that involve very young women, and that's simply not fair to those girls. That would *not be ethical* nonmonagamy.

There have been other people who thought like this. H.G. Wells, for one. Yes, the author of, "The Time Machine" and, "The War of the Worlds." This is according to the book, "A Man of Parts," by David Lodge, and reviewed by Michael Dirda in the *Washington Post*, September 15, 2011.

> *"A Man of Parts" explores, with great verve, Wells's life-long attempt to honor his own complexity, to be true to himself as a sexual being, a loving family man, a creative artists and an ambitious social thinker. ... Lodge reminds us that Wells consistently argues for a more rationally arranged world. He envisions future Utopias where free love flourishes, where a ménage a quatre is possible, where young women speak openly of their sexual desires.*

In some ways, we're here now. I can't argue with that aspiration, either. Life doesn't suck and then you die; *life resonates with love,* or should anyway. And we humans are sexual creatures, no flippin' doubt about it. As Avery says, "Why fight yourself?" Why fight our nature as wonderful, loving, sexual creatures? Because of fear of other people? Give. Me. A. Break. We're better than that.

So, let's explore the differences between monogamy and nonmonogamy. Again, from the book *Opening Up*:

> *It's no wonder people are so dissatisfied: monogamy sets most people up to fail. The rules of traditional monogamy are clear: you've vowed to be emotionally and sexually exclusive with one person forever. ... But it is more common that people are monogamous not by choice but by default; they believe monogamy is what everyone else is doing, what is expected, and how relationships are supposed to be. In addition, they have grown up with messages about the fairy tale, it has seeped into their consciousness, and they work hard to live up to all*

> the hype. The problem is that those unspoken expectations of monogamy are unrealistic and unattainable.
>
> People who practice nonmonogamy begin from the same premise: one partner cannot meet all their needs and they may want to have sex or a relationship with someone other than their current partner. But instead of hiding it, they bring this fact out in the open. They don't stifle their behavior based on how they're supposed to act. They open the lines of communication. They talk honestly about what they want, face their fears and the fears of others, and figure out a way to pursue their desire without deception. They don't limit themselves to sharing affection, flirting, sex, connection, romance, and love with just one person. They believe strongly that you can have all these things with multiple people and do it in an ethical, responsible way.

Sounds open and loving, doesn't it? Not like some Prince Charming live-happily-ever-after kind of story.

> Nonmonogamous folks reject this myth and acknowledge that no one person can be, or should be expected to be, everything for another. People in open relationships enjoy exploring different dynamics with different people – sexual, emotional, psychological, and spiritual. ...
>
> Many people in monogamous relationships deal with cheating all the time; fear of cheating, the suspicion of cheating, the discovery of cheating, the aftermath of cheating. Nonmonogamous folks recognize that during a lifetime you can and will be attracted to other people even if you are in a wonderful, fulfilling relationship; they make room in their relationship for these attractions rather than allow them to cause anxiety, jealousy, and unreasonable expectations. ... We need to let go of the notion that venturing beyond monogamy is wrong or shameful, or that it calls for us to behave dishonestly. Honesty is crucial to creating and sustaining a positive and fulfilling open relationship.

Similar sentiments are noted in *Sex at Dawn*: "[C]ouples with 'open marriages' generally rate their overall satisfaction (with both their relationship and with their life in general) significantly higher than those in conventional marriages do. Polyamorists have found ways to incorporate additional relationships into their lives without lying to one another and destroying their primary partnership. Like many gay couples, these people recognize that additional relationships need not be taken as indictments of anyone."

How to decide what we can do about all this? Well, we could conduct a NWAL-type analysis of human relationship modes. This time, however, let's not look at several different types. Rather we can use our NWAL principles *in reverse* to determine what the preferred societal relationships would be. Think about where we'd end up if we could choose the most loving social model. You still have to be in the now to begin; not back 50 years, or worried about the next generation.

Start from here: I want to love everyone and have them love me. Everyone should love one another, right? That's the golden rule, as I recall. If people understood love – an appreciation of oneness, i.e., nonduality – then we would have a much more peaceful society than we have now, where everyone's afraid and angry all the time.

But what about relationships? Can you say in your heart that the best and most loving type of human relationship is *just one man and one woman* (or one man and one man, or one woman and one woman)? I don't think so. Family belies that notion. Many people can love one another. Family, friends, teammates, band of brothers. Right now, there are so many people I absolutely love. You can't tell me love is *limited* to two people.

But is it limited to two people from the standpoint of intimacy? I think most people, if there were not societal frowns on it, would be very happy to be intimate with lots of different people. It seems to be the case, no matter how much one loves and is comfortable with one other person sexually, it'd still be nice to *get some strange; be allowed* to love another. And another. And then another. Something new from time to time. I think this begins to look more like a fabulous system if you put one caveat on it; you do have to love the other. Not just infatuation or lust. A deep feeling of *no separation*; unbridled sharing and caring. Unconditional forgiveness. *No Fear!*

The key component of this type of paradigmatic relationship would be difficult for many people. It would mean you'd be completely fine with (one of) your lover(s) loving and being in love with her other lovers at the same time. It means you'd be OK with having your wife, girlfriend, boyfriend, and husband having intercourse with someone else. You'd not only forgive them; you'd be *happy* for them. Pleased they could enjoy the bliss of romance with another person(s). That they could enjoy more spice in life, and that *it didn't make you feel unloved or unlovable!* It didn't make you angry or want to leave.

In other words, even if your man got it on with another woman – perhaps someone you knew, likely a friend – you wouldn't be mad. You wouldn't feel betrayed or scorned; you'd be glad they could experience more of love and life with another person, if they wanted to. And you wouldn't feel trust had been broken; in fact, you'd feel *more trust than ever before*, because it would all be *truth*. Avery Kanfer always reminded me, "The truth will set you free." If the truth is love, there's only good that comes ultimately.

Rachel Rabbit White talks about this. She and her husband were in a monogamous relationship, but had discussed the possibility of other ventures. They viewed cheating not as a relationship killer, but maybe as a short-term bummer that provided another avenue for growth. When they finally did experience sex with another couple, she said it was an *exhilarating* experience. That she felt happy for her husband and that they'd made a good team. She also said it had spiced up their own relationship. www.rachelrabbitwhite.com and http://www.youtube.com/watch?feature=player_embedded&v=2rMZCg5Q-iY#! What's not to like about that?

Some more minor conditions, however, are in order. Have to practice smart and safe sex. To do otherwise is *not* loving. We now can discern paternity with blood tests, so we don't have to wonder who the father is. But children growing up in this type of situation would have the most wonderful gift; several mothers and fathers to love and take care of them, in spite of only two parents biologically. No more deadbeat dads, because support would be expanded and shared.

And sexually transmitted diseases must be managed. This isn't a Vegas swingers club I'm talking about. It's not love to be intimate with someone and hide the fact of any communicable diseases. If you got one, you got it. That will affect your relationships. But these things can be managed; think Magic Johnson.

To me, the ideal human relationship is not monogamy, promiscuity, or a harem. Those all involve fear-based concepts. Marriage is based on being exclusive, not inclusive, and the fear of losing your spouse, afraid of being alone again. Promiscuity isn't fearful in itself, but is based on fear of commitment and lack of trust. Harems are based on control; clearly fear-based from what we've learned. My guess is you'll find many of these relationships where people are very happy. That's great. My only point is that polyamory is another option based on love, trust, and inclusiveness. So, it seems more love-based to me. It's not afraid of changing partners. It teaches great lessons of forgiveness and empathy. And I think that humans should celebrate their sexuality more, *like bonobos*. That wouldn't be awful.

There's another concept that must be discussed, because it's central to improving the human relationship system: compersion. Here's how it's described in *Opening Up*:

> One last quality that cannot go unmentioned in any discussion of open relationships is <u>compersion</u>, a concept that may be new to many readers. Compersion is taking joy in your partner's pleasure or happiness with another partner. For some, compersion has an erotic component: they get turned on watching, imagining, or hearing about their partner's sexual experiences. Some practitioners of polyamory think of compersion as the opposite of jealousy, or at least the antidote to jealousy. Given the problems (and drama) ignited by jealousy, you can see how compersion can go a long way toward creating a foundation for pleasure and generosity in any relationship.

In fact, to me, compersion is not limited to erotic or sexual aspects of relationships. It basically means that you love your partner, or anyone, so much that you only want their happiness, no matter what that means for you. I liken compersion to the feeling I have about my kids; I love them and just want them to be happy. I don't want to control them, I just want to love them. I don't need to be with them all the time or tell them what to do. I just want them to be happy. If all humans could have compersive feelings for other humans, we'd be a lot better off. It's like Big Love for everyone.

So, in addition to those persons who're judged for their sexual behavior and proclivities – i.e., Lesbians, Gays, Bisexuals, and Transgendered people (god bless 'em) – it'd be good to add the Polyamorists. So, we could all go around with LGBTP tee-shirts.

I also think society would be more peaceful if humans were allowed and encouraged to have open relationships and not be stuck in the meme of being exclusive to one person. We wouldn't have so many angry men. The Muslim world could be transformed from being dominated by a bunch of controlling, violent, and sexually repressed dudes to a modern and loving society. When I have been asking in this chapter what men and women can do to bring peace to the world, this is it.

Men and women should be more open and not jealous. It may take some getting used to, but I think it could be the key to peace on the planet. We could be peaceful just like the bonobos. We could develop compersion – true love - for all other people and the planet, and stop fucking up the earth with our human generated environmental and economic degradation. We could just stop being assholes and really, finally, live up to our self-proclaimed greatness.

I'm sure lots of smart folks will criticize this. A lot of religious folks will, too. Heck, all my friends might. Surely, I'm not an expert, and it admittedly can seem pretty simplistic or naive. People will pull out the Bible and centuries of tradition and argue their points. Hey, look where that's got us.

Remember, in all this, there's no right or wrong.

The most difficult thing about many of these concepts is, they're not typical. Hardly anyone uses *love and fear* as measuring sticks for how life *can be better*. I believe down to my toes that this personal allegory of mine can be useful in bringing peace to the world.

Yours can, too.

If we can make love our north star, no matter where we travel, we'll be OK. So, how do we do it?

We just *change our minds*.

Twenty One | **Of Course We Can**

I know we can. The human race *will* come to understand and know love. How do I know? Because I've changed my mind, about many things. Everyone can and does change their mind *all the time*. We just need to move *collectively* in the direction of love. It's already started happening.

Problem is, we don't have forever. Each of us *will* die. The human species will perish from the earth. The earth will one day stop spinning around the sun, or the sun will die out or maybe explode. We know this. How long do we have? James Lovelock says in *Gaia, The Practical Science of Planetary Medicine*: "The Earth, as a home for life, is now 3.8 billion years old and with a lifespan of not more than 4.5 billion years." In human terms, if living to 80 years old, the earth is now almost 68! Not much longer. And I don't think we're treating old Mother Earth so well.

Lovelock again, "Through the ceaseless activity of living organisms, conditions on the planet have been kept favourable for life's occupancy for the past 3.8 billion years. Any species that adversely affects the environment, making it less favourable for its progeny, will ultimately be cast out, just as surely as will those weaker members of a species who fail to pass the evolutionary fitness test." This test is not about strength, though, it's about understanding. Eckhart Tolle says in *A New Earth*:

> *Responding to a radical crisis that threatens our very survival – this is humanity's challenge now. … A significant portion of the earth's population will soon recognize, if they haven't already done so, that humanity is now faced with a stark choice – evolve or die.*

This is serious. What are we going to do? An awful lot of people just check out. Watch TV all the time. Live for sports. Strive for money. Bury their heads in technology. So many humans *assume* it's always gonna be the same, so why worry? This is reality, you know - wars, killing, hate, anger, and all that.

But I don't think sitting around and disengaging from life is the best path. For one, watching too much television can make people unhappy. The *Washington Post* reported on November 23, 2008, in an article written by Donna St. George, that, "Unhappy People Watch More TV Than Happy People, U-Md. Researcher Finds." Of course, we all watch TV, and learn a lot from it. I watch the weather report in the mornings, and news/talk shows in the evening for an hour or so, sometimes more. I like to watch my sports teams and a couple shows every week. My favorite is *Real Time* with Bill Maher. But the article said that:

> *Happy people spend more free hours socializing, reading and participating in religious activities, while unhappy people watch 30% more television, according to new research on American life. In a study that is among the first to compare daily free-time activities with perceptions of personal contentment, researchers found that television hours were elevated for people who described themselves as "not too happy." On average, the down-and-out reported an extra 5.6 hours of tube time a week, compared to their happier counterparts.*

Sitting on our butts won't get us where we need to go, and actually can kill. As reported in the *Washington Post* on August 3, 2010, "Too Much Sitting May Hasten Death," "The more of [the subjects'] leisure time they spent sitting – doing such things as watching television, reading and driving – the more likely people were to have died, especially from cardiovascular disease, even if they also exercised."

You'll also be glad to know that the United States is the world leader in – *percentage of obese people*. About 30% of us 'Mericans are considered obese, versus 24% of the next closest country, Mexico, and much more than the 14% world average. http://www.nationmaster.com/graph/hea_obe-health-obesity.

This isn't worth aspiring to. Shutting off Mother Nature and society doesn't work. Fulfilling a passion is one thing; that's good, whether golf, bridge, music, etc. These are often communal endeavors. Losing oneself in video games in a dark room, to me, seems almost disrespectful to nature. Things like that disconnect us from what's really out there. In fact, addicted gamers (i.e., over 30 hours a week) tend to be more aggressive and antisocial, and are at long-term risk of mental illness. According to a study reported in an article in *Time* magazine, "Lab Report: Health, Science, Medicine," by Alice Park, on January 31, 2011, "pathological gamers were at higher risk of developing conditions such as depression, anxiety and social phobia ... ."

Even our fabulous Baby Boom generation, which accomplished a lot in terms of opening up society, isn't as healthy as we should be. According to an article in the *Washington Post* on April 20, 2007, by Rob Stein, entitled "Baby Boomers Appear to be Less Healthy Than Parents:"

> As the first wave of baby boomers edges toward retirement, a growing body of evidence suggests that they may be the first generation to enter their golden years in worse health than their parents. ... Boomers are healthier in some important ways – they are much less likely to smoke, for example – but large surveys are consistently finding that they tend to describe themselves as less hale and hearty than their forebears did at the same age. They are more likely to report difficulty climbing stairs, getting up from a chair and doing other routine activities, as well as more chronic problems such as high cholesterol, blood pressure and diabetes. ...
>
> The baby boomers were much less likely than their predecessors to describe their health as "excellent" or "very good," and were more likely to report having difficulty with routine activities, such as walking several blocks or lifting 10 pounds. They were also more likely to report pain, drinking and psychiatric problems, and chronic problems such as high blood pressure, high cholesterol and diabetes.

Boy, we're in trouble. A bunch of old, fat, ex-hippie couch potatoes. How will we ever win World War III? But hey, at least we're happy, right? Well, it depends. People with more money tend to be happier. I mean, it's true that:

> Money can't buy happiness. Except that, according to a new study from Princeton University, it sort of can – about $75,000 worth a year. The further a person's household income falls below that level, the unhappier he or she is. But no matter how much more than $75,000 people

Heaven is Everywhere

make, it doesn't bring them any more joy. Time magazine, "The Cost of Happiness," by Belinda Luscombe, September 27, 2010.
Another study said that:

> *Pulling in the big bucks makes people more likely to say they are happy with their lives overall – whether they are young or old, male or female, or living in cities or remote villages, the survey of more than 136,000 people in 132 countries found.*
>
> *But the survey also showed that a key element of what many people consider happiness – positive feelings – is much more strongly affected by factors other than cold, hard cash, such as feeling respected, being in control of your life and having friends and family to rely on in a pinch. ...*
>
> *The findings "are really significant" because "we are finally able to answer the big questions, such as 'What is a good society?'" Shigehiro Oishi, an associate professor of psychology at the University of Virginia, wrote in an email. "If the goal of a society is to raise the daily enjoyment of its citizens, then, it seems critical to devise ways to increase the relational wealth of nations (e.g., stronger social networks)."* "Money Buy Happiness? It's a Down Payment," by Rob Stein, *Washington Post*, July 1, 2010.

I like that, "relational wealth of nations." That's what we should be focusing on. Our collective peace and happiness. Sort of like Bhutan, where they strive to improve the Gross National Happiness quotient. They don't think gross national product and capitalism are the most important things, but rather happiness of the people. And happiness is based in spiritual harmony and respect for the environment. http://www.youtube.com/watch?v=CXJwNSkdTH0. What a quaint thought! Worth changing our minds to, don't you think?

And a fabulous goal. Rather than sit around watching TV, getting fat and being angry, we could be focusing on having a wonderful society together and *having fun as a human family*! Thinking about it, what do we get for our slavish devotion to the capitalistic determinants of success and happiness? A lot of people get depressed and try and find happiness through drugs. Yea, sure, pot, oxycontin, alcohol, you name it. But also, antidepressants; *lots* of antidepressants. Many people need them, that's fine.

My Mom does. But, you know what, some studies say they don't really work. Check these out:

- *Here's some depressing recent news. Antidepressants don't work. What's even more depressing is that the pharmaceutical industry and the Food and Drug Administration (FDA) have deliberately deceived us into believing that they DO work. As a physician, this is frightening to me. Depression is among the most common problems seen in primary-care medicine and soon will be the second leading cause of disability in this country.* Mark Hyman, MD, "Why Antidepressants Don't Work for Treating Depression," *Huffington Post*, April 24, 2010.

- *Schizophrenia patients in the United States desperately need social services and rehabilitation programs because medications alone are doing little to help them improve their quality of life, according to a government study into antipsychotic drugs. The findings mirror what many clinicians have long reported – large numbers of patients come into doctors' offices in crisis and are temporarily stabilized with medications but soon relapse because the drugs have limited effectiveness or intolerable side effects.* "Schizophrenics Need More than Medicating," *Washington Post*, March 1, 2007.

- *Antidepressant medications appear to help only very severely depressed people and work no better than placebos in many patients, British researchers say. ... The researchers found that compared with placebos, these new-generation antidepressant medications did not yield clinically significant improvements in depression in patients who initially had moderate or even severe depression. The study found that significant benefits occurred only in the most severely depressed patients.* "Study Doubts Effectiveness of Antidepressant Drugs," *Rueters*, by Will Dunham and Julie Steenhuysen, February 26, 2008.

What does this have to do with changing our minds? Well, our minds – and how we treat them - are vitally important for our health and welfare.

> A recently released medical study confirms that poor mental health and stress can cause us to age more quickly and get sick faster – that there are actually molecular changes in the body when we are stressed. This probably isn't surprising to most people. Mental health professionals, through their experiences with patients, have long known that the mind plays a major role in the health of the body. "More Than Mind Matters," Allen Lebovits, *Washington Post*, December 27, 2004.

So, then, wouldn't our minds collectively also play important roles in the health of our species? Read this advice column written by Carolyn Hax in the January 4, 2010, *Washington Post* called, "It's So Easy Being Glum," and think of it as applied to humanity:

> Yes, I do think it's easier day-to-day to be unhappy. Meaning, when we're faced with these little decisions about how to perceive something, it's always a little bit easier to blame than it is to celebrate. ... And it's always a little bit easier to put that blame on someone/something else: "Guys are such jerks:" is easier than ... "I expected him to read my mind and be my little puppet, when in fact I didn't pay any attention to what he might have wanted or the ways I might have dismissed his feelings."
>
> In the long run, though, those little easy choices make life so much harder. When you're cumulatively pessimistic and/or fundamentally negative, you're actively choosing to accept a lower allotment of joy. ... Optimism also demands that you greet new people and situations with an open mind, instead of just lumping them into some crazy category of Things You Already Knew. When we prejudge, we close doors, deny opportunities, marinate ourselves in the past.
>
> To have an open mind though, we have to assign ourselves to the role of students in life, and to not knowing the outcome in advance. It's trading the secure (if false) sensation of being wise to everyone and

everything, for the possibility of surprise, be it pleasant or un-. Choosing optimism is choosing vulnerability and humility on an ongoing basis and that's often in conflict with our nature.

Except it's not really in conflict with our nature; it's what our nature has become. It's the human species before we finish going through our mid-life crisis and finally finding ourselves in who we are.

Here's what many people do all their lives, because they chose to be miserable:

> *When you hate what you are doing, complain about your surroundings, curse things that are happening or have happened, or when your internal dialogue consists of should and shouldn'ts, of blaming and accusing, then you are arguing with what is, arguing with that which is always already the case. You are making Life into and enemy and Life says, "War is what you want, and war is what you get." Eckhart Tolle, A New Earth.*

That's what happens. People make life an enemy, and hope for heavenly salvation. And, in spite of the slim possibility of the hereafter as they conceive it, they believe it *with certainty*. When you have an enemy, you get pretty certain about your beliefs; darn sure of the correctness of your opinion and the awfulness of the enemy. Heck, that's been going on forever. What is it about making enemies and being certain of their evil nature that's so appealing? "Harvard psychologist Daniel Gilbert, writing in the *New York Times*, inadvertently may have offered a clue. He was explaining that people are happiest when they're certain. "We don't like *not* knowing, apparently, even when what we know is awful." "The Deadliness of Certainty," by Kathleen Parker, *Washington Post*, May 27, 2009.

Is this based on reality, or what people *want* to believe? Um, the latter, I'd say. "People seem to choose frames of reference that supply them with the answers they want." Department of Human Behavior, by Shankar Vedantam, *Washington Post*, November 19, 2007. In fact, "most people, perhaps all, seem hard-wired to be able to interpret reality to suit their needs. ... During Colonial times, there were even people who managed to convince themselves that slavery was in the best interest of slaves; later on, some maintained that colonialism was in the best interest of the poor countries." Same column, same author, same paper, April 30, 2007.

So, on the one hand, it could be difficult to change minds, because people believe what they want. On the other hand, who really wants to think life sucks and then you die, that life is an enemy, and that other humans (who from all observation, are like us) are bad or evil?

I admit I've developed this frame of reference – my personal allegory – to suit what I want. And I want peace in the world. I want to believe humans are good and loving, because that's what I want to be. *That's what I know I am.*

Now, if humans chose the frame of reference and the outcome they want based on their desires, then why wouldn't they want to think of all humans as good, and this life as perfect? It shouldn't be that hard to change minds away from divisiveness and hate, towards love, right?

Wouldn't that also mean accepting that things some time don't go our way? Friends and loved ones do get sick and die. People get swept away by floods and hurricanes, trees fall on cars and kill the occupants, little babies die of starvation *every day*. I can understand that folks must think, "This situation I find myself in can't be that bad. There must be something better after this life." We can, however, get over these feelings, no matter how bad some think life is. An article in the July-August 2006 issue of *Utne*, titled, "Trauma? Get Over it." by Joseph Hart, said:

> *While [Robert] Shaer was working as the medical director of rehabilitation services at Boulder (Colorado) Community Hospital, he discovered that people suffering from persistent physical diseases like chronic fatigue syndrome and fibromyalgia [long-term body-wide pain and tenderness of joints and soft tissue] which are notoriously difficult to heal, respond well to treatment methods normally reserved for those who suffer from trauma. A patient who had been in a car accident and was suffering from whiplash, for instance, finally found relief after Scaer reenacted the accident. His radical conclusion, articulated in his book* The Body Bears the Burden *(Hayworth Medical, 2001) and* The Trauma Spectrum *(Norton, 2005), is that all chronic ailments and mental illness can be traced to trauma, and that virtually everyone in a modern society is traumatized.*
>
> *Scaer goes on to argue that the very institutions of our culture – schools, courts, and government, even the medical establishments – are traumatizing. "In the legal system, for example, if you're deposed,*

you come out in a shambles. You come our shaken and traumatized because it's so adversarial," he explains. *"The physiology of that experience is identical to a car crash. It's the flight-fight-freeze response."*

Huh. Is that right? Is everyone traumatized? Are you? Am I? I think a lot of people around the world are traumatized, like the Taliban, many Christians, Tea Party folks, North Koreans, maybe some of the Occupy Wall Street folks, and many minorities (as a result of eons of discrimination). Maybe women, too, *and* men? I believe fear is ever so present in the lives of many, and we've studied that extensively here. It's clear also that trauma is induced by fear. If this guy is saying that we're collectively choked with fear, how in the world do we get rid of it? Back to the *Utne* article:

> *Our problem, says Scaer, is that we keep our fears, anxieties, and sadness bottled up inside us." ... [But here is] one of the central paradoxes of trauma: In the midst of the deepest suffering lie the seeds of growth, change, and hope. In 1995 psychologists Richard Tedeschi and Lawrence Calhoun coined the term "posttraumatic growth" to describe the flowers of hope and renewal that can grow from the ruins of a catastrophic event. ... Tedeschi and Calhoun show that people who have survived an astonishing range of trauma – triggered by events such as a death in the family, being held hostage, sexual assault, or medical emergency – all report coming out of the experience with positive results. ...*
>
> *Gina Ross, in fact, calls trauma one of the "four paths to spirituality," along with prayer, meditation, and sexuality. Yet most of us don't look at a tornado or a car crash and see spiritual renewal – our culture, and especially our medical establishment, focus almost exclusively on the negative side of trauma. ... Our cultural goal seems to be ... : not to face trauma and heal it, but to avoid it altogether. Failing to grasp this element of renewal, we are increasingly ruled by fear and anxiety.*
>
> *"The object is not to conquer fear, but to become a connoisseur of fear,"* [Sam] *Keen explains.*
>
> *"We teach our [trapeze] students to identify fear – to be aware of the physical sensations of panic and fear. What happens to them*

> *when they finally do go off the platform is that the anxiety is translated into excitement. What was terror becomes joy."*

This is what I'm saying. We talked a lot about the gargantuan effects of fear, but there's another side. If we recognize how we're affected by fear - our perception of separation – then we can learn love just by *letting go* of the fear. Hey, it doesn't always go swimmingly. Emotions are powerful. But it'd be great to get headed in the right direction. And these problems are enhanced when looked at collectively. According to the *Utne* article, "As a nation, we are paralyzed on the trapeze platform, frozen by fear, unwilling to take a risk in the name of exhilaration." What's the risk? That we will become vulnerable if we love everyone? That love will make us losers? That we'll get poor and be defeated? It's hard to jump off the platform and reach for the trapeze, but we can and must do it.

To keep the fear from being bottled up inside us, we have to recognize and address it. This is analogous to treatments for post traumatic stress disorder. According to an article in the *Washington Post* on October 19, 2007, "Most PTSD Treatments Not Proven Effective," by our friend Shankar Vedantam, *facing fear* is the only effective solution.

> *The majority of treatments for post-traumatic stress disorder that are used to treat thousands of veterans lack rigorous scientific evidence that they are effective, according to a report issued yesterday by a panel of the federal government's top scientists. ... The report did find strong evidence that one particular treatment known as exposure therapy was effective; the technique asks patients to repeatedly reimagine traumatic events as a way to make the events lose their potency. ...*
>
> *Edna B. Foa, a professor of clinical psychology in the department of psychiatry at the University of Pennsylvania, and one of the pioneers in developing exposure therapy as a PTSD treatment said the technique was based on the insight that many victims of trauma do all they can to avoid being reminded of traumatic events. ...*
>
> *Two things happen in the process, Foa said. Patients come to replace actual recollections of trauma with other perceptions – taking on blame and guilt, for example, for being afraid. Second, by avoiding situations, patients can fail to see that much of life is not dangerous – the movie is only fiction. Foa said she has patients recount traumatic*

Of Course We Can

> events aloud with their eyes closed. She records the patient and then has the patient listen to the tape repeatedly. ...
>
> Foa also has patients make lists of situations that trigger anxiety and encourages them to deliberately expose themselves to the least-frightening situations. As people realize that many situations are harmless, Foa said they replace images of self-doubt and helplessness with a more healthy outlook.

So, addressing fear is the best way to overcome it. Overcoming fear is changing one's mind, isn't it? Just like succumbing to fear. Either way, it's a change. Here are some additional angles on the same concept. A study published in the journal *Psychological Science* and reported in *Time* magazine on July 27, 2009, "Finding Your Inner Loser," by John Cloud, said that affirmations don't really work, but that accepting even negative thoughts and feelings can help. If someone says to himself, "I'm good, I'm lovely," that actually can make him feel worse, because he *doesn't* feel that way. But if folks recognize their negative thoughts and feelings - rather than fighting them, - they can address those dastardly mental energies and vanquish them.

Similarly, people who learned to reach out for help from their parents as infants – meaning they understood when they felt bad or upset, and sought support from their mother and father – tend to be more confident and socially competent than those who acted out their anger. Department of Human Behavior, by Shankar Vedantam, *Washington Post*, February 12, 2007. In other words, those not being afraid to reach out to others learn their situations can be remedied.

"Contrary to the popular American myth that those who fend for themselves become strong and independent, the psychological research seems to show exactly the opposite is true: It is the people who are confident enough to reach out to others for help – and to whom help is given – who become truly capable of independence," according to Vedantam.

Meaning, those who admit to their fears can better address them, rather than *sucking it up* all the time. I came across another indication of this in an obituary for Jerilyn Ross in the *Washington Post*, January 10, 2010, titled, "Therapist Overcame Her Phobia, Helped Others Overcome Theirs," by T. Rees Shapiro. Ross was only 63 when she died. (It always makes me sad to read about people who die young.) Anyway, she was a psychotherapist who overcame her fear of heights to help hundreds of other people. She

421

had a radio show in the 1980s, "where she was known as the 'Phobia Lady.' She would give common sense advice to listeners on how to confront and overcome their fears."

> "The phobic person has a fear of fear," Ms. Ross told The Washington Post in 1980. "Phobic people are generally considered to be weak and helpless. That's not the case. Most of them are bright and competent people with one thing ruining their lives."
>
> Many of her clients through the years had a fear of driving over the long and narrow Chesapeake Bay Bridge in Maryland. ... "The Bay Bridge has everything to fear, not just height, ... Water, traffic, claustrophobia – anyone who's prone to suffer from a phobia has a problem with the bridge. It's the feeling of being so close to death. ... A good part of therapy is going over and over the bridge again and again with an escort, ... It's like a roller coaster – the first time it's dark terror, but then after time and time again, it gets boring."
>
> "Learning to live with anxiety is like learning how to get along with the mothers, sisters, fathers, brothers, in-laws, friends, co-workers, supervisors, and assorted colorful and eccentric characters who make our lives worth living," she wrote. "And the moment you understand that you have a living, breathing relationship with your anxiety – a relationship whose qualities and character are of your making – is the moment you free yourself from the tyranny of fear and assert your right to challenge, subdue, and even embrace it."

Ah, ha! Yes, a *living, breathing relationship with anxiety*. That's the fear of death, the fear of being alone, or being separated from love forever. If you put that crap out of your head, there's nothing to worry about. To do that, it comes down to the Now. Always. Letting go of the apprehensions, whatever one thinks they are. Becoming more *mindful* and conscious of the present moment and the love abiding within. And the *reality* of death. Hey, there's a study about that, too.

> *If mindful people are more willing to explore whatever happens in the present, even if uncomfortable, will they show less defensiveness when their sense of self is threatened by a confrontation with their own mortality?*

> *Based on the results of several different experiments, the answer appears to be yes. When reminded about their death and asked to write about what will happen when their bodies decompose (in grisly detail), less mindful people showed an intense dislike for foreigners that mention what's wrong with the United States (pro-U.S. bias), greater prejudice against Black managers who discriminated against a White employee in a promotional decision (pro-White bias), and harsher penalties for social transgressions such as prostitution, marital infidelities, and drug use by physicians that led to surgical mishaps. Across these various situations, mindful people showed a lack of defensiveness toward people that didn't share their worldview. … Mindful people were diplomatic and tolerant regardless of whether they were prompted to think about their slow, systemic decline toward obliteration.*
>
> *So, what do mindful people do that allows them to confront death in a non-defensive manner? What we found was that when asked to deeply contemplate their death, mindful people spend more time writing (as opposed to avoiding) and use more death-related words when reflecting on the experience. This suggests that a greater openness to processing the threat of death allows compassion and fairness to reign. In this laboratory-staged battle, mindfulness alters the power that death holds over us. Pretty cool.* Todd Kashdan, professor of psychology, George Mason University, "Confronting Death with an Open, Mindful Attitude," *Huffington Post*, March 2, 2011.

Damn right. Couldn't have said it better myself. Being in the now, mindful of the present, removes fear. Overcomes death. No surprise about that. Here's another take on the concept of facing death from Dr. Andrew Kneier on November 2, 2010, in an article titled "Accepting Death and Letting Go," on the *Huffington Post* website:

> *[This is] a key remedy to the obstacles of letting go – namely, for patients to be supported in discussing the process of coming to terms with death long before their illness progresses to a terminal stage. When patients receive this support, I have found, they have a much*

> *easier time letting go instead of clinging to hope and pursuing medical treatments that prolong their suffering. ...*
>
> *I must say, I was a bit blown away by the results [of my survey], especially the fact that people appreciated the opportunity to talk about these issues because they were so often encouraged, by family members and friends, to be positive and hopeful instead of being encouraged and supported in being realistic and prepared. It seems that our culture's preoccupation with healing, a fighting spirit and control over physiological realities makes it all the more difficult for people to let go because they receive so little support for doing so. Each culture has its own repertoire of acceptable ways of facing death, and in our culture, it seems more acceptable to "rage against the dying of the light" and fight to the bitter end than to take stock of what your life has been about and to be at peace with your coming death.*

I don't mean not do whatever you can to prolong life; I certainly try. I have my regular annual check-up, and go see the doctor if something's awry. You've heard of a few issues I've had. I don't want to die tomorrow or next week, but I do want to have the attitude of living each day as if it were my last. Though, as Steve Jobs said, one day it will be. And Jobs said if you ask yourself that question – "If this were my last day, would I be happy about it?" and the answer is NO too often, you gotta make a change. Change is not bad, change is the only thing you can count on, says my friend Neal.

Still, being mindful and accepting is important. No matter what happens, if I have peace of mind, then everything is OK. So, the point of this discussion is that it's a good idea to do for oneself what brings the greatest peace of mind, in any situation. That means letting go of fear and accepting what is, and knowing that the love of the Universe is what is. It's all about perspective, and changing one's mind to have a viewpoint that brings the most peace, love, and happiness. There are multitudinous ways to do that, for example:

> *Music has always had the power to make me blissfully happy. So it comes as no surprise when I read that researchers have now proved that listening to your favorite melodies and harmonies can release large amounts of dopamine, a chemical that sends "feel good" signals to the rest of the body and plays a role in motivation and addiction.*

> The small study, published last month in Nature Neuroscience, used brain scans to show that college students released significantly more dopamine when they heard their preferred music (which ranged from Beethoven to Led Zeppelin to the Israeli trance band Infected Mushrooms) as opposed to someone else's tunes. "It's interesting, because music is an abstract sequences of tones – you're not really getting anything for it – but somehow the way the brain is interpreting these tones, you get this intense physiological response, and the most potent reinforcing chemicals in the brain is released, creating a wanting, a desire, a craving, and saying, 'Do this again,'" says the studies lead author, Valorie Salimopor, a neuroscientist a the Montreal Neurological Institute in Canada. ...
>
> This biological buzz may help explain why music has played a key role in almost every culture and why we continue to spend so much money on iPods, better speakers, concert tickets, and the like. "We've Long Known that Music Can Move Us. Now We Know Why," by Carolyn Butler, *Washington Post*, March 1, 2011.

We also talked about how focusing on the little things in life can help us understand oneness better. Allowing music to buoy our spirit is a simple way to remove anxiety. What better antidote to some fearful situations than playing one's favorite tunes? As I write, I've been listening a lot to new age music by Phillip Kanakis, because it's soothing and has the right feel for me. http://www.cdbaby.com/cd/phillipkanakis. It may be that these healing sounds of the Universe can also have more than a transient effect. Like bleaching a shirt or dyeing an Easter egg longer makes the colors deeper, dipping our consciousness into the love again and again, and exhaling fear, can have a cumulative effect.

These are but small examples of ideas for transforming one's consciousness. There are innumerable paths to peace. Here are some more suggestions, the first from James S. Gordon, professor of psychiatry and family medicine at Georgetown Medical School and author of: *Unstuck: Your Guide to the Seven-Stage Journey Out of Depression*, from an article titled "Some Simple Steps for the Stressed-Out," September 29, 2009, *Washington Post*:

> *A middle-aged working-class woman recently came to my medical*

office complaining that her back had "seized up." Her husband had lost both his jobs and was feeling quite disheartened; not long after, her blood pressure had "jumped through the ceiling" and she began sleeping poorly.

Another patient came to see me suffering from crippling anxiety attacks. He had lost the better part of his considerable fortune in the economic collapse. Now he was waking in the middle of each night feeling his chest crushed, unable to breathe, half fearing and half wishing he would die.

I've been practicing psychiatry for 40 years, but I've never seen this much stress and worry about economic well-being and the future. ... In this uncertain time, symptoms of chronic illnesses – hypertension, back pain, diabetes – that were controlled or dormant are erupting. Low-level depression, whose hallmarks are feelings of helplessness and hopelessness, is endemic. Large numbers of people across the country are trying to quiet their anxiety with drugs or drink, or have turned to antidepressants, anti-anxiety medications and sleeping pills. But after decades working not only in Washington but also in war-traumatized populations overseas, I've found there are simple strategies for helping people cope that are easy to learn, practice at home, and in these stressful times, free.

What do you think they are? Take drugs? Complain and criticize other folks? Shoot up some other dudes? Get mad at your spouse? Here are the six things Dr. Gordon suggests:

1. *Begin a simple meditation practice. ... Slow, deep breathing - in through the nose, out through the mouth, with the belly relaxed and soft, and the eyes closed – is a sure "evidence-based" antidote to the stress response that uncertainty provokes.*

2. *Move your body. With the possible exception of talking with a sympathetic, skilled human being, physical exercise may be the single best therapy for depression. ... Exercise is often the first item on my prescription pad.*

3. Reach out to others. Human connection – to family, friends, coworkers in the same boat – is an antidote to the sense of aimlessness and isolation that may come from job loss or unexpected economic insecurity. Social connection also helps prevent the chronic illness that can often follow prolonged stress. ...

4. Find someone who will listen and help you take a realistic look at your situation. When the middle-aged woman with the "seized-up" back came to see me, we discussed her finances as well as her feelings. ... This simple exploratory conversation – and subsequent heart-to-heart with her husband – allowed her to turn aside the cascade of anxious emotions. Her body began to repair itself.

5. Let your imagination help you find healing – and new meaning. ... [P]atients find relief and assistance from imagining their inner "wise guide" to help them find peace, direction and meaning. This may seem kind of strange at first, but it's an ancient process used in many indigenous cultures and is actually pretty easy. ...

6. Speak and act on your own behalf. ... Often speaking up for yourself produces valuable information and clarity ... .

You can see what this means. He's saying we can heal ourselves. This is from an article in *Parade* magazine on March 9, 2008, titled, "Thoughts Can Heal Your Body," by Robert Moss:

> Our thoughts can make us sick, and they can help us get well. That may sound like New Age thinking, but medical research increasingly supports the role played by the mind in physical health.
> The "placebo effect" is an example of how the connection between brain and body works in healing. It has been demonstrated that when a patient believes something will relieve pain, the body actually releases endorphins that do so. ... Medical research has suggested that 30% to 70% of successful treatments may be the result of the patient's belief that the treatment will work."

Ever have a scar heal? How about a sore knee or tendonitis? What about a headache? We take healing so for granted, that we don't realize we can truly fix ourselves. So, we can heal our minds, too.

What are some other things that can lead us from fear to peace; that can bring us happiness? What are other ways we can heal ourselves?

Well, talking – *deeply*. People who spend more time in substantial conversations, rather than small talk, are happier. That's what a study in *Psychological Science* shows. "[F]olks who tend to engage in substantial conversations tend to be happier than those whose talk is mostly light and breezy. ... Deep conversations may actually make people happier." "Happy to be Talking," by Jennifer LaRue Huget, *Washington Post*, March 16, 2010.

Being positive also works. People who are positive, "get less stressed in day-to-day life. When something doesn't quite go to plan, a person with a positive attitude might just deal with it, typically refocus, or even look for another solution. But a person with a negative attitude will typically complain more, get angry or frustrated, and they will expend a lot of energy going over and over in their heads what has happened and how much it is a real inconvenience for them." "Do Positive People Live Longer?" by David R. Hamilton, Ph.D., *Huffington Post*, November 2, 2010. From the *Parade* article above, "There is ample evidence that negative thoughts and feelings can be harmful to the body," says Lorenzo Cohen, director of the Integrative Medicine Program at the M.D. Anderson Cancer Center in Houston.

A *Washington Post* article on January 12, 2010 reported on a study (how 'bout that, another study!) about this:

> A recent study published in the journal Circulation showed that a sunnier outlook on life is associated with a lower risk of heart disease and mortality. The research, which tracked more than 97,000 women older than 50 for eight years, found that optimists were 9 percent less likely to develop heart disease and 14 percent less likely to die from any cause than their pessimistic counterparts. Those with a high degree of "cynical hostility" were 16 percent more likely than all others to die during that same period." "Experts Aren't Positive, But Optimism Might be Good for You," by Carolyn Butler.

Exercising is also crucial. We have to *move our bodies*. One time when my Mom was about 90, we were sitting in her back yard in the sun. She noticed the hose with the sprinkler at the end was out in the yard. Well, my old mother walks over there, bends down, and starts tugging the hose in. I looked at her and thought, "I should help." Then it occurred to me, "If she stops doin' it, she'll stop being able to do it." So, I let her.

It's that way with lots of things. Don't stop moving your body around. This was from, "The Year in Health," in *Time* magazine on December 7, 2009:

> *Staying active is more important than ever for seniors. Adopting such painless habits as walking 30 minutes a day or parking farther from the supermarket can lower the risk of dementia and help relieve chronic pain. It also can be a great mood booster. A 2008 study of people in 80 countries found that after the onset of middle age (40 for U.S. women and 50 for men), people enter the highest risk age group for depression. Gloom leads to lethargy, and lethargy exacerbates the sadness. Getting your blood pumping breaks that cycle.*

That's really true. I try and do some exercise every day. I can't imagine not doing that. "Any exercise, when done with enough vigor and for long enough, helps reduce stress and fuels the brain with chemicals that create a sense of well-being even after the sweating is done, says Michael Lehman, a researcher at the Laboratory of Cellular and Molecular Research at the National Institutes of Health." "Rigorous training can boost fitness for the race of life," by Frank Kritz, *Washington Post*, May 28, 2013. And did you know that "15 minutes of exercise may extend your life," which was the subject of a short article by Rob Stein in the Health Section of the *Washington Post* on August 23, 2011. "Compared with couch potatoes, people who exercised for an average of 92 minutes a week – about 15 minutes a day, assuming one day of rest – were about 14 percent less likely to die for any reason while participating in the study, the researchers reported in the Lancet."

Here's a poignant story about how running saved a young woman's life – saved her from fear - and basically taught her oneness. The author is Danielle Seiss, a *Washington Post* staff writer, who penned, "Running for My Life," published on September 15, 2009.

> *I can't even say for certain when dark thoughts started to take control of my life. But I remember, when I was just 6 years old, crying every day. I didn't sleep at night. When I did, I had nightmares. I stopped*

eating. When I did eat, I often couldn't keep the food down. I felt that something terrible was going to happen to me. I was soon plagued by bad headaches. The condition became severe, and I began to develop paranoid thoughts and panic attacks, though at the time no one, including me, would recognize that's what was going on. ...

Over those early years the darkness came and went. When I hit my teens, though, it hit hard. ... For a while, I lived in a state of overwhelming dread. I would look at a perfectly normal object, or a scene out a window, and feel like I was staring at something horrific. When I could bear it no longer, I started planning my death. At this point, my mother interceded. ... This began a long series of treatments. Like many depression sufferers, at first I was prescribed cognitive, or talk, therapy. When that didn't work, often-mammoth cocktails of medications were added, [which were] either ineffective or mildly effective, or they lost their effectiveness over time ... .

And then I discovered running. Or running and long walking, to be exact. One day, particularly agitated, I fled my house and began walking toward a nearby mountain. ... When I got home I was excited about my discovery – and happy. ... I had walked about 27 miles, and it did more for my emotional state than all the therapy and pills. ... At some point I began adding running to my long-distance walks. ... [Some friends] persuaded me to join the [track] team, which I did because I had begun to realize that I really loved to run. I decided I wanted to be a long distance runner ... [b]ut as the track season approached I became anxious about competing, and I finally quit before the meets started. ... I silently decided I would somehow make it up to myself, and I imagined that someday I'd run the Boston Marathon. ...

There's a lot of experience that confirms my experience of mental health through running. ... There is also evidence that the therapeutic benefits of exercise increase with intensity. ... Now, if I am feeling down, I go for a run. I usually start feeling better almost as I head out the door – in part, I believe, because I am taking charge and doing something. But by mile four, I can actually feel my thinking beginning to change, from negative to positive, as if four miles, or about 30 minutes is some kind of threshold. On longer runs, by about mile

13 or 14, I start to feel a mild euphoria. If I run faster, I'll notice it earlier. ... On really long runs, of 18 to 20 miles or more, the nature of my thoughts go beyond just positive to creative. I start having brainstorms, one after the other, and I begin to feel "one with things," for lack of a better way to describe it. It's like deep meditation in which your personal boundaries open up and you no longer notice where you end and everything else begins.

Blimey! Oneness through running. Notice that fear changes to oneness for her. Of course, don't just go out running 20 miles! She also points out in the article that you can blow a gasket. But the key thing is she found something that took away the fear, and taught her about creativity and being one with everything. That's really cool. She didn't know how to face her fear; so she learned how to face the world, on her terms. It shows – again - we can transform ourselves. There's another nugget here. She found her passion – *outdoors*.

That coolest Doc around, Sanjay Gupta, wrote an article in *Time* magazine on December 8, 2008, entitled, "Slender in the Grass." He says:

A new study found that inner-city kids living in neighborhoods with more green space gained about 13% less weight over a two-year period than kids living amid more concrete and fewer trees. ... The new research, published in the American Journal of Preventive Medicine, isn't the first to associate greenery with better health, but it does get us closer to identify what works and why.

At its most straightforward, a green neighborhood simply means more places for kids to play – which is vital since time spent outdoors is one of the strongest correlates of children's activity levels. But green space is good for the mind too; research by environmental psychologists has shown that it has cognitive benefits for children with attention-deficit/hyperactivity disorder (ADHD). In one study, simply reading outside in a green setting improved kids' symptoms. Exposure to grassy areas has also been linked to lower body mass index among adults.

This is more profound than one might think. It's incredibly important that we "get back to nature." Too many just tune it out; don't even know when the moon's full. In *Parade* magazine on December 13, 2009, an article titled with the question, "Does

Fresh Air Keep You Well?" answers with a resounding, Yes! From a study in the *Journal of Epidemiology and Community Health*, it was determined that, "People living near a green space had lower rates of 15 out of 24 diseases, including asthma, diabetes, intestinal complaints, and back and neck problems. The links were strongest for depression and anxiety. For example, people whose environment was 90% green were significantly less likely to have an anxiety disorder than those living where it was only 10% green."

I mean, nature really rocks! Imagine this, tree hugging is actually good for you! Seriously!

> *Die hard conservatives love to disparage liberals as tree huggers, but it has been recently scientifically validated that hugging trees is actually good for you. Research has shown that you don't even have to touch a tree to get better, you just need to be within its vicinity to have a beneficial effect.*
>
> *In a recently published book, Blinded by Science, the author Matthew Silverstone, proves scientifically that trees do in fact improve many health issues such as Attention Deficit Hyperactivity Disorder (ADHD), concentration levels, reaction times, depression and other forms of mental illness. He even points to research indicating a tree's ability to alleviate headaches in humans seeking relief by communing with trees. Uplift, May 13, 2013.*

As a result of these types of studies, I've read about "doctors around the country who are beginning to prescribe nature in order to prevent (or treat) health problems ranging from heart disease to attention deficit disorder." "Take a Hike and Call Me in the Morning," by Dr. Daphne Miller, *Washington Post*, November 17, 2009. More from the article:

> *Other physicians, from New Hampshire to Texas, are sending their patients out to wade through streams and walk on beaches and trails. Earlier this year the city of Santa Fe, N.M. launched a Prescription Trails program to target the high rates of diabetes in the community. ... Richard Louv, author of the best-selling book "The Last Child in the Woods" and coiner of the term "nature deficit disorder" is all for these prescribing patterns.*

> "I think that physicians can do more [to get people out into nature] than any other professional," he said. Louv's book and web site (www.childrenandnature.org) cite dozens of studies documenting the positive impact that wilderness outings can have on mental and physical health. The fact that the American Academy of Pediatrics has invited Louv to deliver the keynote address at its 2010 annual meeting indicates that the larger medical community is starting to recognize the therapeutic value of time in the woods.

Well, some may say, can't I just put pictures of nature around my house and get the same effect? Uh, no. You can create a wonderful environment for living in your house, but the real effects of nature are outside. A study reported in *Spirituality & Health* magazine, November-December 2008, reported on a study in an article titled, "A Nature Photo Soothes like a Blank Wall."

> In the study, published in the Journal of Environmental Psychology, 90 University of Washington undergrads – 30 in each of three groups – spend an hour peering through a window that overlooked an on-campus fountain, looking at a plasma real-time display of the fountain, or staring at a blank wall. The students were given a low-level stress task to raise their heart rates, and then the researchers measured how fast the students' heart rates returned to normal – a sign of how much their stress was reduced. Those participants who looked out the window at the fountain had a faster recover rate than those who looked at the blank wall. And the longer they looked through the window, the faster their heart rate declined. But, strikingly, watching the plasma display was no more effective in lowering heart rates than staring at the blank wall.

There's a reason for this, I think. When we're being part of nature, we can more easily perceive the oneness than at other times. There's no filter hindering our senses' appreciation. Is it just about that? What else? Here's what Eckhart Tolle has to say in *The Power of Now*:

> Presence is needed to become aware of the beauty, the majesty, the

> *sacredness of nature. Have you ever gazed up into the infinity of space on a clear night, awestruck by the absolute stillness and inconceivable vastness of it? Have you listened, truly listened, to the sound of a mountain stream in the forest? Or to the song of a blackbird at dusk on a quiet summer evening? To become aware of such things, the mind needs to be still. You have to put down for a moment your personal baggage of problems, of past and future, as well as all your knowledge; otherwise, you will see but not see, hear but not hear. Your total presence is required.*

And when I'm present, in the now, in nature, another sentiment always rises. *Gratitude.* How lucky I am to be here looking at the moon and the stars. How fortunate I am to be here at all. How wonderful to be alive and with such great friends. There's *community* at the same time. They go hand in hand. … Ah, friends.

Let me tell you the rest of the story about saying goodbye to our friend Eileen. She was Julie's best friend in kindergarten, and then again in senior year of high school. Julie lived overseas during the interim. They lost track of each other and met again twenty years later at a Blues Week music camp I go to in Elkins, West Virginia. Eileen and Grant, a fabulous musician and harmonica instructor, lived in Seattle. But after she and Julie found each other again, we all became great friends and visited often.

Then, Eileen got the Big 'C.' She fought ovarian cancer for two years, several surgeries, with rounds of chemo. Grant called in the autumn years ago, and said Eileen wanted to come home to Bethesda for Christmas to be with her family. And, by the way, could he play a concert at our house? Turns out, it was also Julie's 50th birthday, but we said, what the heck, who could think of a better present? We didn't know if Eileen was coming home for the last time.

She didn't make it. Grant called the week before the concert and said Eileen was going into hospice. We never saw her again. But Grant arranged for a substitute harmonica player (none other than the legendary Phil Wiggins) and we decided to press on with the concert. That was December 19, 2009. The big blizzard. It ended up that we had to cancel the concert, but only after buying beer, wine, food, and ice cream cakes. Nobody could drive anywhere. Snowmageddon! We were stranded, and despondently thinking of Eileen.

Julie sent out an email to our neighbors – "Got booze and food, shovel yourselves over." Well, they came, and we had a fabulous time. Stayed up singing around the fire until the wee hours. Became wonderful friends with all those neighbors. Gratitude – thankful to be alive. Community – with our wonderful friends. Thinking of what we share to this day brings a tear. We weren't the only ones to experience community in that way.

Sally Quinn wrote a column in the *Washington Post* on February 11, 2010, after we got two more blizzards in one week! It was called, "Snowstorms Bring a Chance to Reflect on What Really Matters:"

> *As I sit here at my desk at home writing today, the blizzard is so strong that the city is paralyzed. Nobody is even venturing outside. The wind is howling, the branches are covered with snow and ice, and the flurries are blowing relentlessly. I have lived here on and off most of my life. I have never seen Washington like this before. First of all, it is beautiful. Secondly, it is peaceful – a pervading sense of calm. It's as if I, for one, have been liberated from my daily struggles. I have simply given it to what is happening around me and accepted it. …*
>
> *In the midst of the first storm, as the snow was in full flurry, I took a walk through Georgetown. There was hardly a person on the streets, barely a car in sight. I was overcome. For the first time in a long time I actually saw the city I call home. I walked down to the Potomac River, along the path to the Kennedy Center, and out onto the balcony and over to the corner, where there were no human tracks. I stood there for what seemed forever, just absorbing everything around me with all of my senses. My face was cold, the smell was fresh, I touched the snow and put it in my mouth and tasted it. I could hear no sounds. …*
>
> *I turned back to the river and watched the ice flows slowly drifting south. It was a transcendent moment – one might call it prayerful – as if I were watching chunks of my life floating by in front of me. I haven't often felt like that in Washington. I couldn't help thinking that this blizzard was, for me at least, not an accident. On some level it was a deliberate moment for all of us to stop and contemplate what our lives are about, what is important, who we want to be. …*
>
> *I felt joy as I trudged home. I was going to a warm house, with big fires and plenty of food and a family that I loved. And yes, to a party. The best kind. What we did have was a house full of people. Our close friend was in the ICU at Georgetown Hospital, having just had a liver*

> transplant. His family of four stayed with us because they needed to be near the hospital and couldn't drive. Two of my husband's granddaughters were staying with us, locked in as well. My son and fiancé, who live next door, were also snowed in, with her friend from Sweden and a roommate.
>
> Several friends on the block have come over for dinner since the snow began. We've had huge pots of stew, spaghetti, and soups. We've had big fires and lots of candles. We've had many bottles of wine. We've had an abundance of love. We have been so fortunate. ... Part of what has made this a special time is to realize, no matter what happens, how grateful we are and should be for what we have.

Darn tootin', Sally. The snow brought us a magical time, too. But you know what? The earth is *always* like that. Always beautiful and peaceful and full of love; it just doesn't seem that way to us. Isn't it all about perspective? And isn't the best perspective to see community with everything? *Time* magazine said this on December 10, 2010, in an article called "Happiness is Other People:"

> Churchgoing people tend to be happier and more satisfied than their areligious peers – but why? A new survey finds that the key to their sanguine dispositions may have more to do with friends than faith. People who attend religious services and have rich networks of church-based friends reported more life satisfaction than those who go to church but have few or no friends in the congregation.

It's true. We don't need a religion or dogma to be filled with peace, gratitude, and love. Secular community works, though many of the religious literalists are afraid it won't. We can climb out of the doldrums brought on by excessive focus on money and rates of growth. In an article titled, "The Taboo Cure for Our National Gloom: Live a Little!" in the *Huffington Post*, by Joe Robinson, on November 30, 2010, he said:

> The University of Colorado's Leaf Van Boven has shown that experiences make us happier than material items, since they can't be compared to anyone else's experience and form the positive memories that tell us we like our lives. Thomas DeLeire, at the University of

> Wisconsin, found that only one of the nine categories of consumption he measured was linked to an increase in happiness: leisure purchases. Recreation is so good at "re-creating" mindsets that there is a whole field of health devoted to it: recreation therapy, which builds self-worth and positive mood for people with disabilities, through activities such as horseback riding or wheelchair basketball.
>
> We don't just need an economic recovery; we need a psychological one – a national program of recreation therapy to lift spirits, restore our sense of competence and increase capacity for enjoyment, the proven outcome of recreation participation. The research shows that leisure experiences are far from the trivial sideshow we think they are. They provide a critical line of defense against the setback of life; buffering stress and building self-determination and social connection, which satisfy core needs. Recreation is medicine, only a lot cheaper and more fun than the stuff at the drugstore. ...
>
> Americans are lousy at R&R even in good times. Participation in slow-pitch softball is down 30 percent, beach volleyball by 26 percent since 2000, according to the Sporting Goods Manufacturers Association. U.S. Census data shows that 78% of American over the age of 30 don't get any regular exercise. Just 14 percent of Americans take vacations of two weeks or longer, according to a Harris survey.

People! We've got to do better than that. And to really think outside the box, why don't we just change the whole structure of society and work two days a week and take off five? Why not? Don't we have enough? How much stuff, money, and bling do we need? It's ALL about changing our minds. Like the Occupy Wall Street Crowd. That's not a political movement; it's a spiritual movement!

According to an article titled, "Occupy D.C. has a plan, To stay in the public's face," by Courtland Milloy, January 18, 2012: "We are trying to create a community where love is valued more than money," one of the demonstrators said. "We're not sure how, but we're trying." That's a really excellent goal; much better than trying not to pay taxes.

Joe Robinson finished his article above, by saying: "Let's beat fear by stepping into the center of life, where we may find a silver lining in hard times – a new understanding of where true value lies: in the friends, family, and, yes, recreation, that gets us through. We may be down, but not out. We've got life."

That's the way I feel. We can rally this great country – great world – of ours with a collective can-do attitude. Why let these dumb human "problems" get us down. So many of them are of our own making and come about just because fear sidetracks people. If we accept who, what, and where we are, let go of the negative stuff, have fun, and be open to nature, anything is possible. Why?

Because the feelings of peace and happiness we're learning are contagious. I've told you fear, anger, and loneliness can be transmitted; well, so can love, peace, and happiness. A December 5, 2008, *Washington Post*, article titled, "Happiness Can Spread Among People Like a Contagion, Study Finds," by Rob Stein, says:

> Happiness is contagious, spreading among friends, neighbors, siblings and spouses like the flu, according to a large study that for the first time shows how emotion can ripple through clusters of people who may not even know each other. The study of more than 4,700 people who were followed over 20 years found that those who are happy or become happy boost the chances that someone they know will be happy. The power of happiness, moreover, can span another degree of separation, elevating the mood of that person's husband, wife, brother, sister, friend or next-door neighbor.
>
> "You would think that your emotional state would depend on your own choices and actions and experience," said Nicholas A. Christakis, a medical sociologist at Harvard University who helped conduct the study published online today by BMJ, a British Medical Journal. "But it also depends on the choices and actions and experiences of other people, including people to whom you are not directly connected. Happiness is contagious."
>
> One person's happiness can affect another's for as much as a year, the researchers found, and while unhappiness can also spread from person to person, the "infectiousness" of that emotion appears to be far weaker.

This same study was reported in *Time* magazine on December 22, 2008, which said in an article titled, "The Happiness Effect," by Alice Park, that "emotions can pass among a network of people up to three degrees of separation away, so your joy may, to a larger extent than you realize, be determined by how cheerful your friends' friends' friends are, even if some of the people in this chain are total strangers to you."

This shows that the peace of mind we have individually can transmit to society more broadly. We can collectively bring peace to the world if enough people "get it." This is similar to what the TM'ers called the "Maharishi Effect." They said that, if one percent of the population meditated using Transcendental Meditation, the rest of the people would feel the effects of that in terms of being more peaceful, successful, etc. I don't know if that's true. *But I believe that if more of us started truly believing we could have peace in the world, it would be like a tsunami and spread over the land.*

It's like with singing and playing music. When in a group singing a song, often I can still sing along if I don't know all the words, just because everyone is singing. Same with playing drums, for example. I've been in drum circles and worried about getting off the beat, but it's almost impossible. Your hand automatically slams down on the drum when it's supposed to, as if by magic, or some underlying energy (ya think?)

The other point is that – while fear and its offspring seem very contagious – they're not as strong as peace. Just like light defeats the dark, peace defeats fear. For example, Dr. Dean Ornish changed his focus in dealing with heart patients.

> *Doctors had been trying to motivate patients mainly with the fear of death, he says, and that simply wasn't working. For a few weeks after a heart attack, patients were scared enough to do whatever their doctors said. But death was just too frightening to think about, so their denial would return, and they'd go back to their old ways. ...*
>
> *So instead of trying to motivate them with the "fear of dying," Ornish reframes the issue. He inspires a new vision of the "joy of living" – convincing them they can feel better, not just live longer. That means enjoying the things that make life pleasurable, like making love or even taking long walks without the pain caused by their disease. "Joy is a more powerful motivator than fear," he says.* Fast Company's Greatest Hits, *Edited by Mark Vamos and David Lidsky.*

We just need a critical mass. And I think we're getting there.

> *[W]hat we are doing here is part of a profound transformation that is taking place in the collective consciousness of the planet and beyond;*

> the awakening of consciousness from the dream of matter, form, and separation. The ending of time. We are breaking mind patterns that have dominated human life for eons. Mind patterns that have created unimaginable suffering on a vast scale. I am not using the word evil. It is more helpful to call it unconsciousness or insanity. ...
>
> The doing and the happening is in fact a single process; because you are one with the totality of consciousness, you cannot separate the two. But there is no guarantee that humans will make it. The process isn't inevitable or automatic. Your cooperation is an essential part of it. However you look at it, it is a quantum leap in the evolution of consciousness, as well as our only chance of survival as a race. Eckhart Tolle, *The Power of Now*.

Yea, it depends on us; on us changing our minds. One problem is highlighted in the book, *The Path of Least Resistance*, by Robert Fritz. Many people undermine their own success, because they're afraid to succeed. He says, "The belief that you are unable to have what you want creates a tension that is resolved by not getting what you want." How true that is. I know people who think this way, and probably in my life I've eluded success because I was afraid of it or didn't think I deserved it. Isn't that the situation with the entire human race? Don't we seem plagued by our preconceptions of who we are? What obstacle is there, really, to peace? We have to press on believing, we can have peace on earth.

There are lots of evolutionary thinkers who believe this. James Lovelock is one, and here's how he thinks we should learn to behave – or suffer the consequences - in his book, *Gaia: The Practical Science of Planetary Medicine*:

> Could we, by some act of common will, change our natures and become proper stewards, gentle gardeners taking care of all the natural life of our planet? I think that we are full of hubris even to ask such a question, or to think of our job descriptions as stewards of the Earth. We are all too plainly failing even to manage ourselves and our own institutions. ...
>
> I would suggest that our real role as stewards of the Earth is more like that of that proud trades union functionary, the shop steward. We are not managers or masters of the Earth, we are just shop stewards,

> like workers chose, because of our intelligence, as representatives for the others, the rest of life of our planet. Our union represents the bacteria, the fungi, and the slime moulds as well as the nouveau riche fish, birds, and animals and the landed establishment of noble trees and their lesser plants. Indeed all living things are members of our union and they are angry at the diabolical liberties taken with their planet and their lives by people.
>
> People should be living in union with the other members, not exploiting them and their habitats. A planetary physician observing the misery we inflict upon them and upon ourselves would support the shop steward and warn that we must learn to live with the Earth in partnership. Otherwise the rest of creation will, as part of Gaia, unconsciously move the Earth itself to a new state, one where we humans many no longer be welcome.

Don't we have enough reasons to change our mind toward union and oneness with all things? I think we do. Whose responsibility is it to do this? Mine. And yours. Ours.

As Al Gore said in *Our Choice*, "The only meaningful and effective solutions to the climate crisis involve massive changes in human behavior and thinking … ." And the only real solution to the crisis of fear among humans is the same – changing what we think about us and our world.

It still comes down to one mind at a time. Each of us can change the world. It just depends on how we look at it. Right brain, or left brain. Love or fear.

Dr. Jill Bolte Taylor was a scientist studying the brain, when she had a debilitating stroke. But what she realized was that, the right hemisphere of the brain is about the present moment, peace, love, boundariless, and wholeness. The left brain is just about me – separate. Only thinking about the past and future. I'd urge you to watch this: http://www.ted.com/talks/jill_bolte_taylor_s_powerful_stroke_of_insight.html

Jill says, if she could find and understand nirvana through her right side brain, then we all can. She believes, having lived through oneness, we all can choose who we are and how we want to be.

There's one more perspective I'd like to share. It comes from the ancient Hawaiian teaching of Ho'oponopono. I don't know that much about it, and can't even pronounce the word correctly, but I learned about it from a colleague who told me about the book, *Zero Limits*, by Joe Vitale and Ihaleakala Hew Len, PhD (I mentioned this book a while back). Dr. Hew Len espouses sort of a modern version of Ho'oponopono, as interpreted by a teacher named Morrnah. I'll quote some excerpts from the book:

> *Dr. Hew Len explained that everything you seek and everything you experience – everything – is inside you. If you want to change anything, you do it inside, not outside. The whole idea is total responsibility. There's no one to blame. It's all you. ...*
>
> *"Have you ever noticed that whenever you have a problem, you are there?" he asked. "It's all about 100 percent responsibility for everything. No exceptions. There's no loophole that lets you off the hook for something you don't like. You're responsible for all of it – all."*

Well, Dr. Hen Lew, that sounds sort of depressing. All the problems in the world are totally my responsibility. Yikes!

> *"Ho'oponopono is really very simple. For the ancient Hawaiians, all problems begin as thought. But having a thought is not the problem. So, what's the problem? The problem is that all our thoughts are imbued with painful memories – memories of persons, places, or things."*

Hmm, but wouldn't painful memories really be fear?

> *To do ho'oponopono, you don't have to know what the problem or error is. All you have to do is notice any problem you are experiencing physically, mentally, emotionally, whatever. Once you notice, your responsibility is to immediately begin to clean, to say, 'I'm sorry. Please forgive me.'*

Joe Vitale then said, "Apparently, Morrnah, and now Dr. Hew Len, felt that by asking for forgiveness we cleared the path for healing to be manifest. What was blocking our well-being was nothing more than lack of love. Forgiveness opened the door to allow it back in."

> *"But how do I get that across to people – that we are each 100 percent responsible for problems?" [Dr. Hew Len] asked. "If you want to solve*

> *a problem, work on yourself. If the problem is with another person, for example, just ask yourself, 'What's going on in me that's causing this person to bug me? People only show up in your life to bug you! 'I'm sorry for whatever's going on. Please forgive me.'" ...*
>
> *He went on to explain that at heart we are all pure, with no programs or memories or even inspirations. That's the zero state. There are no limits there. ...*
>
> *Complete responsibility means accepting it all – even the people who enter your life and their problems, because their problems are your problems. They are in your life, and if you take full responsibility for your life, then you have to take full responsibility for what they are experiencing, too. ...*
>
> *But once you accept it, the next question is how to transform yourself so the rest of the world changes, too.*
>
> *The only sure way is with "I love you." That's the code that unlocks the healing. But you use it on you, not on others. Their problem is your problem, remember, so working on them won't help you. They don't need healing, you do. You have to heal yourself. You are the source of all experiences.*

This might seem crazy, but look at it this way. It's saying that, if everything is your problem, you also can fix it. And you fix it with Love, Forgiveness, and Gratitude. The words to say are, "I love you. I'm sorry. Please forgive me. Thank you."

It is quite empowering to believe that you can – I can – solve the problems of the world. And if the world is not at peace, I can solve that problem. I can bring love to the situations in my life.

And you know what else? This is consistent with the theories of quantum physics that say the observer creates the object, creates the reality. So, maybe it's not so loony as one might think. Joe Vitale again:

> *This may be the hardest part of ho'oponopono to understand. There's nothing out there. It's all in you. Whatever you experience is all inside yourself.*

So, if I experience love and peace inside myself, then that's reality. And in some respects, it doesn't matter whether any of this is right. If I can achieve peace and love for myself, then I really have achieved the goal. Nice.

— ••• —

Ok, folks. That's it. I've said my piece.

I don't know how much longer I have on this planet, but dang-gummit, I've tried in this book to make the case for peace and love as best I know how. It doesn't seem that complicated.

You step outside; preferably with friend(s) you love. Then you just appreciate life. We're not here that long. So short, yet so special.

My Mom's tired of living; my Dad's no more. My thoughts of them are only about love.

Look at that sky! Saturn is right below the Moon tonight. I can't imagine not ever seeing that again. But someday I won't.

So, I relish looking up at the stars (with Dad) whenever I can. And feeling that oneness.

Can we take the feelings of love we have at such moments, bring them into our daily lives to conquer fear, and create the paradise *we all want*?

Yes, we can. We really can.

The last words of this book, are from John Lennon:

You may say I'm a dreamer,
But I'm not the only one,
I hope someday you'll join us,
And the world will live as one.

— *Finis* —

"Everything in this book may be wrong."

- Richard Bach,
Illusions; The Adventures of a Reluctant Messiah

Gratitude

I'd like to especially thank a number of people who helped me with this book. Mainly, the Brookmont Arts Alliance. You were always there to provide excellent and loving feedback. I appreciate our sessions tremendously. In particular, thanks to Peter Ainslie, Jody Bolz, Janet Wittenberg, Alice Covington, Adrienne Hand, Raye Leith, Sharon Schultz, Jane Callen, Louisa Jagger, Mary Kearney, and my dear wife Julie Littell.

About the **Author**

Jefferson Glassie has been studying and writing about concepts of peace for most of his adult life. It's sort of an obsession. His main goals in life are to have peace of mind and bring peace to the earth. He's still working at it.

In addition to writing, he's an attorney in Washington, DC representing nonprofit organizations and associations. He has written and spoken about nonprofit legal topics for way too long. He also is co-chair of his firm's Well-Being program, and reportedly has served as the Chief Spiritual Dude of the Harmonic Unity Musical Ministry ("HUMM").

Jeff lives in Bethesda, Maryland with his wife, Julie Littell. He loves his children and his most favorite thing is to sing and play music with his friends.

Heaven is Everywhere is his magnum opus thus far. His other spiritual books include *Peace and Forgiveness* and *Poems of Peace and Forgiveness*.

Other tools for peace from Peace Evolutions, LLC:

Peace and Forgiveness
by Jefferson Glassie
ISBN 0-9753837-0-1, 112 pages, $14.95
This life is our perfection, says the author. Who could imagine any heaven more perfect than this earth, with butterflies, snowflakes, and mountain tops? Though we are all peace and love, man has fears that cause war, anger, hate, and everything that isn't love. Letting go of fear – forgiving - brings peace. If we learn this, we can change the world.

Also available:
Double Audio CD read by the author
ISBN 0-9753837-1-X, $14.95

Poems of Peace and Forgiveness
by Jefferson Glassie
ISBN 0-9753837-2-8, 72 pages, $12.95
With photographs by the author
This book captures the concepts from Glassie's book, Peace and Forgiveness. These beautiful poems explain there's no right or wrong, no evil or sin, in the Universe. Everything that's not love is just based on fear. Glassie teaches the lessons of forgiveness that can lead to peace of mind, and peace in our society. We are all one, in perfection.

Fonging for the Soul
by Erasmus Caffery
ISBN 0-9753837-3-6, 78 pages, $14.95
Gathering with others, tapping on an oven rack attached to strings tied to fingers that are stuck in your ears, listening to primal sounds. Fonging brings us to together in laughter, and is much more sane than war. This book explains how to fong. It's very simple and you can do it with anyone.
By understanding the simultaneous silliness and splendor of life, we learn to create a better and more peaceful world through inanity. With many helpful illustrations, because you'll need them.

Songs of Peace and Forgiveness
ISBN 0-9753837-4-4, $16.98
Featuring original and public domain songs by Gaye Adegbalola, Scott Ainslie, Roddy Barnes, Eleanor Ellis (on a Bill Ellis song), Andra Faye and the Mighty Good Men, Grant Dermody and Frank Fotusky, Allen Holmes and Alison Radcliffe, Kelley Hunt (on a Jim Ritchey song), Ray Kaminsky, Mark Kinniburgh, MSG – The Acoustic Blues Trio, Jesse Palidofsky, and Alex Radus. The most unique blues CD you've ever heard. It will make your heart soar. Proceeds go to help preserve the famous "Barbershop" in Washington, DC run by the Archie Edwards Blues Heritage Foundation, winner of The Blues Foundation's 2005 Keeping The Blues Alive (KBA) Award.

Other tools for peace from Peace Evolutions, LLC:

**My Love Affair with an Island:
The History of the Jefferson Islands Club
and St. Catherine's Island**
by Jefferson Glassie
ISBN 0-9753837-5-2, 128 pages, $20.00
With photographs
This book tells the history of the famous Jefferson Islands Club, called the "Playground of Presidents," which was the private island retreat for Presidents including Franklin Roosevelt and Harry Truman as well as many Senators and Congressmen. With many humorous anecdotes and comments, Glassie recounts the history of both Poplar Islands where the Club was initially located and St. Catherine's Island, mixing in tales of politicians and watermen, along with the harm caused by erosion and the gradual degradation of the health of the Bay.

Rest in the Knowing
by Lynda Allen
ISBN 0-9753837-6-0, 88 pages, $15.00
With photographs by the author
Prepare for a personal journey from darkness to light. Lynda Allen's poems guide you along a path which reflects life's twists and turns. About mid-way, you will find yourself in a familiar spot, at 'The Corner of Trust and Doubt'. Will you stop or turn back to the dark? Or lift the veil to the light of the 'Waking World'?

Illumine
by Lynda Allen
ISBN-13: 978-0975383797, 79 pages, $15.00
This is Lynda Allen's second book of poems, which continues her spiritual and poetic journey. Using the gorgeous flower photographs of Joyce Tenneson as her inspiration, Lynda touches the soul with her beautiful words and spirit. A wonderful, uplifting yet realistic look at life from a fabulous author and spiritual leader.

peaceEvolutions LLC

ORDER FORM

Fax orders to (301) 263-9280 with completed order form.
Email orders by logging on to www.peace-evolutions.com
Telephone orders by calling (301) 263-9282.
Postal orders may be sent to: **Peace Evolutions, LLC**
P.O. Box 458-31, Glen Echo, MD 20812-0458

Please send the following:

Peace and Forgiveness, book	$14.95 each	quantity: _____
Peace and Forgiveness, audio CD	$14.95 each	quantity: _____
Poems of Peace and Forgiveness, book	$12.95 each	quantity: _____
Songs of Peace and Forgiveness, CD	$16.98 each	quantity: _____
Fonging for the Soul	$14.95 each	quantity: _____
My Love Affair With An Island	$20.00 each	quantity: _____
Rest in the Knowing	$15.00 each	quantity: _____
Illumine	$15.00 each	quantity: _____
Heaven is Everywhere	$20.00 each	quantity: _____

We will honor all requests for full refund on returned items.

Please send more free information on:
❏ presentations   ❏ other publications and information

Name: _____

Address: _____

City: _____ State: _____ Zip: _____

Telephone: _____

Email address: _____

Sales tax: Please add 5.00% for products shipped to Maryland addresses.

Shipping and handling:
United States: $5.00 for first book/CD and $2.00 for each additional item. International: $7.00 for first book and $5.00 for each additional item.

Payment:
❏ Check   or   Credit Card
❏ Visa ❏ Master Card ❏ Discover ❏ American Express

Card number: _____ Exp. Date: _____

Name on Card: _____